FAO Fisheries Report No. 352 Supplement
FAO Rapport sur les pêches N° 352 Supplement

FIRI/R352(Suppl.)

**D'YOUVILLE COLLEGE
LIBRARY**

Papers presented at the
FAO/UNEP MEETING ON
THE EFFECTS OF POLLUTION ON MARINE ECOSYSTEMS
Blanes, Spain, 7-11 October 1985

Communications présentées à la
REUNION FAO/PNUE SUR LES EFFETS DE LA POLLUTION
SUR LES ECOSYSTEMES MARINS
Blanes, Espagne, 7-11 octobre 1985

FOOD AND AGRICULTURE ORGANIZATION OF THE UNITED NATIONS
ORGANISATION DES NATIONS UNIES POUR L'ALIMENTATION ET L'AGRICULTURE
Rome 1987

The designations employed and the presentation of material in this publication do not imply the expression of any opinion whatsoever on the part of the Food and Agriculture Organization of the United Nations and of the United Nations Environment Programme concerning the legal status of any country, territory, city or area or of its authorities, or concerning the delimitation of its frontiers or boundaries.

Les appellations employées dans cette publication et la présentation des données qui y figurent n'impliquent de la part de l'Organisation des Nations Unies pour l'alimentation et l'agriculture aucune prise de position quant au statut juridique des pays, territoires, villes ou zones, ou de leurs autorités, ni quant au tracé de leurs frontières ou limites.

M-45
ISBN 92-5-002297-2

The copyright in this book is vested in the Food and Agriculture Organization of the United Nations and in the United Nations Environment Programme. Applications for permission to reproduce this book, in whole or in part, by any method or process, should be addressed, with a statement of the purpose and extent of the reproduction desired, to the Director, Publications Division, Food and Agriculture Organization of the United Nations, Via delle Terme di Caracalla, 00100 Rome, Italy.

Reproduction interdite, en tout ou en partie, par quelque procédé que ce soit, sans l'autorisation écrite de l'Organisation des Nations Unies pour l'alimentation et l'agriculture, détentrice des droits avec le Programme des Nations Unies pour l'environnement. Adresser une demande motivée au Directeur de la Division des publications, Organisation des Nations Unies pour l'alimentation et l'agriculture, Via delle Terme di Caracalla, 00100 Rome, Italie, en indiquant les passages ou illustrations en cause.

© FAO and UNEP 1987

PREPARATION OF THIS DOCUMENT / PREPARATION DE CE DOCUMENT

This document was prepared as part of a cooperative project of the United Nations Environment Programme, entitled

Mediterranean Action Plan: Support to the Implementation of MED POL - Phase II,

with the Food and Agriculture Organization of the United Nations,
the United Nations Educational, Scientific and Cultural Organization,
the Intergovernmental Oceanographic Commission,
the World Health Organization,
the World Meteorological Organization and
the International Atomic Energy Agency
as cooperating agencies.

Ce document a été préparé dans le cadre d'un projet du Programme des Nations Unies pour l'environnement, intitulé

Plan d'Action pour la Méditerranée: soutien à la mise en oeuvre de MED POL - Phase II,

avec l'Organisation des Nations Unies pour l'alimentation et l'agriculture,
l'Organisation des Nations Unies pour l'éducation, la science et la culture,
la Commission océanographique intergouvernementale,
l'Organisation mondiale de la santé,
l'Organisation météorologique mondiale et
l'Agence internationale de l'énergie atomique
comme agences de coopération.

Distribution:

FAO Fisheries Department/Département
  des pêches de la FAO
FAO Regional Fisheries Officers/
  Fonctionnaires régionaux des pêches
  de la FAO
Selector GFCM/Sélecteur CGPM
UNEP/PNUE

For bibliographic purposes this document should be cited as follows / La référence bibliographique de ce document doit être donnée ainsi:

FAO/UNEP, FAO/PNUE, Papers presented at the
1987 FAO/UNEP Meeting on the effects of pollution on marine ecosystems. Blanes, Spain, 7-11 October 1985. Communications présentées à la réunion FAO/PNUE sur les effets de la pollution sur les écosystèmes marins. Blanes, Espagne, 7-11 octobre 1985. FAO Fish.Rep./FAO Rapp.Pêches, (352)Suppl.:279 p.

DEFINITION OF MARINE POLLUTION

Pollution of the marine environment means: "The introduction by man, directly or indirectly, of substances or energy into the marine environment (including estuaries) which results in such deleterious effects as harm to living resources, hazards to human health, hindrance to marine activities including fishing, impairment of quality for use of sea water and reduction of amenities".

IMO/FAO/Unesco/WMO/WHO/IAEA/UN/UNEP Joint Group of Experts on the Scientific Aspects of Marine Pollution (GESAMP)

DEFINITION DE POLLUTION DU MILIEU MARIN

On entend par pollution du milieu marin: "L'introduction, directe ou indirecte, par l'homme, de substances ou d'énergie dans le milieu marin (y compris les estuaires) lorsqu'elle a des effets nuisibles tels que dommage aux ressources biologiques, risques pour la santé de l'homme, entrave aux activités maritimes, y compris la pêche, altération de la qualité de l'eau de mer du point de vue de son utilisation et dégradation de valeurs d'agrément."

Groupe mixte d'experts OMI/FAO/Unesco/OMM/OMS/AIEA/ONU/PNUE chargé d'étudier les aspects scientifiques de la pollution des mers (GESAMP).

PREFACE

The Long-term Programme for Pollution Monitoring and Research in the Mediterranean Sea (MED POL - Phase II), which is the scientific/technical component of the Mediterranean Action Plan, is basically divided into two groups of activities, namely Monitoring and Research. The Research component is divided into twelve topics one of which is concerned with the ecosystem modifications in areas influenced by pollutants (research activity 'I').

The Meeting on the Effects of Pollution on Marine Ecosystems (Blanes, Spain, 7-11 October 1985) was jointly convened by FAO and UNEP in the framework of research activity 'I'. One of the objectives of the meeting was to provide a forum for Mediterranean scientists to present their work on the subject. Many of the authors of the papers are principal investigators of research projects carried out in the framework of the above activity. The papers were not reviewed before presentation. Furthermore, no attempt was made to group the papers by subject and they appear here in alphabetical order of the senior author's name.

The views expressed in the papers are those of the authors and do not necessarily represent the views of either FAO or UNEP.

Final editing and compilation of this volume was done by the staff of the FAO Fishery Resources and Environment Division, particularly Mr. G.P. Gabrielides. Ms Mary Rollo and Ms Vanta Papapanagiotou were responsible for the typing and Ms Gloria Soave for correcting the references.

PREFACE

Le programme à Long-Terme de Surveillance Continue et de Recherche en Méditerranée (MED POL - Phase II), qui est la composante scientifique/technique du Plan d'action pour la Méditerranée, est divisé à la base en deux groupes d'activités, à savoir la Surveillance Continue et la Recherche. La composante Recherche est divisée en douze sujets parmi lesquels, un s'occupe des modifications des écosystèmes dans des endroits influencés par des polluants (activité de Recherche 'I').

La Réunion sur les Effets de la Pollution sur les Ecosystèmes Marins (Blanes, Espagne, 7-11 octobre 1985), a été convoquée conjointement par la FAO et le PNUE dans le cadre de l'activité de recherche 'I'. Un des objectifs de la réunion a été de donner aux scientifiques Méditerranéens l'occasion de présenter leur travail sur le sujet. Plusieurs auteurs des communications sont des investigateurs principaux des projets de recherche réalisés dans le cadre de l'activité mentionnée ci-dessus.

Les communications n'ont pas été révisées avant la présentation. De plus, aucune tentative n'a été faite pour regrouper les communications par sujet et elles sont présentées ici par ordre alphabétique suivant le nom de l'auteur principal.

Les vues exprimées dans les communications sont celles des auteurs et ne correspondent pas nécessairement à celles de la FAO ou du PNUE.

La mise en forme définitive et la préparation de ce document ont été effectuées par le personnel de la Division des ressources halieutiques et de l'environnement de la FAO, et plus particulièrement par M. G.P. Gabrielides. Mme Mary Rollo et Mlle Vanta Papapanagiotou ont été responsables pour la frappe et Mme Gloria Soave pour la correction des références.

CONTENTS/TABLE DES MATIERES

	Page
EVALUATING THE EFFECTS OF POLLUTION ON NATURAL MARINE ECOSYSTEMS - SOME OUTSTANDING PROBLEMS OF BIOLOGICAL SURVEILLANCE TECHNIQUES by P.D. Abel	1
ETUDE DES PEUPLEMENTS BENTHIQUES D'UN MILIEU PERTURBE: LE PORT D'ALGER par A. Bakalem et J.C. Romano	27
ETUDE COMPARATIVE DES MODIFICATIONS DES ECOSYSTEMES DES EAUX COTIERES DU GOLFE ET DU PORT DE PATRAS SOUMISES A L'INFLUENCE DES POLLUANTS par J. Castritsi-Catharios, G. Ganias et D. Panagopoulos	44
RECOVERY OF A SOFT BOTTOM COMMUNITY AFTER EXTENSIVE DREDGING. I: MOLLUSCA by M.C. Curini-Galletti	54
LES EFFETS DE LA POLLUTION SUR LES PEUPLEMENTS BENTHIQUES DE SUBSTRATS ROCHEUX DU PORT D'AUGUSTA (SICILE ORIENTALE) par S.I. Di Geronimo	64
OCCURRENCE OF TINTINNIDS IN TWO POLLUTED AREAS OF ALEXANDRIA COAST by M.M. Dorgham	76
EFFECTS OF POLLUTANTS ON MARINE COMMUNITIES AND ECOSYSTEMS IN LIMASSOL BAY by M. Hadjichristophorou and A. Demetropoulos	84
FISH DISEASES, AN INDEX OF WATER POLLUTION : A REVIEW by Y. Halim, M. Faisal and I. Ahmed	97
ENVIRONMENTAL CONDITIONS IN ABU-KIR BAY, EAST OF ALEXANDRIA, DOWNSTREAM FROM "EL-TABIA" EFFLUENT. IMPACT ON THE FISH ASSOCIATIONS IN THE COASTAL ZONE by Y. Halim, H.H. Saleh and A. Salim	105
CHANGES IN THE BENTHIC COMMUNITIES DUE TO VARIOUS POLLUTANTS IN IZMIR BAY (TURKEY) by A. Kocatas, Z. Ergen and T. Katagan	112
MODIFICATIONS DE L'ECOSYSTEME PLANCTONIQUE PAR LA POLLUTION DES EAUX COTIERES LIBANAISES par S. LAKKIS et R. Zeidane	123
SEASONAL FLUCTUATIONS OF A POLYCHAETE COMMUNITY IN A SMALL BRACKISH ENVIRONMENT (MICROHABITAT) by C. Lardicci and A. Castelli	160
ECOLOGICAL OBSERVATIONS OF A LOCALLY LIMITED SUMMER BLOOM by I. Marasovic and T. Pucher-Petkovic	167
LEVELS OF GENETIC DIVERSITY AND RESISTANCE TO POLLUTION IN MARINE ORGANISMS by E. Nevo, R. Noy, B. Lavie and S. Muchtar	175
COMPARISON OF THE EFFECTS OF ORGANIC POLLUTION AND POLLUTION BY MINING WASTE IN GREEK WATERS by A. Nicolaidou, A. Zenetos, C. Bogdanos and D. Papadopoulos	183

	Page
ETUDE DE LA DISTRIBUTION ET DE LA PHENOLOGIE DE L'ANGIOSPERME MARINE *Posidonia oceanica* par P. Panayotidis et A.V. Catsiki	192
THE IMPACT OF POLLUTION ON THE COPEPOD COMMUNITY OF THE KASTELA BAY by D. Regner	201
EFFECTS OF POLLUTION ON FISH POPULATIONS IN EGYPTIAN WATERS by H.H. Saleh and A.F. El Karashily	216
L'EXPERIMENTATION *in situ* EN TANT QUE CRITERE D'EVALUATION DES EFFETS DE LA POLLUTION: ETUDE DES MECANISMES EVOLUTIFS ET DU REPEUPLEMENT DE SEDIMENTS POLLUES par G. Stora, A. Arnoux et C. Diana	229
ACTION DES POLLUANTS SUR LES ECOSYSTEMES LITTORAUX MEDITERRANEENS par N. Vicente	252
REPORT ON BOTTOM FAUNA IN TWO NORTHERN ADRIATIC AREAS PRESUMED TO BE INFLUENCED BY INPUTS by D. Zavodnik and J. Vidakovic	263

EVALUATING THE EFFECTS OF POLLUTION ON NATURAL
MARINE ECOSYSTEMS - SOME OUTSTANDING PROBLEMS
OF BIOLOGICAL SURVEILLANCE TECHNIQUES

by

P.D. ABEL*
Biology Department
Sunderland Polytechnic
Sunderland SR1 3SD
England

1. INTRODUCTION

The study of ecosystem processes is essentially the study of the rates of biological processes at the individual, population and community levels of organisation. Estimation of the rates of reproduction and mortality, of energy consumption and expenditure, of growth and production, and of energy transfer between trophic levels, all are necessary. In turn, these estimates can only be made on the basis of accurate knowledge of population densities, and of many other ecophysiological data. While ecosystem processes can reasonably be studied in simple, small-scale ecosystems (including experimental ecosystems), such studies tell us only what can happen, not what actually does happen, in any real ecosystem. For real ecosystems, the information required to make accurate statements about ecosystem processes does not exist, or is itself based on gross approximations or unjustifiable assumptions. Even the accurate estimation of polulation density is astonishingly difficult. For many years, I offered my students an exercise requiring them to estimate the number of plastic beads buried in a large, shallow tray of sand. For teaching purposes, the exercise has the advantage that the estimate obtained can be compared with the true value. Many students were satisfied if their confidence limits embraced the true value, but the more discerning ones realised that even in this simple system, the most sophisticated methods which could practicably be applied in the field yielded answers only within $\pm$ 30-40% of the true value. What confidence, then, can we have in any estimate of the fish population, or plankton biomass, of a large ecosystem such as the Mediterranean sea?

If we could reliably show that, as a result of pollution, the population of a species was reduced by 20%, or the energy flow from one trophic level to another were altered by 20%, we should be proud indeed of our technical prowess. However, the fact is that we cannot reliably measure such changes and are unlikely to be able to do so in the foreseeable future. For these reasons (which will become clearer later in the text), biologists do not, in practice, seek to determine the effects of pollutants on ecosystems. We do talk in such terms, as the title of this paper and indeed of the whole conference indicates. But we must recognise that in practice, we are involved with measuring the effect of pollutants on ecological communities and community structure.

The study of communities has the advantage that with available methods of sampling and data analysis it is possible to draw reasonably sound conclusions which are of practical use. Nevertheless, not all components of the marine community are equally accessible and amenable to study. Reasons for this include the availability of sampling techniques and taxonomic expertise. A community can be considered as a set of interacting sub-communities, and for reasons which will be mentioned later, the benthic invertebrate community, particularly of coastal areas, is generally considered the most suitable assemblage of organisms for studying the effects of pollution in the field. This discussion will be concerned primarily with the benthic invertebrate community, although much of the material included is relevant to other components of the marine ecosystem as a whole.

---

* This work was completed during a period of sabbatical leave at the Department of Applied Hydrobiology, Agricultural University of Athens, Greece

The changes which may take place in a community subject to pollution are summarised in Fig. 1. Such changes, if they are detected, indicate that the physical and chemical environment may have altered in a significant way, the community changes representing the response of the biota to the new environmental conditions. Such changes do not, of course, necessarily indicate that pollution has occurred, since they may be the result of natural processes. Nevertheless it has long been recognised that regular surveillance of the biota of aquatic habitats is a useful technique in monitoring the effects of pollution.

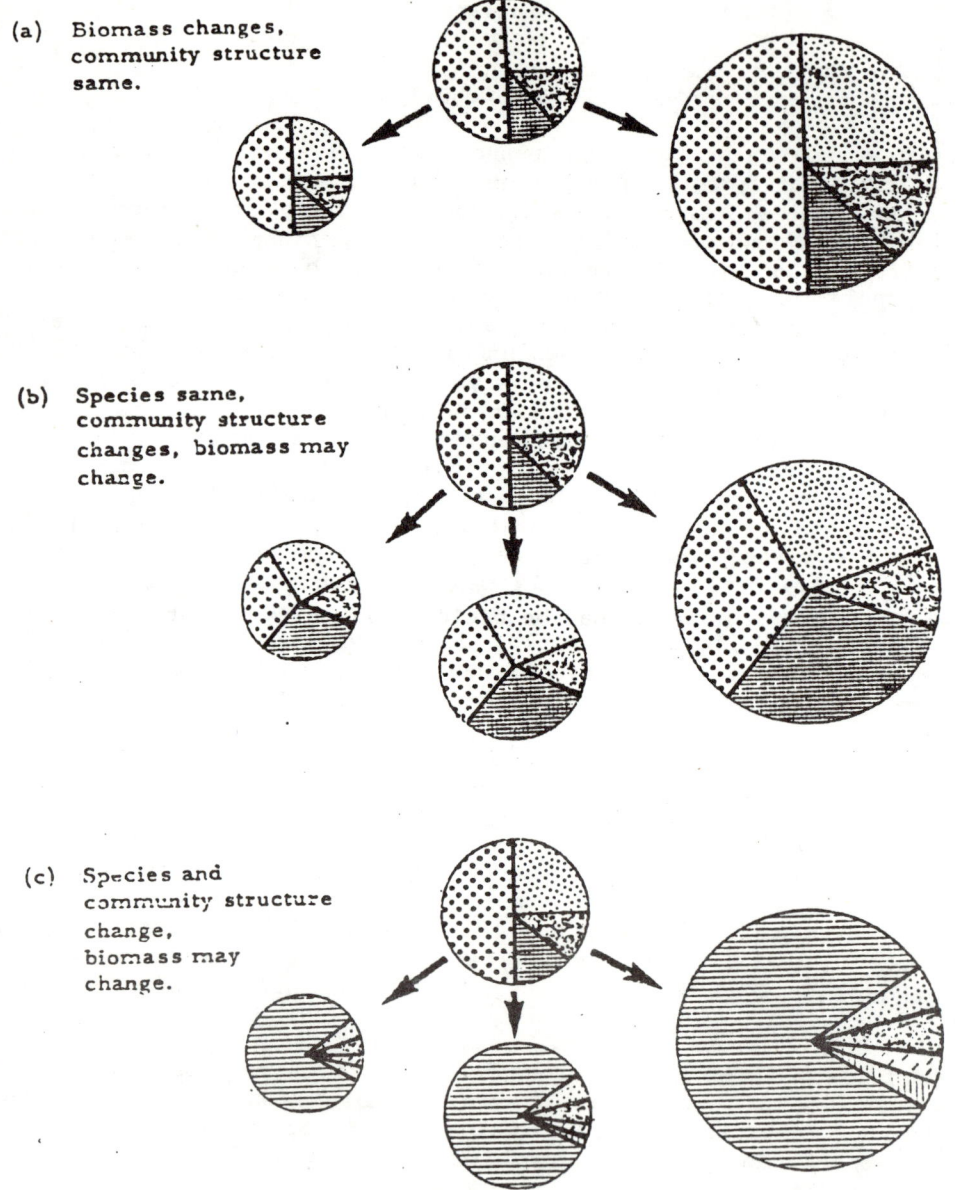

Figure 1. Diagrams illustrating the possible responses of a community to an environmental change, including the incidence of pollution. From Hellawell (1978).

The advantages of biological surveillance over chemical surveillance are simply stated. First, a chemical analysis relates only to the moment at which the sample was taken. Since pollutant loads vary rapidly and unpredictably, it is very easy to draw false conclusions from chemical data (Fig. 2). Since living organisms respond to the totality of their environment as they have experienced it during their lifetimes, theirs is an integrated response to the conditions which prevail in their environment over a longer period. Secondly, a chemical analysis detects only those pollutants which are specifically sought, so the presence of an unsuspected pollutant will not be detected. Thirdly, a chemical analysis tells us nothing about the effect which the pollutant exerts on the biological community.

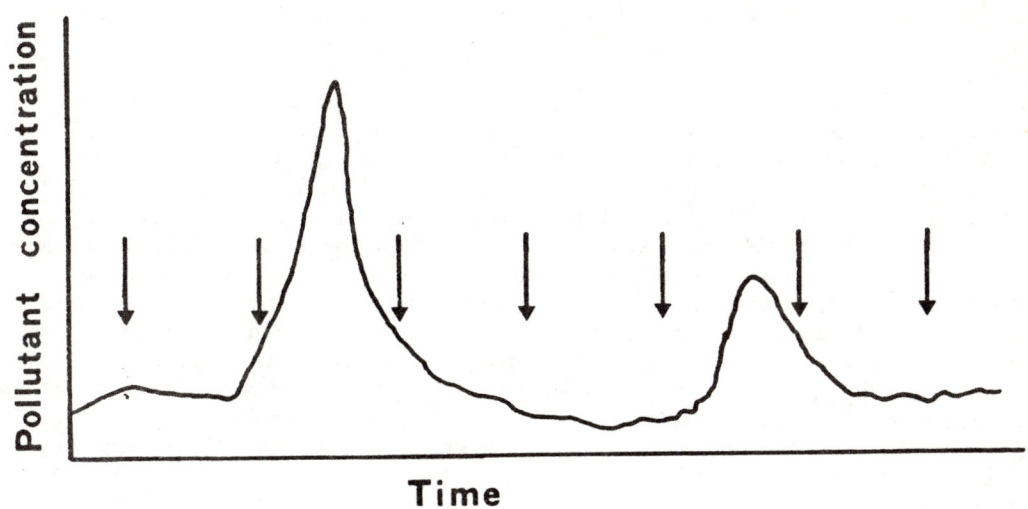

Figure 2. Pollutant concentrations frequently fluctuate rapidly, depending upon the strength and pattern of release of effluents, and the extent of rainfall and river discharge. Even very frequent chemical sampling (arrows) tends to miss peak concentrations which may be biologically significant.

The study of pollutant effects in marine ecosystems is a relatively recent activity compared to the study of freshwater systems, which has a history dating from nearly 100 years ago. Principally this is because the effects of pollution in fresh water, and their consequences for public health, became obvious much earlier. In principle, the answers sought from the study of freshwater and marine systems are the same. Also, there are practical requirements which are common to all ecological investigations, e.g. relating to sampling methodology, taxonomy, and the correct analysis and interpretation of data. But since biological monitoring of freshwater has a much longer history than its marine counterpart, many aspects of the subject have received fuller and more detailed study than is the case for marine systems. Indeed, the theory and practice of biological monitoring in fresh water is well established and is the subject of a number of authoritative reviews (Alabaster, 1977; Hellawell, 1977, 1978; Dickson and Cairns, 1978; James and Evison, 1979; Mathews et al., 1982; Herricks and Cairns, 1982; Pascoe and Edwards, 1984). In contrast, although biological surveillance techniques have been widely and successfully employed in the marine environment, marine biologists have still to agree on appropriate methodologies. It is therefore informative to consider what the marine biologist can learn from the experience of his colleagues working on freshwater systems, and I believe this is important for three reasons. Firstly, it is clearly wasteful to rediscover principles and methodologies which are already known, and to repeat mistakes which have already been made. Secondly, and on the other hand, the uncritical application to the marine environment of ideas and approaches which are appropriate in fresh water may lead to error, since there are important differences between marine and freshwater habitats and the characteristics of their respective biota. Thirdly, a comparison of these differences, and their implications for monitoring programmes, will focus attention on gaps in our knowledge and help to direct research activity where it is most needed.

The purpose of this paper is therefore to look at some key elements of the theory and practice of biological monitoring and to consider the implications of applying them, in modified or unmodified form, to the monitoring of the marine environment. Perkins

(1979) in an earlier discussion pointed out that in fresh waters, biological monitoring has been successfully applied in rivers, but less so in lakes, and suggests some possible reasons. Insofar as lakes and the sea more closely resemble each other than they do rivers, it is important to consider how far the principles of monitoring can be applied to the marine environment.

2. INDICATOR ORGANISMS

It is fundamental to ecology that an organism cannot survive indefinitely in an environment that does not meet its physical, chemical and nutritional requirements. Thus the presence of a particular species, particularly if it is abundant, indicates that its requirements are being met. Its absence, however, does not necessarily indicate the converse: one species may, for example, be competitively excluded by another. Nevertheless within certain limitations the presence, absence or relative abundance of species may be used as indicators of environmental quality. In its broadest sense, therefore, the term "indicator organism" can be used of any member of the fauna or flora of a habitat. Some organisms, however, have such wide tolerances of different environmental conditions that their distribution and abundance may tell us little about their environment. Some authors therefore reserve the term "indicator organism" to those species which have fairly narrow and specific environmental tolerances, so they will show a marked response to quite small changes in environmental quality. If the environmental factors which are commonly limiting to the species are known, its presence will be indicative of a specific environmental condition. Thus, in freshwater, nymphs of Plecoptera are intolerant of reduced dissolved oxygen levels. Their presence therefore indicates that the water is well oxygenated. Their absence does not necessarily indicate that the water is not well-oxygenated, but their absence from waters where they might normally be expected at least suggests the need for further investigation. In practice, it is rare that enough is known of the ecology and physiology of a species for its presence or absence to be indicative of a specific environmental condition. There are, furthermore, very few organisms whose presence specifically indicates that the water is polluted. An obvious exception is the presence of coliform bacteria, indicating faecal contamination. These, of course, indicate the presence of pollution but give no information concerning its ecological significance. Several organisms are frequently associated with polluted conditions - in freshwater, tubificid worms and the "sewage fungus" Sphaerotilus natans are examples, and the marine polychaete Capitella capitata, a cosmopolitan species, appears to be associated with sewage discharges to the sea (Reish, 1972). However, such species may be found naturally in non-polluted environments. In the present discussion, indicator organisms are considered to be "those which, by their presence and abundance, provide some indication, either qualitatively or quantitatively or both, of the prevailing environmental conditions" (Hellawell, 1978).

Ideally all members of a community should be considered as potential indicators of environmental quality and included in biological monitoring programmes. In practice, groups such as bacteria, plankton, algae, protozoa, macroinvertebrates and fish require such different sampling methods and taxonomic skills that most investigators choose only one such group. Normally this is satisfactory except for the most extensive and detailed research programmes. The advantages and disadvantages of various groups have been summarised for freshwater by Hellawell (1977, 1978) and for the marine environment by Stirn (1981). Benthic macroinvertebrates show most of the required characteristics and are the most widely-used group both in marine and freshwater studies. However, it is important to consider some characteristics of marine organisms and of the marine environment which influence the validity of the indicator organism concept.

Two requirements of an indicator organism are, relatively limited mobility, and relatively long life cycles. Thus the indicator organisms, by their presence or abundance, will reflect the environmental conditions of the place where they are situated, over their entire life up to the point of sampling. However, most marine benthic invertebrates have a planktonic phase in the life cycle. Thus the distribution and abundance of the adults may reflect events which took place at some previous time in a different location, during the larval stage or at a crucial time of metamorphosis. Together with certain other features of marine invertebrate life histories, this could have profound consequences for the usefulness of these organisms as indicators.

Most freshwater invertebrates have more or less annual life cycles, though some have two or more generations per year and some have overlapping generations in a life-span of 2 years or so. Perhaps for this reason (and because depopulated stretches of river are quickly colonised by downstream drift of organisms) in rivers even after catastrophic pollution incidents, the "normal" flora and fauna tends to be re-established within a matter of months. In contrast, many marine species appear to have longer life spans, and there is evidence that in marine communities a single event can have profound consequences whose effects, because of the long life-span of species and the nature of interspecific interactions, can persist for a very long time. Lewis (1972) followed population changes of four littoral species on an unpolluted English shore over 4 years. A combination of climatic events and biotic interactions resulted in repeated large fluctuations in population density, in some cases of more than 10-fold. A major determinant of the marked community changes which occurred during the period of observation was the success or failure of planktonic stages in settling and metamorphosis. Perkins (1979) refers to examples where single events had consequences which persisted for several years. Thus it appears that in marine systems, single events, often of natural origin, can cause long-term community changes which persist long after the disturbing influence has ceased, and that the use of marine organisms as indicators of environmental quality may give a quite misleading picture.

Populations of marine organism are also known to undergo long-term fluctuations in abundance. Perkins (1979) again cites several examples. Therefore the collection of data on presence, absence or abundance of species is of little value unless it extends over several years, possibly decades. A cautionary example for Mediterranean workers is that of <u>Acanthaster planci</u>, the "Crown of Thorns" starfish, which attracted much attention in the Pacific Ocean 10-15 years ago. Its increased abundance and its resultant effect on coral reef ecosystems was widely attributed to pollution or other human interference. However it is quite likely that periodic increases in population density are a natural feature of its ecology (Moore, 1978). Sea-urchins and jellyfish are currently under examination in the Mediterranean for similar reasons, and it is important not to lose sight of the possibility that we may be dealing with a natural phenomenon.

I do not mean to suggest that the concept of indicator organisms is invalid or inappropriate in marine systems, but is is clear that the concept cannot simply be transferred from freshwater without further fundamental research. The fact that the idea has worked so successfully in rivers may be due to a fortuitous peculiarity of the river ecosystem - perhaps its simplicity - but it is certainly in part due also to the much greater knowledge which exists about rivers and their biota. In interpreting the data from biological surveys, we implicitly or explicitly compare what we have found with what we expected to find. Our expectations depend upon knowledge of the previous conditions of the habitat under study, on contemporary knowledge of physically-similar and geographically adjacent habitats, or at worst on a broad but detailed general knowledge of the ecophysiology of our indicator organisms. For marine organisms such detailed knowledge does not yet exist, and is urgently required. Perkins (1979) refers to an example which illustrates the need for more background knowledge of the ecology of marine indicator organisms. Observations of the benthic community along a transect away from a marine outfall showed that the diversity and abundance of species increased with distance from the outfall. The obvious conclusion, that the outfall was exerting a significant polluting effect, was however erroneous. Subsequent studies on a physically-similar environment showed that the improverished community near the pollution discharge was due to natural, physical disturbances caused by wave action on the substratum, and that the pollutant discharge had no detectable effect.

Several areas of further research can thus readily be identified as important.

a) Long-term studies of benthic community structure in unpolluted areas, particularly with a view to determining the effects of natural processes on variations in community structure and in population densities of selected indicator organisms.

b) Studies on the effects of local physical conditions (current, climatic factors, substratum characteristics) on the distribution and abundance of benthic invertebrates.

c) Long-term studies on the benthic community structure of areas known to be polluted, to improve knowledge of the response of individual species and communities to specific forms of pollution.

d) Detailed studies of the ecophysiology and population ecology of potentially useful indicator species.

In addition, taxonomic studies, including training programmes and the development of reliable keys, are highly desirable.

In the absence of the above, the application of the indicator organism concept to marine studies is likely to remain of limited value.

3. SAMPLING METHODS

The validity of any ecological investigation depends crucially upon the sampling technique and strategy adopted at the outset. Many factors influence the design of a sampling programme. These include: the objectives of the investigation and the type of analysis to which the data are to be subjected; the physical characteristics of the habitat to be sampled; the characteristics of the organisms to be sampled - their size, habits, abundance, and patterns of distribution; and constraints on the human, physical and financial resources available for the investigation. Frequently, then, a sampling programme is devised on the basis of a series of compromises between what the investigator would like to do, and what it is possible to do. In order to make sensible decisions about sampling - in particular, how to maximise the quantity and quality of the information obtained for a given sampling effort - it is necessary to understand something of sampling theory and practice, including statistical considerations. There are objective criteria for making such decisions, which are frequently ignored in favour of arbitrary decisions on such matters as sample size, number and distribution. As a result, much expensive effort is wasted or, worse, inaccurate or wrong information is published which serves only to confuse and mislead. The editor of one well-known journal recently drew attention to the low quality of many submitted manuscripts, and cited inadequate sampling as a major reason for rejection (Clark, 1982; Hartley, 1982) discussed a number of common failings of many investigations, and made some specific recommendations. An alternative approach, which is appropriate within the framework of MEDPOL, is to adopt standardised sampling techniques, such as described by Stirn (1981).

Standardisation of sampling techniques has two principal advantages. Firstly, inexperienced teams, by following the standard methods, can avoid error. Secondly, results from different laboratories are more easily comparable. It also has dangers. Inexperienced workers, through lack of understanding of the rationale underlying the methods, may be tempted to omit or modify procedures which appear to them of small significance, but which in reality can decisively affect the results. On the other hand, the existence of standard methods frequently inhibits the development and use of alternatives which may be superior, or more appropriate in particular circumstances. Therefore there is no substitute for a widespread and thorough understanding of the principles of sampling theory and practice, and steps should be taken to achieve this among all laboratories and personnel.

It is beyond the scope of this paper to cover the subject fully. Several sources of guidance are available, either specifically for marine studies or directly relevant to marine studies, including Holme and McIntyre (1984), Dybern et al. (1976), Elliot (1977) and Hellawell (1978). This discussion is confined to a selection of points which provides examples of the kinds of difficulties faced when a sampling programme is devised.

The design of the equipment used for sampling can greatly influence the results. Stirn (1981) recommends a modified Van Veen grab for sampling of benthic invertebrates, but many different designs are in use and even grabs of similar designs from different manufacturers may perform differently (Ankar et al.,1979). Elliot and Drake (1981) examined the performance of seven different designs of benthic grab. In their experiments, known numbers plastic pellets, representing animals, were placed in large tanks containing water with different types of substratum. They found that the efficiency of the grabs (expressed as the number of "animals" caught as a percentage of those available to be caught) was generally low, and varied widely between designs and in different kinds of substratum. Thus it appears that estimates of population density based on grab samples must be interpreted with great caution, if indeed they are of any use at all.

Extraction of the invertebrates from benthic grab samples usually involves some form of sieving process. Stirn (1981) recommends the use of a 1 mm mesh sieve as a basic tool, but suggests the additional use of a 0.5 mm mesh. Hartley (1982) also recommends the use of a 1 mm sieve as a standard in pollution studies. In his fuller discussion of the topic, he points out that the use of the smaller sieve may in practice result in a decrease in accuracy. The smaller size of mesh generally retains largely juvenile forms which are frequently unidentifiable. Ankar et al., (1979) found that differences in extraction procedure was a major source of variation in results between two laboratories. Therefore it is arguable that a more closely specified procedure for sorting and extracting marine benthic samples is desirable. It is disappointing to note that, among the papers presented at this meeting, in only a small minority of cases have investigators used a 1 mm sieve, in spite of well-known recommendations for the adoption of this standard.

A recurrent problem is the number of replicate samples required from each station. Clearly the answer depends upon the purposes of the study and the type of data analysis to be employed. Stirn (1981) merely states that "a minimum of 0.3 to 0.5 m^2 of surface should be sampled as an integral of 3 to 5 grab samples taken at random", but further suggests that "an adequate number of samples per station be checked by a species-area test".

For quantitative sampling, the number of samples required to estimate population density is given approximately by the formula

$$n = \left(\frac{st}{D\bar{x}}\right)^2 \qquad \text{(Elliot, 1977)}$$

where n = the number of samples required, $\bar{x}$ and s are the mean, and its standard deviation, respectively of the number of individuals per sample caught in a pilot survey.

t = the value of Student's t for the required level of confidence.

D = the index of precision, i.e. the ratio of the standard error to the arithmetic mean expressed as a decimal.

Thus for example, if the mean number of individuals per sample is 10 and the standard deviation is 5, to obtain an estimate of the population density which lies within $\pm$ 10% of the true value with 95% confidence,

$$n \simeq \left(\frac{5 \times 2}{0.1 \times 10}\right)^2 = 100 \text{ samples}$$

Unfortunately, the pattern of distribution of many benthic invertebrate species seems to be such that very large numbers of samples are required for reliable estimates of population density. Even to estimate population densities to within $\pm$ 20% or $\pm$ 40% of their true values may require several hundred samples (Hellawell 1977, 1978). Hartley (1982) gives some data from studies in the North Sea which are reproduced in Table 1. The number of samples required to estimate population density to within $\pm$ 20% of its true value was small ($<$5) when major taxonomic groupings were considered (e.g. all polychaetes, all crustaceans, all molluscs). However, this level of taxonomic analysis is clearly inadequate for most purposes. The substitution of, say, one polychaete species by another is precisely the sort of community change which biological monitoring seeks to detect. It is therefore futile to consider only major taxonomic groupings. If we wish to estimate population densities of individual species, however, Table I shows that the number of replicates required is much larger, and probably 10 to 20 is a more reasonable figure.

For quantitative sampling, the most important criterion is that the sample should contain at least one representative of each species present in the habitat. Since most species in a community are rare, it is the need to sample the rarer species which determines the sampling effort required.

Table I

Comparison of the number (n) of replicate 0.1 m^2 samples required to give standard error of counts per taxon equal to 20% of the mean. Data from survey of Forties oilfield (Hartley, 1979), given by Hartley (1982). The number of samples required (n) is generally <5 for gross taxonomic groupings (polychaetes, molluscs, crustaceans). However for individual species the number of replicate samples required is much larger. Note that only the more abundant species have been considered. For rare species, the number of samples required may be much larger.

	Σx	$\bar{x}$	s^2	n	Range and mean of values of (n) for numerically dominant species*	
					Range	Mean
Station 15						
Polychaetes	420	84.0	278.5	0.99	0.6-18.6	5.51
Crustaceans	58	11.6	17.3	3.21		
Molluscs	132	26.4	11.3	0.41		
Station 16						
Polychaetes	407	81.4	419.3	1.58		
Crustaceans	59	11.8	11.2	2.01		
Molluscs	157	31.4	31.3	0.79		
Station 17						
Polychaetes	431	86.2	237.7	0.92	2.7-96.1	20.68
Crustaceans	66	13.2	16.2	2.32		
Molluscs	113	22.6	26.3	1.29		
Station 18						
Polychaetes	341	68.2	17.2	0.09		
Crustaceans	62	12.4	32.3	5.25	1.5-14.9	6.00
Molluscs	127	25.4	43.3	1.68		

* Regarded as those top ranked species contributing to the first 50% of the number of individuals at each station (10-12 taxa for these stations).

An important characteristic of a qualitative sampling technique is the rate of taxon accretion, i.e. the relationship between sampling effort and the number of taxa caught. Obviously it is important to ensure that a quantitative sample contains a reasonably high proportion of the species actually present. To determine the extent of the sampling effort required to achieve this, a taxon accretion curve (Fig. 3) should be plotted. Stirn (1981) uses the term 'species-area test' for this process. The correct procedure is to take a large number of samples, and determine the mean number of taxa recovered, with its confidence limits. Samples are then considered together in randomly-chosen pairs, threes, fours and so on. (A simple computer program can be used to calculate the values for all possible permutations of two, three, four, etc). A simpler method is to plot the cumulative number of taxa recovered against the number of samples taken, but this method can give misleading results owing to the possibility that the earlier samples may, by chance, contain abnormally high or abnormally low numbers of taxa. The resulting curve (Fig. 3), shows the point at which large increments in sampling effort yield few additional taxa, and the optimum sampling effort can thus be determined.

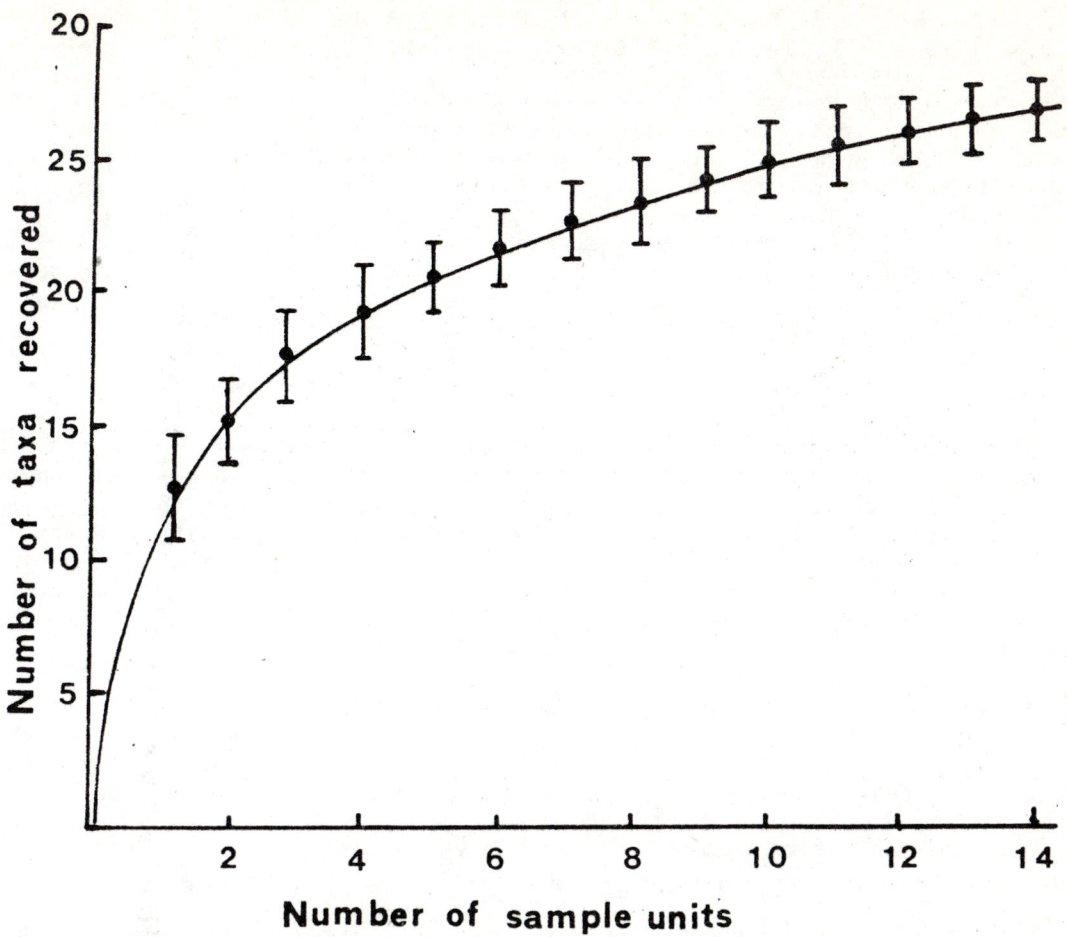

Figure 3. Taxon accretion curve showing the mean number of invertebrate species recovered with increasing sampling effort. Vertical lines are 95% confidence limits.

As with many aspects of sampling, this process is tedious but essential. If, for example, the results are to be analysed by any form of diversity, similarity, or biotic index (see below), it is essential that all, or nearly all, of the species actually present in the habitat are actually sampled. It is easy to show, for example, that if only 50% or 70% of the species actually present in the habitat are recorded, two sites which are in fact identical may produce a similarity coefficient of only 30-50%, thus rendering the correct interpretation of the results impossible. It is a common weakness of many ecological investigations to use sophisticated analytical techniques on inherently poor data, as if mathematical manipulation could compensate for bad experimental design or inadequate sampling effort.

These limited examples suggest that many aspects of sampling methodology require further discussion, with a view to establishing more specific guidelines which should be rigidly adhered to. Insofar as a degree of standardisation already exists in the MEDPOL programme, it probably serves only to inhibit the development of superior methodology without increasing the degree of comparability between contributions from different laboratories.

4. DATA ANALYSIS AND INTERPRETATION

The interpretation of biological survey data is essentially a series of comparisons - spatial, temporal, or both - and a variety of methods of data analysis are available to facilitate the process. One difference between marine survey data and that from freshwater is immediately apparent. For practical purposes, a river is usually considered as a one-dimensional system, i.e. having length but negligible width or depth. Thus, comparisons of communities above and at varying distances below a pollutant source usually allow any effects of pollution on the receiving water fauna to be readily detected. The relatively strong and unidirectional flow of rivers, and the accretion of unpolluted water along the length of the river, tend to accentuate differences between communities at different points along the river. Marine habitats, in contrast, are at least 2-dimensional (e.g. for benthic surveys), and may be 3-dimensional if the pelagic biota are to be considered. As was argued earlier, observations at a single point in time are likely to be of limited value, if not positively misleading. Therefore the additional dimension of time is likely to be involved in many surveys. Thus the analysis of marine data is intrinsically more complex than that of freshwater data. It follows that the data required for meaningful comparisons (and consequently the design of the sampling programme which produces it) and the methods of interpretation of the data need also to be more complex.

Consider a single sample site at a single point in time. To know if pollution is affecting the community, we need to know whether that community is "normal" or not. This can only be done by making temporal or spatial comparisons - is the community at this site the same as it was on some previous occasion? Or is it similar to physically comparable sites at different locations? Clearly in order to answer these questions we need to know the range of temporal and spatial variation in community characteristics which may be considered "normal". To the extent that this knowledge is lacking, no matter how sophisticated our analytical techniques, a reasonable interpretation is impossible. Therefore a major need is for basic research on the range of temporal and spatial variation in the community characteristcs of <u>unpolluted</u> habitats.

Assuming this knowledge were available, there are several bases on which such comparisons can be made. One approach, which is widely used in freshwater, is simply to examine the data for the presence, absence, and abundance of certain indicator organisms, but for reasons which have already been discussed this approach is not yet feasible for marine studies. Therefore there is a need for further research on a range of marine species and communities, particularly in relation to their ecological responses to varying types and levels of pollution.

Given that the data from a single site at a single point in time take the form of a list of species and their relative abundance, and that comparisons between data sets have to be made in three or four dimensions, some means of condensing the data to manageable proportions is essential. It is also important because although an experienced biologist may be capable of interpreting complex ecological data sets by inspection, other specialists with whom he must co-operate (chemists, engineers, administrators) cannot be expected to do so. Therefore the raw data is frequently condensed by using it to compute one or more of several numerical indices or coefficients, of which four types can be distinguished - pollution indices, biotic indices, diversity indices, and similarity indices or coefficients. The applications of each of these will now be considered in turn.

<u>Pollution indices</u>

Pollution indices were developed for freshwater studies and have reached a high degree of sophistication in many central and Eastern European countries (Sladecek, 1979). They are essentially developments of the descriptive Saprobien system of Kolkwitz and Marsson (1909), and are based on the fact that in rivers subject to organic pollution, communities downstream of the pollutant input show a regular and more or less predictable sequence of changes in the presence and abundance of indicator species.

A typical pollution index is that of Pantle and Buck (1955). Organisms present in a sample are given an s-score and an h-score. The h-score is an index of relative abundance: organisms are scored 1,3 or 5 depending upon whether they are rare, frequent or abundant respectively. The s-score relates to the saprobic zone of which the species are characteristic - a score of 1,2,3 or 4 is assigned to each species depending upon whether it is typically found in oligosaprobic, α-mesosaprobic, β-mesosaprobic or polysaprobic zones respectively. The saprobity index is given by

$$S = \frac{\Sigma sh}{\Sigma h}$$

and typically varies from less than 1.0 for unpolluted waters to 4.0 for heavily organically polluted waters. Hellawell (1978) and Washington (1984) give several other examples of pollution indices.

The application of pollution indices to marine studies is obviously hampered by the lack of knowledge of the distribution patterns of most species in relation to pollution, but there is no reason in principle why they should not work successfully. The nearest approach to a pollution index which has been used in marine studies is the Nematode/Copepod ratio. It is frequently found that the ratio of nematodes to copepods in a sample of benthic fauna (meiofauna) alters with the intensity of organic pollution, and the use of this ratio as an index of pollution was suggested by Rafaelli and Mason (1981). The idea has not yet been fully tested. Amjad and Gray (1983) found that it produced satisfactory results in Oslo fjord, but in the Firth of Forth (Shiells and Anderson, 1985), the ratio varied along pollution gradients in an inconsistent manner.

Biotic indices

Biotic indices are empirically-derived and are based on two observed characteristics of the communities inhabiting polluted waters. Firstly, they generally contain fewer species than communities from comparable unpolluted waters; and secondly, as the degree of pollution increases species will tend to be selectively removed in order of their relative susceptibility to that form of pollution. The Trent Biotic Index (Woodiwiss, 1964) is most commonly used, and is reproduced in Table II. To derive the index value, each specimen in the sample is identified to the level at which it can be assigned to one of the groups listed in the lower part of the table. It can be seen that in most cases fairly limited taxonomic expertise is required. When the number of groups in the sample is known, the table is entered at the appropriate column. The line in the table at which the correct index value is given is determined by the presence of the highest-ranking indicator group (Column 1). Thus if seven groups are present, of which the highest-ranking member is _Gammarus_, the TBI value is V. If seven groups are present, but one is a species of Plecoptera, the TBI score is VII. Hellawell (1978) and Mason (1981) give worked examples showing the derivation of TBI values from raw data. There are several similar biotic indices including that of Chandler (1970), which takes abundance into account and is taxonomically more sophisticated. However, sufficient information to construct a workable biotic index for marine systems does not yet exist, though again in principle such indices should work if enough were known about the responses of marine organisms to polluted conditions. Sanders' Rarefaction Method (1968), which is described and recommended by Stirn (1981) can be considered as a form of biotic index (Washington, 1984), since although it is claimed to be a measure of diversity it is based on the polychaete/bivalve fraction of the community which Sanders considered to be indicator organisms. However several authors have pointed to certain ambiguities in the method (Washington, 1984), and it is by no means clear that all authors have applied the method in the same manner.

Diversity indices

Diversity indices were developed by theoretical ecologists who were interested in such questions as the relationship between stability and diversity in ecosystems. Washington (1984) lists no fewer than eighteen different diversity indices which have been used in water pollution studies (Fig. 4). Such a variety of diversity indices has arisen because ecosystem diversity is not easily defined, and therefore can be measured in several different ways, depending on how the concept is defined. Some simple examples will illustrate the point.

A simple measure of diversity is the number of species present in the community. However, it could be argued that this is an inadequate measure. For example, consider two communities of 1000 organisms. Community A contains one individual each of species a,b, and c, and 997 of d; community B contains 250 individuals each of a,b,c and d. Although each community contains 4 species, nearly all the individuals in A belong to the same species. Arguably, therefore, community B is a more diverse one than A. A numerical value derived from the number of species in the community, and the distribution of individuals between those species, can be used as an index of community

Table II

The Trent Biotic index, an example of a simple but reasonably effective index used in fresh water biological monitoring. For explanation, see text. (Woodiwiss, 1964)

Clean			Number of groups present				
			0-1	2-5	6-10	11-15	16
				Biotic index			
Organisms in order of tendency to disappear as degree of pollution increases	Plecoptera nymphs present	More than one species	-	VII	VIII	IX	X
		one species only	-	VI	VII	VIII	IX
	Ephemeroptera nymphs present	More than one species*		VI	VII	VIII	IX
		One species only*		V	VI	VII	VIII
	Trichoptera larvae present	More than one species+		V	VI	VII	VIII
		one species only+	IV	IV	V	VI	VII
	Gammarus present	All above species absent	III	IV	V	VI	VII
	Asellus present	All above species absent	II	III	IV	V	VI
	Tubificid worms and/or red chironomid larvae present	All above species absent	I	II	III	IV	--
	All above types absent	Some organisms such as Eristalis tenax not requiring dissolved oxygen may be present	0	I	II	--	--
Polluted							

*Baetis rhodani (Ephem.) is counted in this section for the purpose of classification.
+Baetis rhodani excluded

Groups : The term "Group" here denotes the limit of identification which can be reached without resorting to lengthy techniques. Thus the Groups are as follows: Each known species of Plathelminthes (Flatworms). Annelida (worms) excluding Genus Nais (worms). In each known species of Hirudinae (leeches). Each known species of Mollusca (snails). Each known species of Crustacea (log-louse, shrimps). Each known species of Plecoptera (stone-fly). Each known genus of Ephemeroptera (may-fly) excluding Baetis rhodani. Baetis rhodani (may-fly). Each family of Trichoptera (caddis-fly). Each species of Neuroptera larvae (alder-fly). Family Chironomidae (midge larvae) except Chironomus thummi (blood worms). Family Simulidae (black-fly larvae). Each known species of other fly larvae. Each known species of Coleoptera (beetles and beetle larvae). Each known species of Hydracarina (water-mites).

---

diversity. The application of diversity indices in water quality monitoring is discussed by Wilhm and Dorris (1968), Hellawell (1977, 1978), Kaesler et al., (1978) and Washington (1984). Diversity indices differ from one another in, for example, the relative weighting given to the number of species and to the distribution of individuals between species (the "evenness" component). Some indices are based on specific assumptions of community structure, and many are not independent of sample size. Washington (1984) divides diversity indices into eight groups, according to the basis on which they are derived (Fig. 4). Having considered each in detail, he concluded that all were unsuitable for application to aquatic ecosystems except for Simpson's D (1949), Hurlbert's PIE (1971), indices based on the theory of runs (Cairns et al., 1968 ; Keefe and Bergerson, 1977) and McIntosh's M (1967). The most widely used indices in practice are those based on information theory, such as the Shannon-Weaver index (Shannon and Weaver, 1949). However, as Washington (1984) points out, the biological relevance of such indices has been widely doubted in recent years.

(1) **Simpson's Index**

Simpson's D
where $D = \dfrac{\sum_{i=1}^{s} n_i(n_i-1)}{n(n-1)}$ (1949)

(2) **Relatives of species number**

Kothé's species deficit (1962)

$$\dfrac{A_1 - A_x}{A_1} \times 100$$

Odum's species per thousand individuals (1960)

(3) **Guesses by data fitting**

Gleanson's index $D = \dfrac{S}{\ln N}$ (1922)

Margalef's index $D = \dfrac{S-1}{\ln N}$ (1958)

Menhinick's index $D = \dfrac{S}{\sqrt{N}}$ (1964)

(4) **Curve fitting approach**

Motomura's geometric series,

$y = A_c^{(x-1)}$ (Whittaker 1965)

Fisher's α,
where $S_1 = \alpha \ln\left(1 + \dfrac{N}{\alpha}\right)$ (1943)

The modified Yules "characteristic"

$\dfrac{M_1^2}{M_2 - M_1} = \dfrac{n^2}{\sum n(n-1)}$ (Williams, 1964)

Preston's log-normal "a"
where $y = y_0 \exp(-aR)^2$ (1948)

(5) **Information theory**

$$\text{Brillouins } H = \frac{1}{N} \ln \frac{N!}{\prod_{i=1}^{s} N_i!}$$ (1951)

$$\text{Shannon's } H' = -\sum_{i=1}^{s} \frac{n_i}{n} \ln \frac{n_i}{n}$$ (Shannon and Weaver, 1949)

$$\text{Evenness } E = \frac{H'}{H'_{max}}$$

$$\text{Redundancy } R = \frac{H'_{max} - H'}{H'_{max} - H_{min}}$$ (Patten, 1962)

(6) $$\text{Hurlbert's PIE} = \left(\frac{N}{N-1}\right)\left(1 - \sum_{i=1}^{s} p_i^2\right)$$ (1971)

(7) McIntosh's "Ecological distance" relative

$$\text{McIntosh's } M = \frac{n - \sqrt{\sum_{i=1}^{s} n_i^2}}{n - \sqrt{n}}$$ (1967)

(8) Theory of runs

$$\text{Cairns SCI} = \overline{DI}_1$$ (Cairns et al., 1968)

$$\overline{DI}_1 = \frac{\sum \frac{\text{no. runs}}{\text{no. specimens}}}{\text{no. times done to be statistically significant}}$$

$$\text{Keefe's TU} = 1 - \left(\frac{n}{n-1}\right)\left\{\sum_{i=1}^{K} p_i^2 - \frac{1}{n}\right\}$$ (Keefe and Bergerson, 1977)

★ See Washington (1984) for a discussion of Evenness.

List of terms

S = the number of species in either a "sample" or a "population"

K = number of taxa in either "sample" or a "population"

N = the number of individuals in a population or community

N_1 = the number of individuals in species i of a population or community

n = the number of individuals in a sample from a population

n_i = the number of individuals in a species i of a sample from a population
$p_i = n_i/n$ = the fraction of a sample of individuals belonging to species i
$\pi_i = N_i/N$ = the fraction of a population of individuals belonging to species i

Symbols of indices from the literature have been changed to conform to the above.

Figure 4. Diversity indices which are widely used in aquatic ecosystem studies, arranged in eight groups according to their derivation. After Washington (1984).

Similarity indices

Calculating the degree of similarity between samples, like the calculation of diversity, can be done in various ways depending upon how similarity is defined. Thus there are a variety of similarity coefficients available for use.

One measure of similarity is the number of species common to both, as in Sorensen's (1948) coefficient:

$$S = \frac{2c}{a+b}$$

where a = the number of taxa in community a
b = the number of taxa in community b
c = the number of taxa in common to both.

If we have two samples, A and B, each containing 2 species, x and y, then according to Sorensen's coefficient they are 100% similar. However, if sample A contains 99 individuals of x and one of y, and sample B contains one of x and 99 of y, they are clearly biologically different, and Sorensen's coefficient is a misleading figure. Clearly there are advantages in using coefficients which take into account the relative abundance of species in the sample, and this can be done in various ways. Sneath and Sokal (1973) give a comprehensive list of similarity indices. Of the large number available, relatively few have been applied in aquatic studies and it is not clear which indices are generally preferable (Washington, 1984). They have been more widely used by terrestrial ecologists. Stirn (1981) recommends Jaccard's (1908) coefficient and Sanders' (1960) index of affinity, but these recommendations do not appear to be based on a full consideration of all available indices. Jaccard's index is purely qualitative.

Analysis of survey data by similarity indices offers some important advantages, and may in fact eventually be found superior to analysis by diversity indices. Comparisons may be made simultaneously in space and time, each site or sample being compared in turn with every other site or sample. The following simple example illustrates the technique, and is based on unpublished data obtained by Chabrzyk and myself.

Samples of benthic invertebrates were taken from each of nine consecutive stretches of the river East Allen in Northern England. Sites AA, B, D, F and H are riffles (stretches of turbulent water) and sites A, C, E and G are pools. Two similarity coefficients were used, Sorensen's (1948) and Raabe's (1952). The latter is one which takes into account the relative abundance of the species in the sample. Every possible pair of sites was compared, and the results displayed as a matrix (Fig. 5). Boxes in the matrix can be shaded differentially, each pattern of shading corresponding to a different level of similarity. In some cases, visual inspection of the matrix allows any patterns to be detected. For example, in Fig. 5 it is immediately obvious that site G has a low similarity with any other site, which suggests that there is some unusual feature of this site which deserves further investigation. More detailed examination of the matrix reveals other interesting features.

Stirn (1981) recommends this approach for marine studies (under the name "trellis method") and there is no doubt that the approach is a very promising one. Its principal disadvantage is that the calculations involved are very tedious, particularly for large data sets. However, they can easily be programmed into a modern microcomputer of modest capacity.

In fact, further treatment of the data greatly assists interpretation. The process of average-linkage cluster analysis (Williams, 1971; Pielou, 1984) allows a dendrogram to be constructed which displays visually the varying degrees of affinity between the sample stations (Fig. 6). Site G is now clearly seen to be dissimilar to all other sites, with a similarity to them of less that 40%. Two major groupings are apparent now - sites AA, D, B, and F, with a similarity of at least 70%, and sites E, H, A and C with a similarity of just under 60%. All the sites in the first group are riffles, and would be expected to have a high similarity. However, another interesting result is that riffle site H has a high affinity with the pool sites E, A and C. Thus this analysis has drawn attention to two stations, out of the nine sampled, which have unusual or unexpected biological characteristics, and which perhaps require further investigation.

Fig. 7 shows the corresponding dendrogram based on the Sorensen coefficient of similarity. Here the pattern is somewhat different. Again, riffle sites AA, B, D and F form a group of high affinity, but the relationship of the "anomalous" sites G and H with the remaining sites is more complex, and perhaps less obviously anomalous. As argued above, the Sorensen coefficient is a purely qualitative one and arguably inferior to the Raabe one. The example is included to demonstrate the fact that the choice of coefficient can greatly influence the outcome of the analysis, and that sophisticated data analysis is not a substitute for sound biological reasoning, but an adjunct to it. Ideally, several coefficients should be used and the results compared before any definite conclusion is reached. Also, there are several different methods of cluster analysis (Williams, 1971; Pielou, 1984).

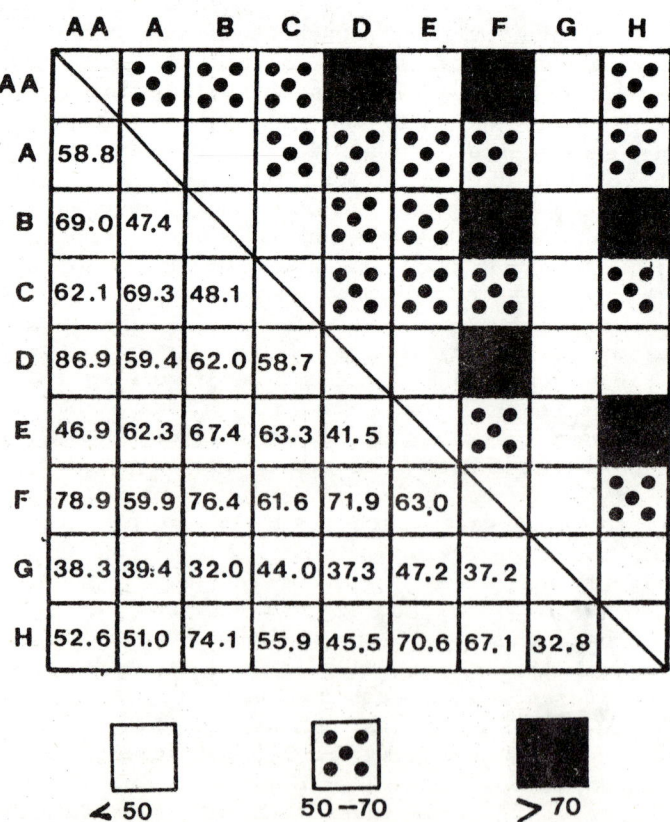

Figure 5. Matrix of Raabe similarity coefficients for 9 alternating pool and riffle stretches on the river East Allen, used to generate the dendrogram in Fig. 6.

A further level of sophistication is to estimate the degrees of probability with which the sites are similar (or different). This is necessary because most samples do not include all the species present in the habitat. Thus two sites which are in fact identical may produce a similarity coefficient of 90%, 80%, 70%, 60% or even less, depending upon the efficiency of sampling. At what level of similarity, therefore, do we conclude that two sites are significantly different? The question can be answered objectively by using, for example, Kendall's rank correlation coefficient (Kendall, 1962) in place of the more conventional indices of similarity. Examples are given by Hellawell (1978). A further example of this approach is shown in Figs. 8 and 9, which relate to data drawn from a survey of the invertebrate communities at 23 stations in a polluted river system.

The system comprises two confluent rivers, and their tributaries, which drain adjacent valleys. Stations numbered with Roman numerals are situated on one, unpolluted, river; stations numbered in Arabic numerals are situated on the other river which is polluted in certain parts. The similarity matrix (Fig. 8) immediately indicates certain stations which have a low similarity to the remaining stations. Construction of the dendrogram by average-linkage cluster analysis (Fig. 9) allows a more detailed interpretation. In interpreting the dendrogram, we are particularly interested in those sampling stations which appear in an unexpected position. It must be remembered, however, that the appearance of a station in an unexpected place does not in itself indicate that the station is polluted.

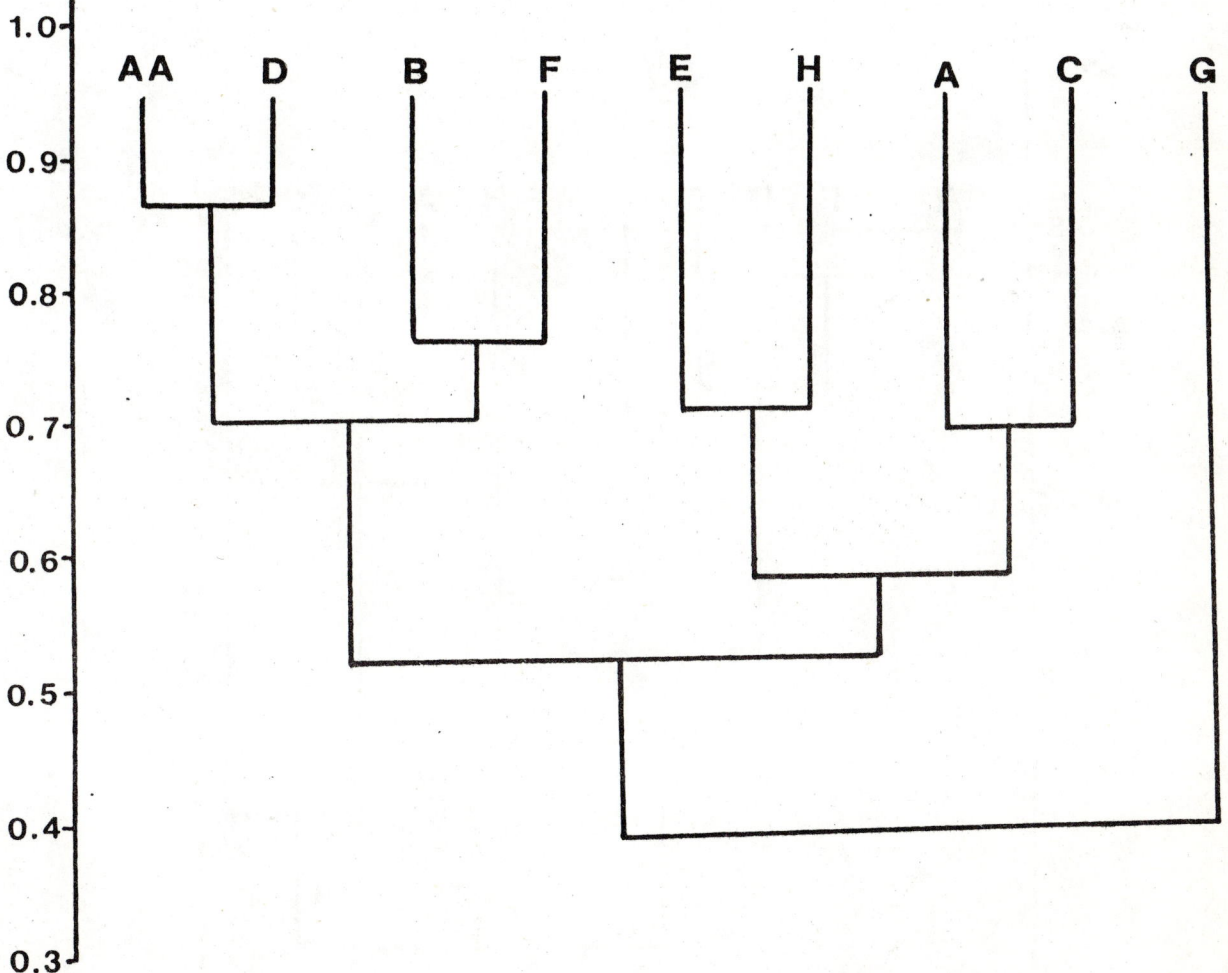

Figure 6. Average-linkage cluster dendrogram for nine sampling stations on the river East Allen (see text), based on Raabe's coefficient of similarity.

At the level of Kendall's _tau_ which indicates significant dissimilarity between stations, seven groups, labelled A-G, are apparent (Fig. 9). Groups A and B each consist of only one station, IX and 10 respectively, which are thus both dissimilar from each other and from any other station. These sites are, in fact, the headwaters of the two rivers, where the physical conditions tend to promote unstable and highly variable communities, so this grouping is not unexpected. Group D contains four stations: 9, 9a and 9+ are all physically similar and close to one another. The fourth station in the group, 6a, is a small tributary with similar physical properties to 9, 9a and 9+, so this grouping is again in accordance with expectations. Group A comprises 13 stations which include most of the downstream sites on both rivers. Note, however, that within this group two large sub-groups occur, comprising stations 1,2,3,4 and 5 on one river and stations III, V, Va, V+ and VI on the other. Thus stations on the same river tend, as expected, to be more closely similar to one another than to stations on another river. The groupings which are anomalous are therefore C, E and F, comprising stations 8, 6+ and 7, and 6 respectively. These groups have low similarity to one another and to any other groups, without any obvious physical explanation. Chemical analysis of the water at these stations revealed that it contained abnormally high levels of zinc (Abel and Green, 1981; Green, 1984). Thus the use of this technique does allow polluted sites to be identified from a large data matrix, and would appear to show great promise for application to marine benthic studies.

Its use is not easy however. The calculations require extensive computing facilities, which in turn requires that certain adjustments be made to the raw data,

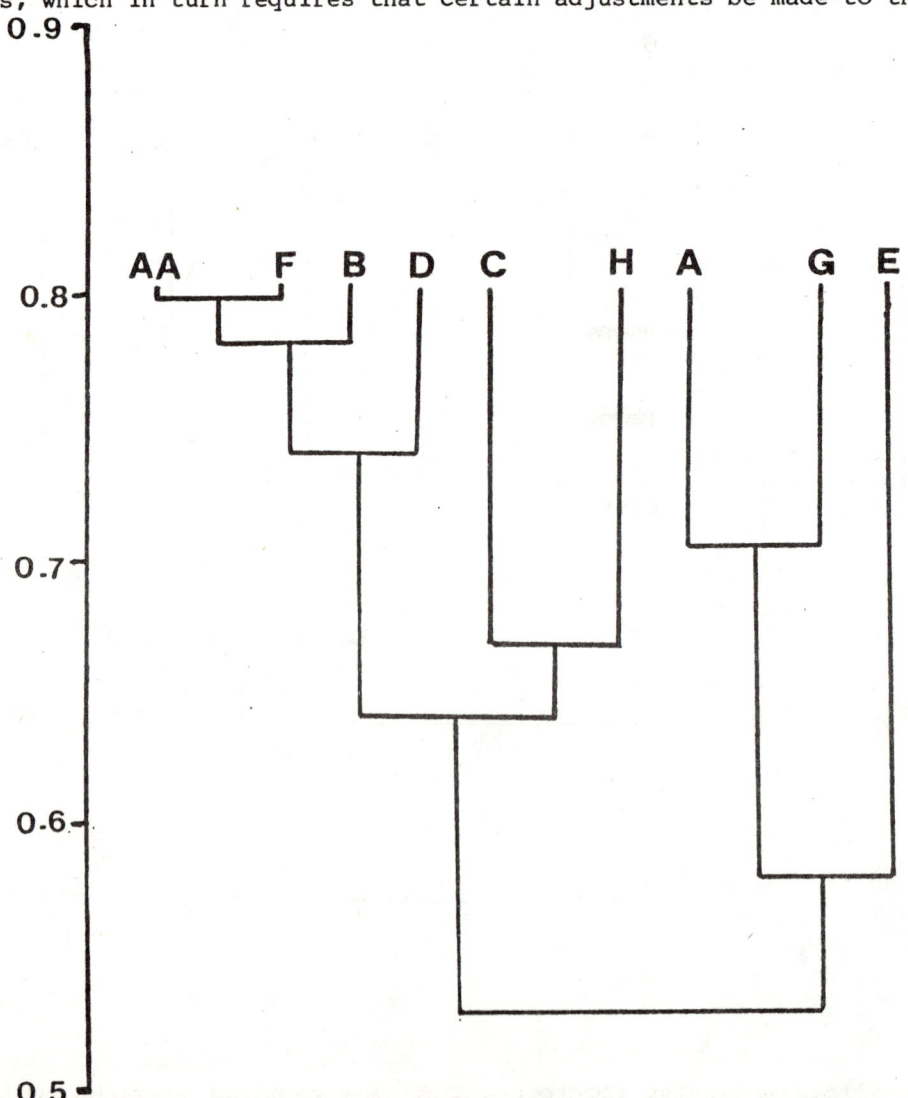

Figure 7. Average-linkage cluster dendrogram for nine sampling stations on the river East Allen (see text), based on Sorensen's similarity coefficient.

without which computer programs produce erroneous or nonsensical results. The correct use of the method requires a detailed understanding of the mathematical principles which underlie it. While it cannot therefore be recommended as a routine method in the Mediterranean at the present time, its evaluation is recommended where suitable facilities and expertise exist. By comparison, the use of simple similarity coefficients, rather that of rank correlation coefficients, is of course more straightforward.

The analysis of data by similarity indices, although relatively recent in its application to aquatic ecosystems, is nevertheless a promising approach. It does not allow a particular habitat to be diagnosed as "polluted", but it does allow a large number of spatial and temporal comparisons to be made simultaneously. It draws attention to habitats which have unusual biological characteristics by comparison with the majority of habitats samples, and thus allows further research to be directed where it will be most profitable.

Some general comments on indices

Not all of the available index methods can be discussed in detail within a single paper. Those which have been mentioned are only some of those which are appropriate,

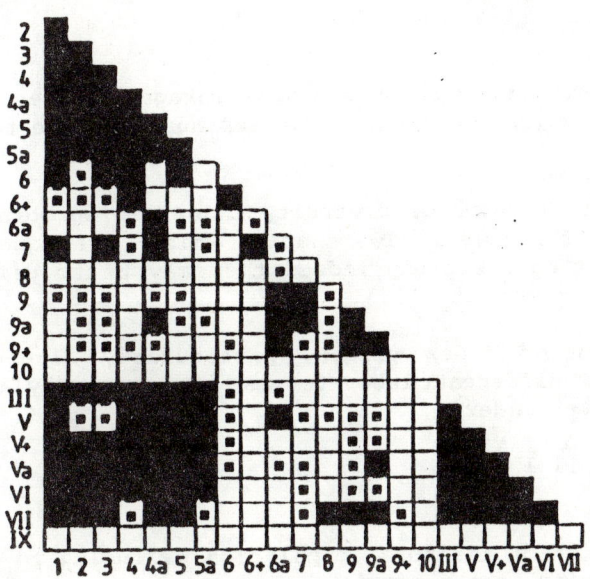

Figure 8. Matrix of Kendall's rank correlation coefficient for sites on the Allen river system (Green, 1984). See text for explanation.

and it is important that no single method be relied upon. There is, unfortunately, a tendency to invest indices with a scientific validity which they do not necessarily possess, and it is necessary to stress some of their limitations.

First, it must be realised that the process of calculating an index results in the loss of most of the information represented by the raw data. The data contained in one sample typically consists of a list of species, tabulated to show the relative abundance of each. Such a table clearly contains many separate pieces of information. To reduce this to a single number, by calculating an index value, results in just one piece of information, the index value itself. Thus to rely solely on the index value is to reject most of the information which has been collected. That the consequences of this lead to error is easily demonstrated. Consider the Trent biotic index (Table II). It can be seen that a score of VII can be obtained in at least eight different ways. A sample consisting of two different stoneflies is scored identically with one containing no stoneflies or mayflies, one caddis larva and ten other groups. Consequently the index is often insensitive to major differences in community characteristics.

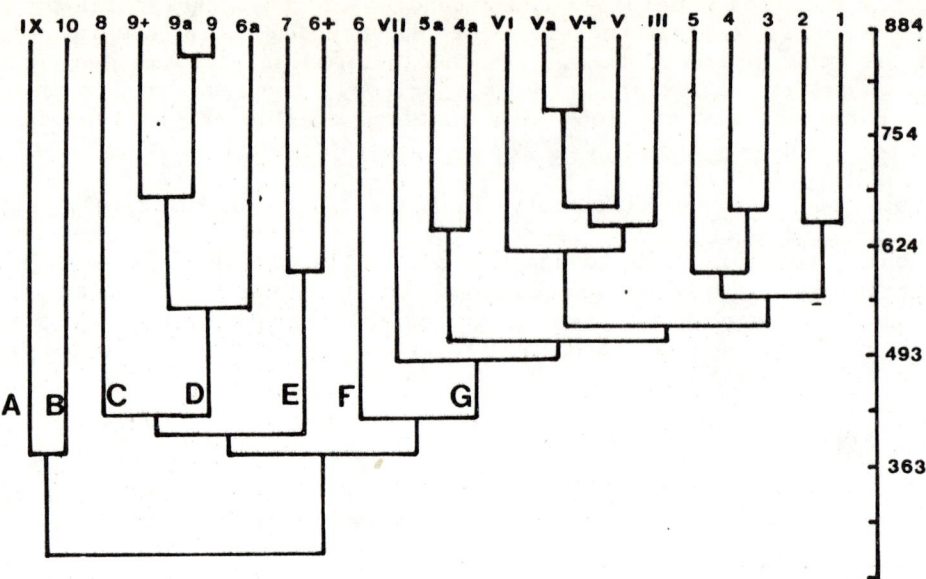

Figure 9. The dendrogram obtained by average-linkage cluster analysis of the matrix shown in Fig. 8 (Green, 1984). For explanation, see text.

Similarly, a particular value of diversity index can be obtained from any of a large number of quite different samples. Two samples which have equal diversity index values may have quite different composition, indeed they may be so different biologically that they have no species in common at all.

Secondly, comparison of index values, particularly diversity index values, from different sources (i.e. different laboratories) is extremely dangerous. Consider the widely used Shannon-Weaver index,

$$H = \sum_{i=1}^{s} P_i \log P_i$$

where Pi is the proportion of the total population belonging to the ith species and s is the total number of species in the sample.

Some authors use $\log_2$, some use natural logarithms, and some use $\log_{10}$. Provided the same base is always used, no problem arises. However, most authors do not even state which base they have used, and therefore comparisons between results from different laboratories are meaningless.

A further difficulty arises through the use of different levels of taxonomic analysis. In theory every individual specimen is identified to the species. In practice this is rare, either because adequate keys do not exist or because individual laboratories lack suitably trained personnel. The extent of simple taxonomic error is, of course, unknown, but it is clear that different laboratories indulge to a varying, and usually undisclosed, extent in taxonomic "lumping", which again renders inter-laboratory comparisons meaningless. Even within a single laboratory, during a single research programme, personnel changes can result in differing taxonomic standards being applied during a single investigation.

Thirdly, there is no absolute relation between a particular index value and a particular level of pollution, and nor is there any reason why there should be. There is no number which can be said to indicate that a particular habitat is, or is not polluted.

This does not mean that index values are useless in pollution studies. It does mean, however, that the index value is not the end-point of the data analysis, as many people seem to think. Index values themselves require biological interpretation; they are simply an aid to the analysis of the raw data, and should always be interpreted by referring back to the raw data before any conclusion is reached.

5. TOXICOLOGICAL RESEARCH AND BIOLOGICAL SURVEILLANCE

For many years, biological surveillance and toxicology existed as two quite distinct aspects of water pollution research, frequently carried out by different people in different places. This situation is mirrored in the MEDPOL programme, where the two types of research are represented by different "activities" in terms of the organisation of the plan. Further, in many organisations research into bioaccumulation of pollutants is the responsibility of a third group, frequently chemists.

In recent years, however, the toxicological and ecological approaches to pollution problems have become more closely integrated. Indeed, some recent authors (Buikema et al., 1982; Pascoe and Edwards, 1984) have included both toxicological work and ecological surveillance as essential components of a more broadly-defined activity, biological monitoring. In my view there is a strong case to be made for adopting this approach within the framework of the Mediterranean Action Plan.

The reasons for these developments are easy to understand. Ecologists who detect changes in communities subject to pollution can make few specific recommendations for improvement if they cannot identify the pollutant responsible. Since most polluted waters receive not one but several different pollutants, some toxicological information is required in order to interpret the ecological data and to make recommendations for environmental improvement. Equally, the toxicologists, who are today able to use quite subtle criteria of sublethal toxicity, require some ecological insight in order to assess the significance of their findings, particularly when they are studying novel pollutants, or are attempting to predict the effects of pollutants in novel habitats. Whereas, twenty years ago, ecological and toxicological techniques were so insensitive that it was more or less impracticable to use them in direct conjunction, today this is not the case. Two examples will show how the development of toxicological techniques now offers the opportunity directly to study pollutant effects in field populations.

Metallothioneins are a group of low-molecular-weight proteins which are produced in the tissues of both vertebrates and invertebrates exposed to heavy metals (Vallee, 1979; Webb, 1979; Olafson et al., 1979; Roesidaji, 1980; Brady, 1982). The concentrations which are effective in inducing metallothioneins are comparable to those found in polluted waters, and the measurement of metallothionein levels in wild fish populations has already been used as an indicator of heavy metal pollution in lakes (Roch et al., 1982).

Genotoxic techniques are based on the finding that in virtually all organisms, exposure to poisons causes alterations in the DNA of their cells, e.g. increases in the frequency of DNA strand breaks. The technique has been used to demonstrate sublethal toxic effects in marine organisms (Zahn et al., 1981, 1982, 1983; Hoehn-Bentz et al., 1983; Kurelec et al., 1983) and is extremely sensitive. Indeed it is so sensitive that many constituents of normal foods have been found to have an effect, and a major practical difficulty may be that many laboratories in the Mediterranean are not sufficiently clean to allow reliable results to be obtained. Problems related to contamination by airborne dust and pollutants, and tobacco smoke, may require the establishment of specialised clean laboratories.

Such techniques, although perhaps unfamiliar to many pollution biologists, are not in fact intrinsically difficult and are routinely used in well-equipped biological laboratories all over the world. They have already been recognised as of potential value to the MED POL programme (FAO/UNEP, 1985). Their existence offers the possibility that pollutant effects in marine populations can perhaps be detected most easily by such biochemical means, rather than by purely ecological techniques. Certainly there should be much closer collaboration between ecologists and toxicologists, and clearly toxicological and ecological investigations can no longer be considered as separate activities. To the extent that artificial interdisciplinary boundaries and organisational frameworks inhibit the application of appropriate combinations of investigative techniques, they must be discouraged.

6. CONCLUSIONS

Although there have been many successful examples of the application of biological surveillance techniques in the study of pollutant effects in marine systems, there are several areas in which further research is required.

The concept of indicator organisms is in principle equally valid in marine and freshwater habitats, but further work is required on the ecology, distribution, ecophysiology, life histories and population ecology of marine species which are potentially useful as indicators.

There is evidence that marine populations and communities undergo long-term changes which are due to natural phenomena. Thus the detection of a sudden or gradual change does not necessarily indicate the presence of pollution. It is suggested that long-term studies of unpolluted areas are urgently required, in order to define the nature and extent of those changes which can be considered "normal". Only changes outside this range can be considered as being due to pollution. It follows that there is at present no possibility of determining, on the basis of a single survey or set of observations, whether a particular community is affected by pollution, except in the case of the most severe effects. Long-term studies (extending at least over several years) are essential.

Such long-term studies are also required on habitats and communities known to be polluted, preferably in comparison with parallel studies on physically-similar unpolluted sites, in order to improve knowledge of the response of communities and of individual species to different types of pollution, and the range of variability of such responses. Such information is also required to refine the concept of indicator species in the marine ecosystem, and in time may allow the development of empirically-based biotic and pollution indices.

Diversity and similarity indices are promising techniques of data analysis. A wide variety of diversity indices are in use, but the validity and biological relevance of many, including some of the most popular ones, has recently been questioned. Similarity indices have only recently been applied to pollution studies, but deserve further investigation as they allow habitats with unusual biological characteristics to be identified, provided that appropriate techniques of sampling, taxonomy and data analysis are applied. It must be stressed, however, that the calculation of index values involves the loss of considerable information. Indices, of whatever kind, must not be regarded as the end-point of the data analysis and interpretation, but simply as an aid to the interpretation of the raw data. They are not a substitute for the exercise of sound scientific judgement.

Consideration of selected aspects of sampling theory and practice indicates that apparently minor variations in sampling procedure can markedly influence the results, and inadequate sampling procedures entirely invalidate them. In the short term, more precise standardisation of technique is desirable. In the longer term there is no substitute for increased awareness of the principles of sampling theory and practice. Care should be taken, however, that standardisation does not inhibit the development and application of improved techniques where they are appropriate and where sufficient expertise exists.

Closer integration of ecological and toxicological research activities is recommended, in view of recent technical developments which offer the possibility that pollutant effects on natural populations may be more readily detected by biochemical rather than ecological techniques.

7. ACKNOWLEDGEMENTS

I am grateful to the Director and staff of the National Centre for Marine Research, Athens, for the use of library facilities.

8. REFERENCES

Abel, P.D. and D.W.J. Green, Ecological and toxicological studies on the invertebrate
1981   fauna of two rivers in the northern Pennine orefield. In Heavy metals in
       northern England: environmental and biological aspects, edited by P.J. Say
       and B.A. Whitton, Durham, University of Durham, Botany Department, pp.109-22

Alabaster, J.S., (ed) Biological monitoring of inland fisheries. London, Applied Science
1977   Publishers, 226 p.

Amjad, S. and J.S. Gray, Use of the nematode-copepod ratio as an index of organic
1983   pollution. Mar.Pollut.Bull., 14:178-81

Ankar, S. et al., Methods for studying benthic macrofauna. An intercalibration
1979   exercise between two laboratories in the Baltic Sea. Finn.Mar.Res.,
       (246):147-60

Brady, F.O., The physiological function of metallothionein. Trends Biochem.Sci., 7:
1982   143-5

Brillouin, L., Maxwell's Demon cannot operate: information and entropy, I and II.
1951   J.Appl.Phys., 22:334-43

Buikema, A.L., B.R. Niederlehner and J. Cairns, Biological monitoring. Part 4.
1982   Toxicity testing. Water Res., 16:239-62

Cairns, J., et al., The sequential comparison index - a simplified method for
1968   non-biologists to estimate relative differences in biological diversity in
       stream pollution studies. J.Wat.Pollut.Control Fed., 40:1607-13

Chandler, J.R., A biological approach to water quality management. Water
1970   Pollut.Control, 69:415-22

Clark, R.B., Environmental science and all that. Mar.Pollut.Bull., 13:335-6
1982

Dickson, K.L., and J. Cairns, (eds) Biological data in water pollution assessment:
1978   quantitative and statistical analyses. ASTM Spec.Tech.Publ.,(652):184 p.

Dybern, B.I., H. Ackefors, and R. Elmgren, Recommendations on methods for marine
1976   biological studies in the Baltic sea. Balt.Mar.Biol.Publ., (1):1-98

Elliot, J.M., Some methods for the statistical analysis of samples of benthic
1977   invertebrates. Sci.Publ.Freshwat.Biol.Assoc.Ambleside, (25):160 p.

Elliot, J.M. and C.M. Drake, A comparative study of seven grabs used for sampling
1981   benthic macro-invertebrates. Freshwat.Biol., 11:99-120

FAO/UNEP, Report of the FAO/UNEP Meeting on the toxicity and bioaccumulation of selected
1985   substances in marine organisms. Rovinj, Yugoslavia, 5-9 November 1984. FAO
       Fish.Rep., (334):22 p.

Fisher, R.A., A.S. Corbet and C.B. Williams, The relation between the number of
1943   species and the number of individuals in a random sample of an animal
       population. J.Anim.Ecol., 12:42-58

Gleanson, H.A., On the relation between species and area. Ecology, 3:158
1922

Green, D.W.J., Ecological and toxicological studies on the invertebrate fauna of
1984 metalliferous streams. Ph.D. Thesis (CNAA), Sunderland Polytechnic, England, 220 p.

Hartley, J.P., Biological monitoring of the seabed in the Forties oilfield. In
1979 Proceedings of a conference on ecological damage assessment. November 1979, Arlington, Virginia, Society of Petrol Ind.Biol. pp 215-53

_____ Methods for monitoring offshore macrobenthos. Mar.Pollut.Bull., 13:150-4
1982

Hellawell, J.M., Biological surveillance and water quality monitoring. In Biological
1977 monitoring of inland fisheries, edited by J.S. Alabaster. London, Applied Science Publishers, pp.69-88

_____ Biological surveillance of rivers: biological monitoring handbook.
1978 Medmenham, U.K., Water Research Centre, 332 p.

Hoehn-Bentz J, B. Kurelec and R.K. Zahn, Fast ephemeral DNA damage upon BaP
1983 injection. Sci.Total Environ., 32:13-27

Holme, N.A., and A.D. McIntyre, (eds) Methods for the study of marine benthos.
1984 IBP Handb., (16):350 p. 2nd ed.

Herricks, E.E. and J. Cairns, Biological monitoring Part 3. Receiving system
1982 methodology based on community structure. Water Res., 16:141-54

Hurlbert, S.H., The concept of species diversity: a critique and alternative
1971 parameters. Ecology, 52:577-86

Jaccard, P., Nouvelles recherches sur la distribution florale. Bull.Soc.Vaud.Sci.Nat.
1908 44:223-69

James, A. and L. Evison, (eds), Biological indicators of water quality. Chichester,
1979 U.K., John Wiley, 605 p.

Kaesler, R.L., E.E. Herricks and J.S. Crossman, Use of indices of diversity in stream
1978 surveys. In Biological data in water pollution assessment: quantitative and statistical analysis, edited by K.L. Dickson and J. Cairns. ASTM Spec.Tech.Publ., (652):92-112

Keefe, T.J., and E.R. Bergerson, A simple diversity index based on the theory of runs.
1977 Water Res., 11:689-91

Kendall, M.G., Rank correlation methods. London, Griffin and Co., 199 p.
1962

Kolkwitz, R. and M. Marsson, Okologie der Tierischen Saprobien. Int.Rev.Gesamt.
1909 Hydrobiol., 2:125-52

Kothe, P., Der "Artenfehlbetrag", ein einfaches Gütekriterium und seine Anwendung bei
1962 biologischen Vorfluteruntersuchungen. Dtsch.Gewässerkund.Mitt., 6:60-5

Kurelec, B., et al., Sea water chlorination: creation of mutagenic by-products. Rapp.
1983 P.-V. Réun.CIESM, 28(7):135-6

Lewis, J.R., Problems and approaches to baseline studies in coastal communities. In
1972 Marine pollution and sea life, edited by M. Ruivo. West Byfleet, Surrey, U.K., Fishing News Books for FAO, pp.401-4

McIntosh, R.P., An index of diversity and the relation of certain concepts to
1967 diversity. Ecology, 48:392-404

Margalef, R., Information theory in ecology. Gen.Syst., 3:36-71
1958

Mason, C.F., The biology of freshwater pollution. London, Longman, 250 p.
1981

Mathews, R.A., et al., Biological monitoring. Part 2A. Receiving system funtional
1982   methods, relationships and indices. Water Res., 16:129-40

Menhinick, E.P., A comparison of some species-individuals diversity indices applied to
1964   samples of field insects. Ecology, 45:859-61

Moore, R.J., Is Acanthaster planci an r-strategist? Nature,Lond., 271: 56-7
1978

Odum, H.T., J.E. Cantlon and L.S. Kornicker, An organisational hierarchy postulate for
1960   the interpretation of species-individuals distribution, species entropy and
ecosystem evolution and the meaning of a species-variety index. Ecology,
41:395-99

Olafson, R.W., R.G. Sim and K.J. Boto, Isolation and chemical characterisation of
1979   heavy-metal-binding protein metallothionein from marine invertebrates.
Comp.Biochem.Physiol.(B Comp.Physiol.), 62:407-16

Pantle, R. and H. Buck, Die Biologische Überwachung der Gësasser und Darstellung der
1955   Ergebnisse. Gas-und Wasserfach, 96:604

Pascoe, D. and R.W. Edwards, (eds), Freshwater biological monitoring. Proceedings of a
1984   specialised conference, Cardiff, U.K., of the International Association on
Water Pollution and Control. Oxford, Pergamon Press, 167 p.

Patten, B.C., Species diversity in net plankton of Raritan Bay. J.Mar.Res., 20:57-75
1962

Perkins, E.J., The effects of marine discharges on the ecology of coastal waters. In
1979   biological indicators of water quality, edited by A. James and L. Evison.
Chichester, U.K., John Wiley, Chapter 12, pp.1-42

Pielou E.C., The Interpretation of ecological data. New York, John Wiley and Sons,
1984   263 p.

Preston, F.W., The commonness and the rarity of species. Ecology, 29:254-83
1948

Raabe, E.W., Über den "Affinitätswert" in der Planzensoziologie. Vegetatio, Haag,
1952   4:53-68

Rafaelli, D.G. and C.F. Mason, Pollution monitoring with meiofauna, using the ratio of
1981   nematodes to copepods. Mar.Pollut.Bull., 12:158-63

Reish, D.J., The use of marine invertebrates as indicators of varying degrees of
1972   pollution. In Marine pollution and sea life, edited by M. Ruivo. West
Byfleet, Surrey. Fishing News Books for FAO, pp 203-7

Roch, M., et al., Hepatic metallothionein in rainbow trout (Salmo gairdneri) as an
1982   indicator of metal pollution in the Campbell river system.
Can.J.Fish.Aquat.Sci., 39 (12):1596-601

Roesidaji, G., The significance of low molecular weight, metallothionein-like proteins
1980   in marine invertebrates: current status. Mar.Environ.Res., 4:167-79

Sanders, H.L., Benthic studies in Buzzard's Bay. 3. The structure of the soft-bottom
1960   community. Limnol.Oceanogr., 5:138-53

_____ Marine benthic diversity: a comparative study. Am.Nat., 102:243-82
1968

Shannon, C.E. and W. Weaver, The mathematical theory of communication. Urbana,
1949 Illinois, University of Illinois Press. pp 82-3, 104-7

Shiells, G.M. and K.J. Anderson, Pollution monitoring using the nematode/copepod ratio -
1985 a practical application. Mar.Pollut.Bull., 16:62-8

Simpson, E.H., Measurement of diversity. Nature,Lond., 163:688
1949

Sladecek, V., Continental systems for the assessment of river water quality. In
1979 Biological indicators of water quality, edited by A. James and L. Evison.
Chichester, U.K., John Wiley, Chapter 3, pp.1-32

Sneath, P.H.A. and R.R. Sokal, Numerical taxonomy. San Francisco, W.H. Freeman,
1973 pp. 141-5

Sörensen, T., A method of establishing groups of equal amplitude in plant sociology
1948 based on similarity of species content, and its application to analyses of
the vegetation on Danish commons. Biol.Skr., (5):1-34

Stirn, J., Manual of methods in marine environmental research. Part 8. Ecological
1981 assessment of pollutant effects. (Guidelines for the FAO(GFCM)/UNEP Joint
Co-ordinated Project on Pollution in the Mediterranean). FAO Fish.Tech.Pap.
(209): 70 p. Issued also in French

Vallee, B.L., Metallothionein: historical review and perspectives. In
1979 Metallothioneins, edited by J.H.R. Kagi and M. Nordberg. Basel, Birkauser
Verlag, p.76 et seq.

Washington, H.G., Diversity, biotic and similarity indices. A review with special
1984 reference to aquatic ecosystems. Water Res., 18:653-94

Webb, M., The Metallothioneins. In The chemistry, biochemistry and biology of
1979 cadmium, edited by M. Webb. Amsterdam, Elsevier North-Holland, pp.195-266

Whittaker, R.H., Dominance and diversity in land plant communities. Science,Wash.,
1965 147:250-60

Wilhm, J.L. and T.C. Dorris, Biological parameters for water quality criteria.
1968 BioScience, 18:477-81

Williams, C.B., Patterns in the balance of nature, and related problems in quantitative
1964 ecology. New York, Academic Press, pp.14-31, 147-92

Williams, W.T., Principles of clustering. Annu.Rev.Ecol.Syst., 2:303-26
1971

Woodiwiss, F., The Biological system of stream classification used by the Trent River
1964 Board. Chem.Ind., 11:443-7

Zahn, R.K., et al., Assessing consequences of marine pollution by hydrocarbons using
1981 sponges as model organisms. Sci.Total Environ., 20:147-69

_____ The effect of benzo (a) pyrene on sponges as model organisms in marine
1982 pollution. Chemico-Biol.Interactions, 39:205-20

_____ DNA damage by PAH and repair in a marine sponge. Sci.Total Environ., 26:
1983 137-56

ETUDE DES PEUPLEMENTS BENTHIQUES D'UN MILIEU PERTURBE:
LE PORT D'ALGER

par

A. BAKALEM et J.C. ROMANO
ISMAL, BP 90, Alger - 1er Novembre, Algérie

1. INTRODUCTION

Au cours de ces dernières années les études des modifications affectant les peuplements benthiques de fonds meubles dues à l'influence de la pollution par matières organiques ont connu un grand essor depuis les premiers travaux de Reish(1959).

En méditerranée grâce aux travaux d'un certain nombre de chercheurs (Bellan, 1967; Romano, 1973; Bellan et al., 1980) les peuplements de tels milieux perturbés sont bien connus; plus récemment Pearson et Rosenberg(1978) ont fait le point des connaissances sur les peuplements des milieux soumis aux apports importants en matières organiques. Glemarec et Hily(1981) et Hily(1983) ont également étudié ce type de milieu des côtes bretonnes.

Les complexes portuaires, milieux semi-fermés, constituent des zones privilégiées pour l'étude des modifications dues à la pollution affectant les peuplements benthiques. Dans ce sens, nous avons entrepris l'étude, depuis novembre 1981, du port d'Alger.

2. LE MILIEU

Le port d'Alger est un vaste ensemble portuaire (180 hectares), aux activités très diverses.

Cet ensemble portuaire est divisé en deux parties:

- le port de commerce : partie la plus importante outre les activités specifiques à un port de commerce, il existe d'autres activités dues à un certain nombre de petites unités industrielles (savonnerie, huilerie, centrale thermique...) installées dans cette partie du port.

- le port de pêche.

Le port d'Alger reçoit les eaux usées d'une grande partie de la ville d'Alger (2 millions d'habitants), et de toutes les industries portuaires.

L'augmentation constante des flux de pollution, liée au développement urbain de l'agglomération algéroise et aux activités croissantes de son port, entraîne au niveau du port d'Alger, milieu semi-fermé, une détérioration du milieu dont les répercussions tant sur le plan qualitatif que quantitatif se font nettement sentir sur la macrofaune des fonds sédimentaires.

3. MATERIEL ET METHODES

C'est dans le port de pêche que nous avons effectué pendant une année (novembre 81-novembre 1982) des prélèvements mensuels au niveau d'une station. La profondeur de la station prospectée est de 7.5 m.

Chaque prélèvement mensuel représente une surface prélevée de 0.5 m^2 qui correspond à 6 coups de benne "Orange-Peel". Le sédiment prélevé est de nature sablo-vaseux auquel s'ajoute une fraction gravier.

4. RESULTATS

4.1 Les Mollusques: (Fig. 1, 2, 3 et 4)

Au total 50 espèces de Mollusques, soit 32 espèces de Lamellibranches, 18 de

Gastéropodes, ont été inventoriées. Le maximum d'espèces (21) a été enregistré en automne (novembre 1981) et au printemps (mai 1982), et le minimum (10 espèces) en hiver.

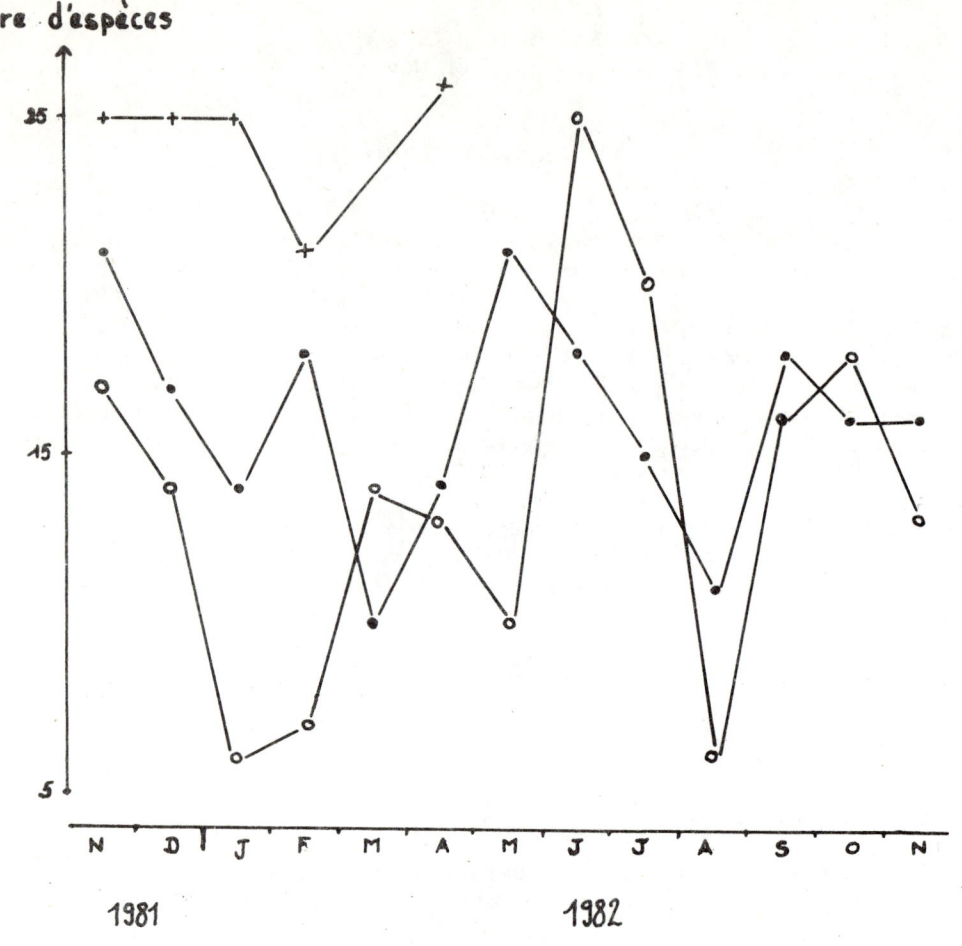

Figure 1. Evolution du nombre d'espèces des Mollusques (—●—), des Crustacés (—o—), et des Polychètes (—x—)

Les variations annuelles de la densité des Mollusques mettent en évidence des pics saisonniers:

- un pic printanier avec une densité maximale de 810 individus/m^2 en mai;

- des pics estival et automnal, mais moins importants que le précédent;

- et également une période hivernale où les Mollusques sont quantitativement peu abondants dans le milieu.

Les fortes densités enregistrées au printemps, en été et en automne sont dues à l'abondance des Bivalves, par contre en hiver les Gastéropodes sont assez bien représentés dans le milieu sur le plan quantitatif.

La prépondérance des Lamellibranches est due à un certain nombre d'espèces: Abra alba, Corbula gibba, Venerupis aureus, V. rhomboïdes et Cardium exigum ayant des fréquences, des densités et des dominances élevées. A ce groupe d'espèces principales nous pouvons ajouter des espèces telles que Dosinia lupinus et Venus verrucosa dont les fréquences sont élevées mais les dominances faibles.

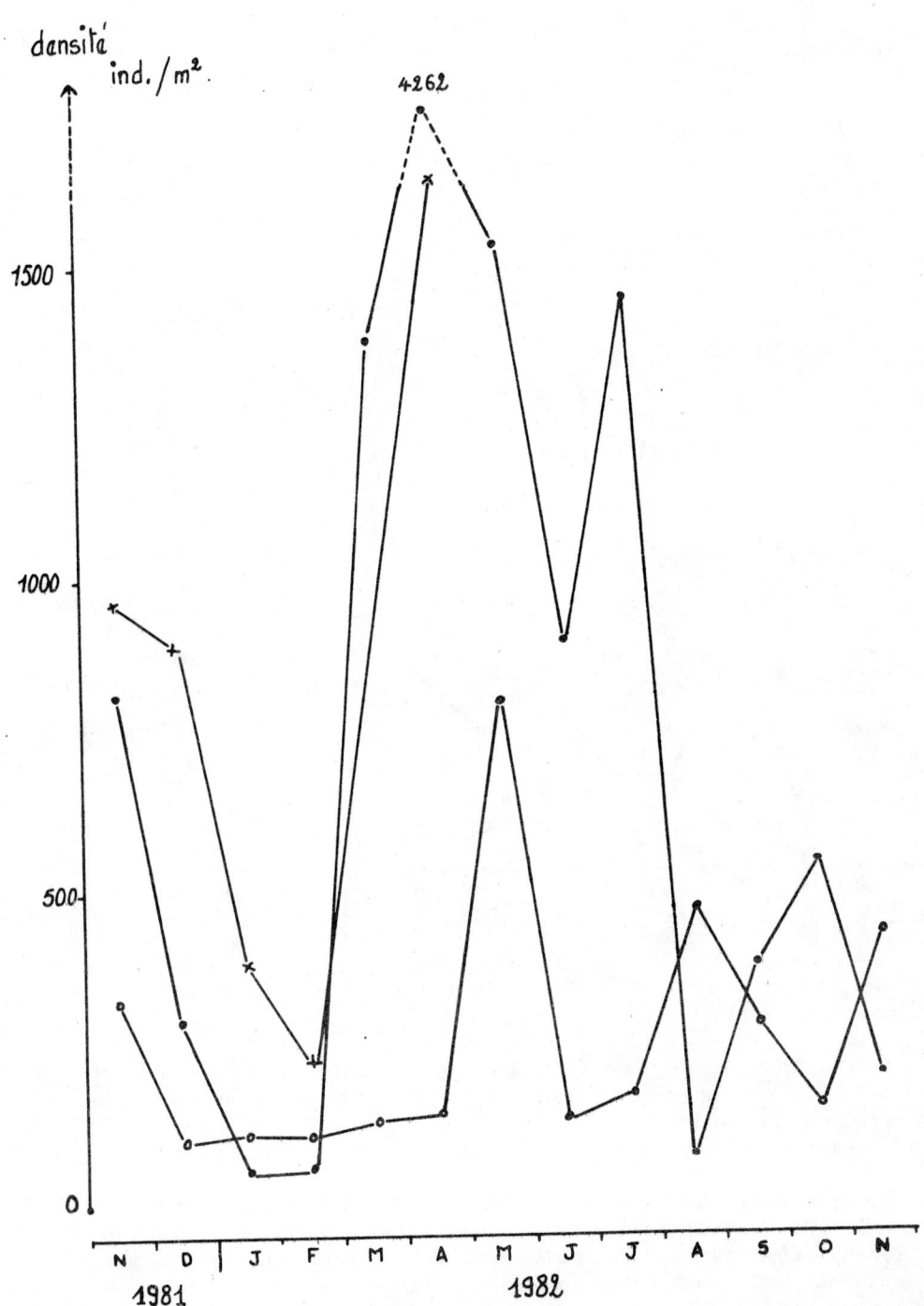

Figure 2. Evolution des densités des Mollusques (—o—), des Crustacés (—•—), et des Polychètes (—x—)

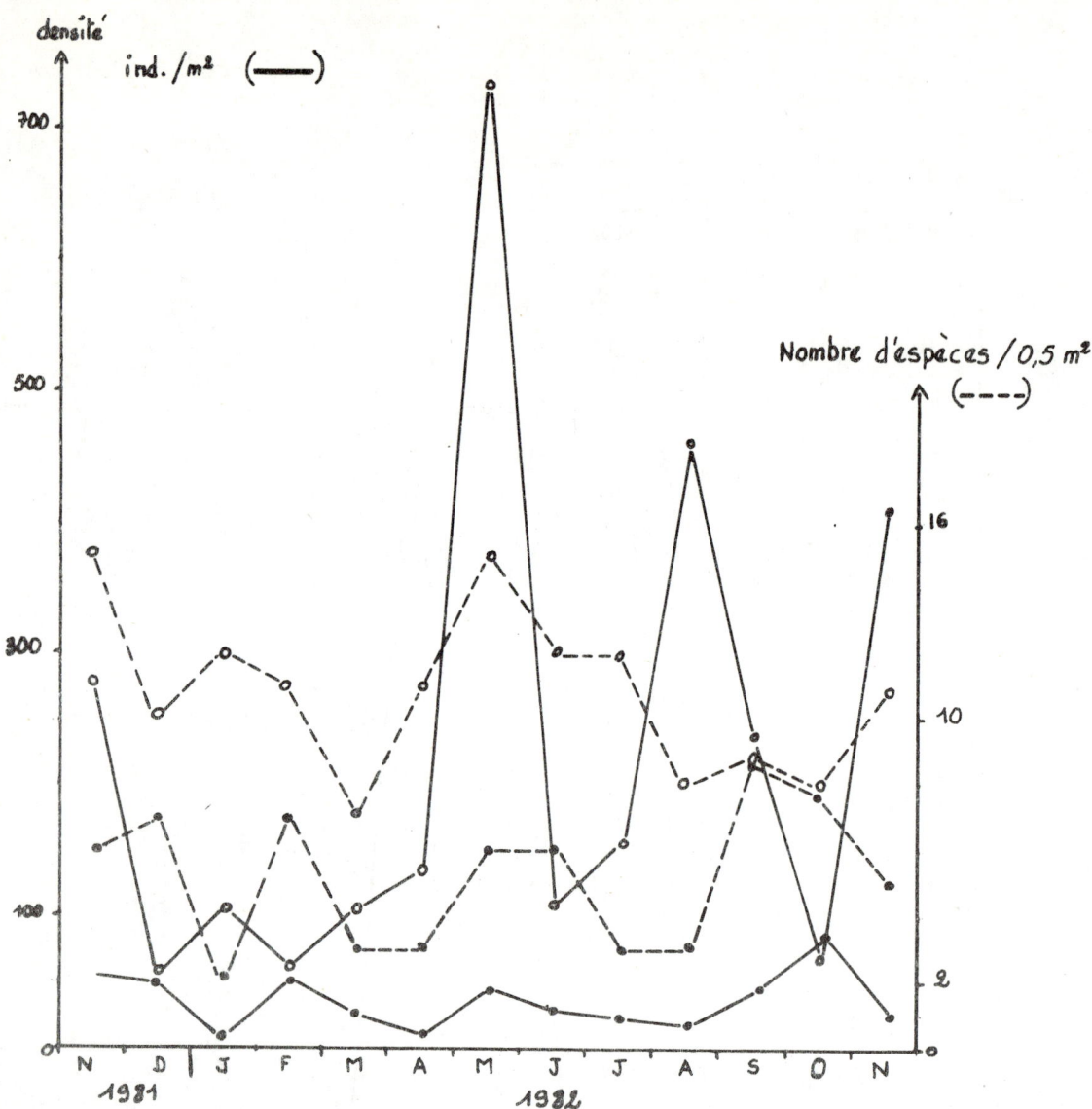

Figure 3. Evolution du nombre d'espèces et des densités des Lamellibranches (o) et des Gastéropodes (•).

Par contre, pour les Gastéropodes seule une espèce: <u>Nassa reticulata</u> est bien représentée au sein du groupe des Mollusques. Cependant, il est à noter des espèces comme <u>Murex trunculus</u> (Fréquence = 92.30%) et <u>Cerithium vulgatum</u> (Fréquence = 69.23%) présentes régulièrement dans les prélèvements mais dont les dominances sont faibles.

<u>Abra alba</u> est considérée par Picard(1965) et Bellan(1967) comme une espèce vasicole tolérante. Hily(1983) la considère comme une espèce tolérant une surcharge du milieu en matières organiques. En baie d'Alger, Bakalem(1979) signale l'abondance d'<u>Abra alba</u> au niveau des fonds sablo-vaseux et vaso-sableux, fonds riches en matières organiques.

Avec une fréquence égale à 100, elle est une espèce constante, son abondance élevée lui conférant une forte dominance. C'est ainsi l'espèce principale ou "leader" du groupe des Mollusques. L'évolution annuelle de la densité d'<u>Abra alba</u> nous permet de relever:

- Les valeurs maximales:

en novembre 1981 où la densité est de 74 individus m^{-2} soit une dominance élevée (22,29%).

au printemps (mars, avril, mai) où les densités et les dominances enregistrées dont les plus élevées du cycle annuel.

en été (juillet, août).

- Les valeurs minimales:

en hiver (décembre, janvier et février). Au début de cette saison, bien que la densité diminue, la dominance de l'espèce au sein des Mollusques reste élevée, ces deux paramètres ne chutent vraiment qu'en février (densité = 2 individus m^{-2} et dominance = 1.75%).

en septembre et octobre c'est-à-dire à la fin de l'été et au début de l'automne.

Les maxima de densité saisonniers: automnal, printanier et estival d'Abra alba correspondent aux périodes de recrutement de l'espèce et aux conditions du milieu favorables à l'espèce. Lors de l'hiver et de la période fin été - début automne, les conditions défavorables du milieu ne permettent pas à Abra alba de se maintenir et de se développer dans le milieu.

Corbula gibba est classée comme espèce à large répartition écologique par Picard (1965). Des travaux plus récents ont permis de mieux préciser les exigences écologiques de cette espèce: Bourcier et al.,(1979) la considèrent comme une espèce indicatrice de la zone subnormale, c'est-à-dire une espèce proliférant dans les milieux où il y a une quantité modérée de matières organiques.

L'évolution annuelle de Corbula gibba est identique à celle d'Abra alba. En hiver: janvier, février, Corbula gibba est mieux représentée dans le milieu qu'Abra alba qui domine le reste des autres mois; cela laisse supposer qu'Abra alba supporte moins bien les conditions défavorables du milieu que Corbula.

Corbula gibba, au sein du groupe des Mollusques, se classe, selon ses densités et dominances, en 2ème position, bien qu'en certains mois (janvier et novembre 1982), elle soit l'espèce principale du peuplement de Mollusques.

Venerupis aureus, Bellan(1967) signale l'abondance relative de cette espèce dans le port de Marseille en des stations appartenant à la zone dite polluée; alors que Picard(1965) considère Venerupis aureus comme caractéristique exclusive de la biocénose SVMC. Elle disparaît du milieu quand les conditions sont très défavorables comme c'est le cas au mois d'avril et octobre. Le cycle annuel de Venerupis aureus suit le même schéma évolutif que celui d'Abra alba et de Corbula gibba. Venerupis rhomboïdes a une évolution annuelle similaire à celle de Venerupis aureus.

Cardium exigum est considérée aussi bien par Picard(1965) que Bellan(1967) comme une espèce indicatrice de pollution; présente dans tous les prélèvements mensuels sans cependant être dominante au sein du groupe des Mollusques sauf en juin où sa dominance (22,06%) est la plus forte de son cycle et aussi du prélèvement de juin.

L'évolution annuelle de ces amellibranches: Abra alba, Corbula gibba, Venerupis aureus et V. rhomboïdes, et Cardium exigum, qui est à l'origine et induit les variations observées au cours du cycle global des Mollusques tant sur les plans des effectifs (abondance, densité) que représentatif (nombre d'espèces, dominance).

Nassa reticulata est la seule espèce de Gastéropodes très abondante. En baie d'Alger, elle se rencontre surtout sur les fonds sableux, et semble être une espèce sabulicole (Bakalem, 1979).

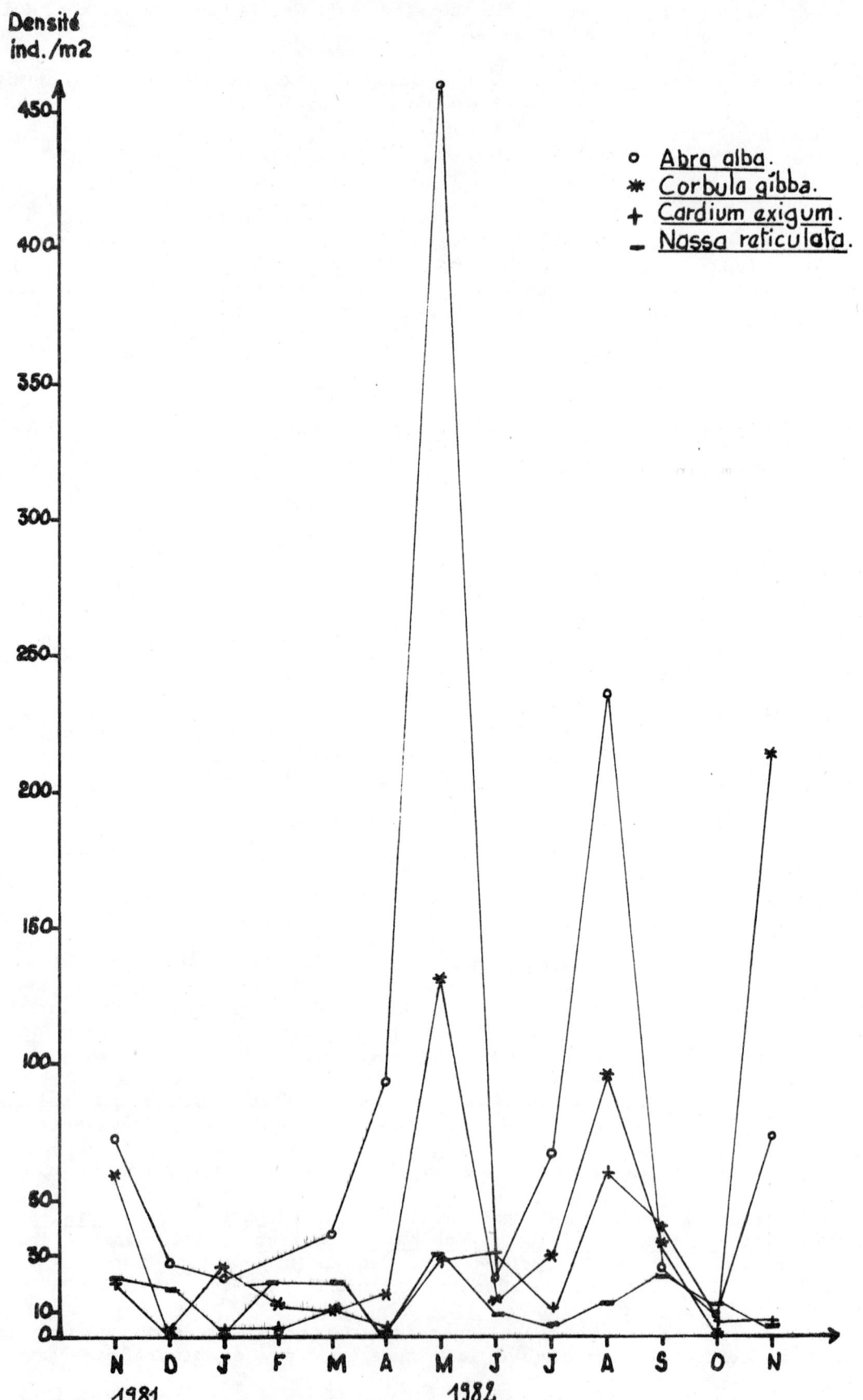

Figure 4. Evolution annuelle des densités des principales espèces de Mollusques

Nous la trouvons dans les prélèvements du port d'Alger, tout au long de l'année.

L'évolution annuelle de Nassa reticulata diffère de celle des Lamellibranches précedemment étudiés. Contrairement à ces espèces Nassa reticulata présente un maximum en hiver: pic hivernal (abstraction faite de janvier) où les effectifs sont élevés et les dominances parmi les plus fortes des Mollusques.

Nous avons également pour ce gastéropode un pic printanier (mai, juin) et un pic en septembre - octobre. Les minima enregistrées pour Nassa reticulata se situent aux mois de janvier, avril et juillet.

L'étude du cycle annuel des Mollusques d'une station du port d'Alger fait ressortir les points suivants:

- richesse qualitative du milieu en Mollusques (50 espèces) comparativement à des milieux ouverts comme la baie d'Alger: 45 espèces (Bakalem, 1981) ou la baie de Bou-Ismaïl: 14 espèces (Bakalem et Romano, 1983) ou à des milieux similaires, Bellan(1967) ne récolte que 28 espèces de Mollusques dans le port de Marseille (France), Ergen et Onen(1983) dans leur étude du port d'Urla (Turquie) n'ont recensé que 32 espèces de Mollusques.

- richesse quantitative du peuplement benthique de la station en Mollusques. Les densités enregistrées en certains mois (november 1981, mai, août, septembre et novembre 1982) sont bien supérieures aux densités maximales trouvées en baie d'Alger (235 individus m^{-2}) et baie de Bou-Ismaïl (Bakalem, 1979; Bakalem et Romano, 1983).

- richesse qualitative et quantitative du peuplement en Mollusques en période estivale où pour un tel milieu fermé les conditions écologiques sont des plus défavorables. Kocatas(1981) a montré que dans un tel milieu, en été, les stations deviennent en général azoïques.

- la dominance au sein du groupe des Mollusques des espèces comme Corbula gibba, Abra alba, tolérant une surcharge du milieu en matières organiques, ou telle que Cardium exigum, indicatrice de pollution.

4.2   Les Crustacés: (Fig. 1, 2, 5, 6, 7 et 8)

Les 59 espèces de Crustacés recensées au cours de l'année se répartissent ainsi: 30 Décapodes, 11 Amphipodes, 8 Isopodes, 4 Cirripèdes et 2 Mysidacées.

Les maxima de densité de Crustacés s'observent au printemps (4262 individus m^{-2} en avril), en été (1450 individus m^{-2} en juillet) et à un degré moindre en automne. Le maximum d'espèces se rencontre en été et en automne.

Les minima tant qualitatif que quantitatif sont enregistrés en hiver et au mois d'août.

Les Décapodes, présents toute l'année dans le milieu, sont le groupe ayant les plus d'espèces (30) dont les maxima sont situés en été et en automne; le restant de l'année correspond aux minima hivernal, printanier et du mois d'août.

Les maxima de densité de Décapodes sont enregistrés en:

- automne: densité maximale en octobre (84 individus m^{-2}).

- juin et juillet où nous observons un pic important avec des valeurs les plus fortes du cycle annuel des Décapodes: 108 et 150 individus m^{-2}.

Les minima de densité se trouvent:

- hiver; janvier: 4 individus m^{-2} ; février: 12 individus m^{-2}.

- au printemps.

- en période estivale: août (24 individus m^{-2}) et septembre (30 ind.m^{-2}).

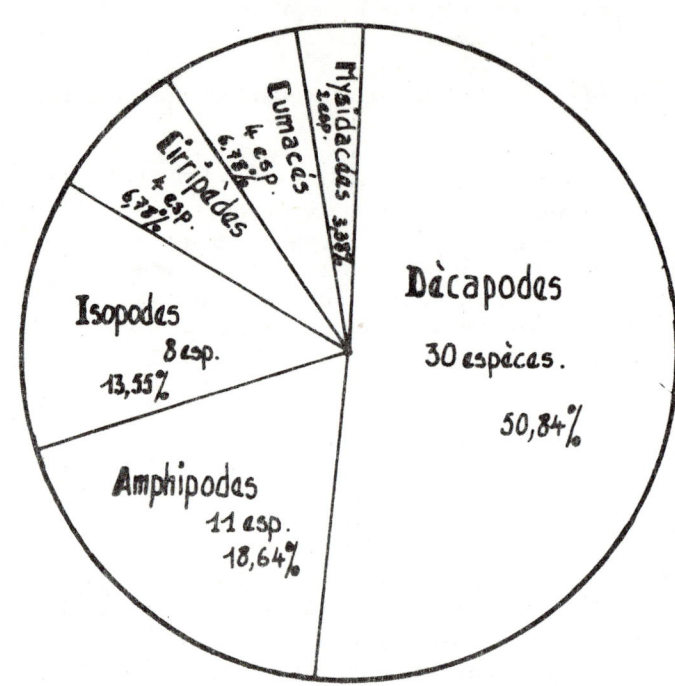

Figure 5. Les différents groupes de Crustacés

Les Amphipodes avec leurs 11 espèces viennent en seconde position, en nombre d'espèces, après les Décapodes. L'aspect de la courbe des variations du nombre d'espèces d'Amphipodes au cours de l'année est identique à celle des Décapodes. Cependant, pour les Amphipodes, en hiver (janvier, février) et en août, il n'y a aucune espèce dans le milieu; le maximum d'espèces (5) est noté en juin et octobre.

Sur le plan quantitatif, il apparaît très nettement deux pics:

- un pic printanier, le plus important débutant en mars et s'achevant en juin. Le maximum de densité de ce pic est 964 individus m^{-2} : cette densité élevée confère au groupe des Amphipodes la dominance la plus forte (65,92%) du peuplement des Crustacés de mois de mai.

- un pic automnal où le maximum d'individus se trouve dans les prélèvements d'octobre (82 individus m^{-2}). Les minima de densités comme les minima du nombre d'espèces sont enregistrés en hiver (0 à 4 individus m^{-2}) et au mois d'août (0 individus m^{-2}).

Les Isopodes, présents toute l'année sont représentés dans le peuplement par 8 espèces. Comme pour les autres groupes, nous distinguons dans les fluctuations de l'abondance des Isopodes des maxima et des minima:

- un maximum fin hiver - début printemps correspondant aux mois de mars, avril et mai. En ces mois, les densités sont très élevées: avril 4158 individus m^{-2}, mars 1306 individus m^{-2} ; de ce fait, la dominance des Isopodes est très nette (dominance = 94,21 et 97,54%).

- un maximum fin printemps - début été (juin-juillet) moins important que le précédent. La densité la plus forte n'est que de 1252 individus m^{-2}.

- un pic automnal avec des densités faibles comparativement à celles des maxima précédents.

- un minimum hivernal (janvier et février) et estival (août) avec des densités faibles, oscillant entre 40 et 48 individus m^{-2}, mais conférant toujours aux Isopodes des dominances élevées (comprises entre 63,49 et 85,75%).

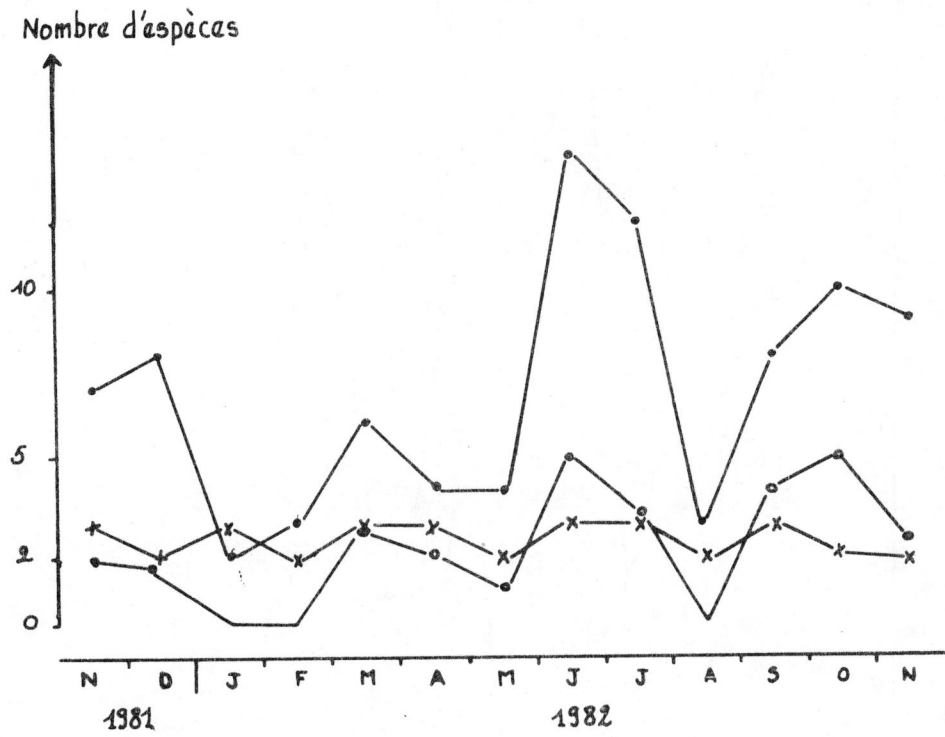

Figure 6. Evolution du nombre d'espèces d'Isopodes (—x—), d'Amphipodes (—o—) et des Décapodes (—•—)

Les trois groupes: Décapodes, Isopodes et Amphipodes sont les principales composantes du peuplement des Crustacés. Les Isopodes, qui sont moins bien représentés qualitativement que les Décapodes et les Amphipodes, sont quantitativement plus importants. Les variations annuelles quantitatives des Crustacés sont surtout la résultante des fluctuations saisonnières des Isopodes, principalement Apseudes africanus et à un degré moindre Apseudes latreilli mediterraneus.

La fréquence d'Apseudes africanus est maximale (F = 100). C'est l'espèce principale ou "leader" du peuplement carcinologique de la station étudiée. Le cycle annuel d'Apseudes africanus sur le plan de l'abondance ou densité, de la dominance est identique à celui des Isopodes. La densité (4 individus m^{-2}) et la dominance (7,14%), les plus faibles ont été enregistrées en janvier. Les fluctuations de la densité d'Apseudes africanus au cours de l'année sont généralement élevées, comme par exemple de mars à avril elle passe de 1300 à 4132 individus m^{-2}. Contrairement à Apseudes africanus, A. latreilli mediterraneus est peu abondant dans les prélèvements: densité maximale 74 individus m^{-2}, d'où ses faibles dominances, sauf en janvier où avec une densité seulement de 42 m^{-2} il est l'espèce principale du peuplement avec une dominance de 75%. En certains mois (novembre 1981, décembre et octobre 1982) Apseudes latreilli mediterraneus est inexistant dans les prélèvements.

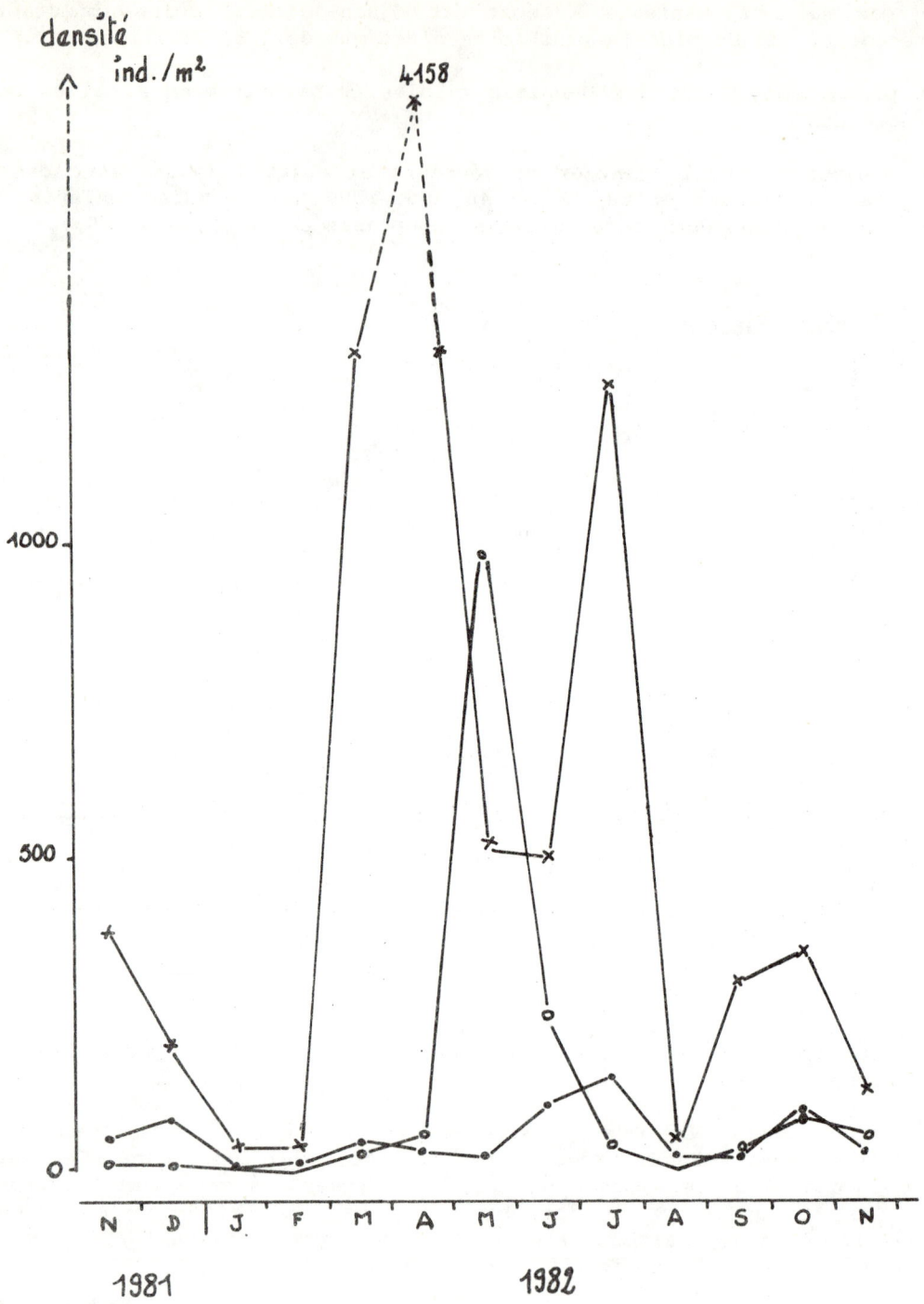

Figure 7. Evolution des densités des Isopodes (—x—), des Amphipodes (—o—), et des Décapodes (—•—)

Les principales espèces d'Amphipodes sont: Corophium acutum et C. acherusicum, et Phtisica marina; cette dernière très abondante et fréquente dans les prélèvements est responsable des fluctuations observées au niveau du cycle annuel des Amphipodes. Le pic printanier des Amphipodes est la résultante d'un "boum" printanier de Phtisica marina qui est alors une des espèces principales des Crustacés (densité = 964 individus m^{-2} et dominance = 65,92%). Phtisica marina comme toute autre espèce d'Amphipodes est totalement absente du milieu en hiver (décembre, janvier, février) et en août, elle est abondante dans le milieu en automne comme l'est également Corophium acutum en octobre.

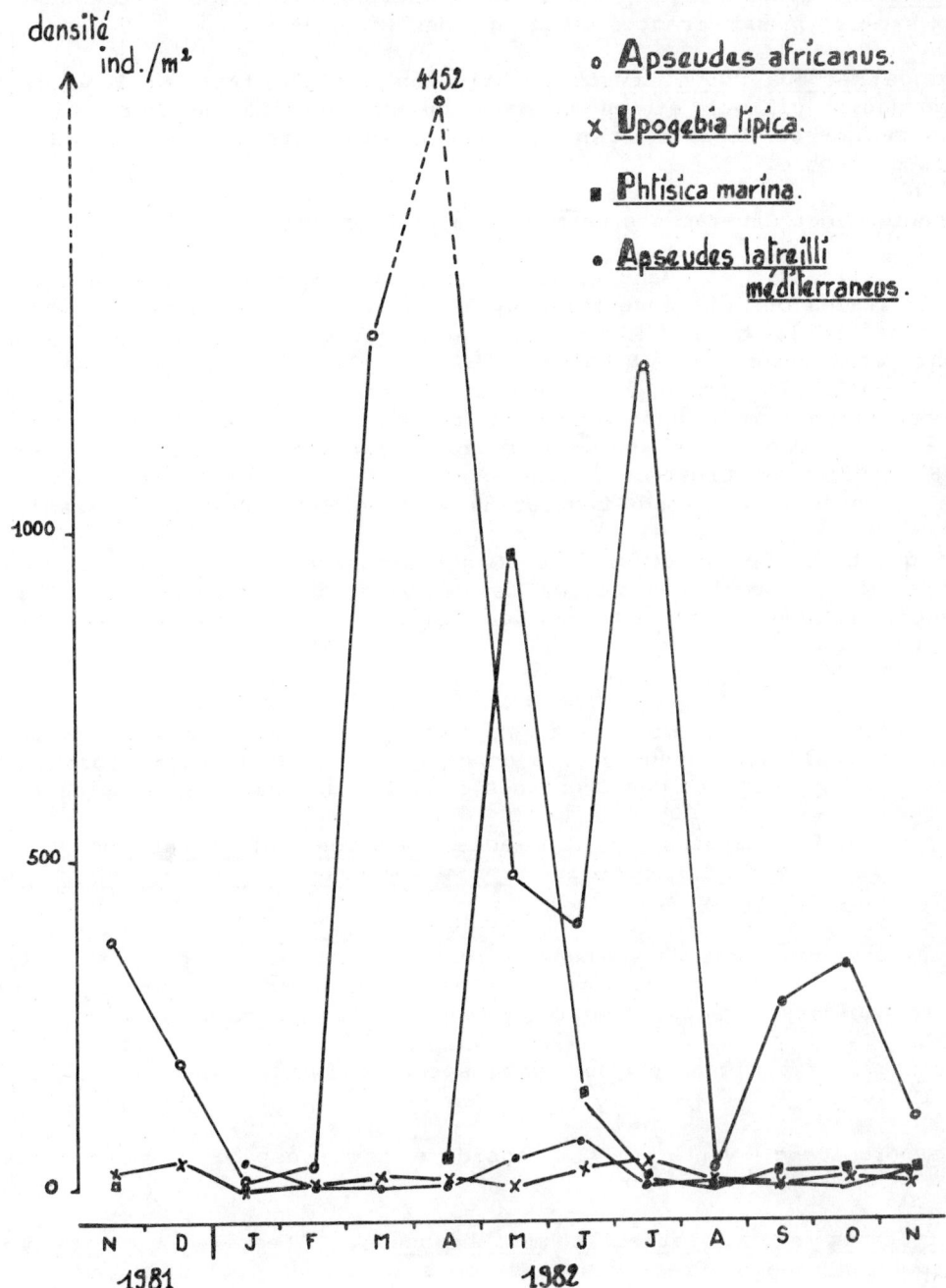

Figure 8. Evolution des densités des principales espèces de Crustacés

Upogebia tipica est l'espèce principale des Décapodes; cependant il existe tout un cortège d'espèces accompagnatrices d'importance secondaire: Pagurus cuanensis, Macropipus arcuatus, Sirpus zariquievi, Galathea bolivari, Galathea strigosa et Pagurus callidus, qui, en certains mois, renforce l'importance du groupe des Décapodes au sein des Crustacés.

Upogebia tipica, bien que ses effectifs restent assez faibles, voit ses dominances augmenter jusqu'à atteindre un maximum en hiver (décembre, février) et en août. Les plus faibles densités de cette espèce se rencontrent au printemps, et plus particulièrement en été et automne.

Medaeus couchi est assez fréquent dans les prélèvements; il constitue avec Upogebia tipica les deux expèces prépondérantes du groupe des Décapodes.

L'abondance de Medaeus couchi reste relativement faible tout au long de l'année, excepté en été (juin-juillet) et en octobre où sa densité augmente ainsi que sa dominance. Les maxima de densité et de dominance enregistrés sont 30 individus m^{-2} et 5,57% au mois d'octobre.

L'étude dynamique des Crustacés a permis de mettre en évidence:

- la richesse qualitative et quantitative de ce peuplement portuaire en Crustacés, 59 espèces de Crustacés ont été inventoriées, alors qu'en milieu ouvert comme les fonds de sables fins de la baie d'Alger, les plus riches en Crustacés, seulement 35 espèces ont été recensées (Bakalem, 1979). Cette richesse qualitative est comparable à celle des fonds de graviers envasés de la baie de Bou-Ismaïl (62 espèces) considérée comme une zone non polluée (Bakalem et Romano, 1983) et également à celle des fonds du port d'Urla (Turquie) où Ergen et Onen (1983) récoltent 61 espèces de Crustacés. Par contre, Bellan (1967) signale la pauvreté qualitative (7 espèces) des fonds meubles du port de Marseille, en Crustacés.

Sur le plan quantitatif, les densités de Crustacés enregistrées dans le port d'Alger sont très élevées par rapport aux maxima de densités de Crustacés trouvées en baie d'Alger (50 individus m^{-2}) et en baie de Bou-Ismaïl (189 individus m^{-2}), (Bakalem et Romano, 1983).

- un développement maximal tant en nombre d'espèces qu'en individus des Crustacés de la fin de l'hiver (février) au début de l'été (juillet); alors qu'en août, la densité et le nombre d'espèces des Crustacés sont à leurs valeurs les plus faibles.

- Les principales espèces de Crustacés: Apseudes africanus et A. Latrielli mediterraneus, Phtisica marina, Upogebia tipica, Medaeus couchi, Pagurus cuanensis, Corophium acutum et C. acherusicum.

La signification écologique de chacune de ces espèces est inconnue ou peu connue. Il semblerait que certaines de ces principales espèces soient caractéristiques des milieux riches en matières organiques ou légèrement pollués, notamment:

Corophium acutum, déjà signalée comme une espèce dominante dans les milieux pollués (Stirn et al., 1975).

Corophium acherusicum: Swartz (1972) considère cette espèce comme tolérante à la pollution.

Apseudes africanus et A. latreilli mediterraneus. L'espèce Apseudes latreilli, trouvée abondamment en rade de Brest dans les zones légèrement polluées par Hily (1983), est considérée par cet auteur comme une espèce tolérante à la pollution.

Upogebia tipica: cette espèce semble avoir beaucoup d'affinité pour les milieux riches en matières organiques.

4.3 Les Polychètes: (Fig. 1, 2, et 9)

Les résultats concernant les Polychètes exposés ici sont fragmentaires, ils ne portent que sur les Polychètes récoltés de novembre 1981 à février 1982 et au mois d'avril 1982.

Un total de 52 espèces a été recensé au cours de ces mois. Le nombre d'espèces récolté mensuellement est à peu près constant (25 espèces) sauf en février où il y a une légère diminution (21 espèces).

Les espèces, quantitativement importantes, appartiennent à la famille des Capitellides et Cirratulidés.

Les fortes densités de Polychètes sont enregistrées en automne et en avril. En janvier et février, période hivernale, les densités sont plus faibles. Cela laisse supposer l'existance d'un pic automnal et printanier et un minimum hivernal.

Les principales espèces de Polychètes sont, par ordre d'importance: Tharyx marioni, Notomastus latericeus, Nereïs caudata, Audouinia tentaculata, Prionospio malmgreni, Hétéromastus filiformis, Polydora antennata et Lumbriconereis latreilli.

Tharyx marioni est l'espèce la plus abondante, sauf en avril, du peuplement de Polychètes, cela fait d'elle l'espèce "leader" des Polychètes.

Picard (1965) la considère comme une espèce sans signification écologique.

Notomastus latericeus est classée par Picard (1965) comme une espèce à large répartition écologique, cela explique sa présence dans un milieu tel que le port d'Alger.

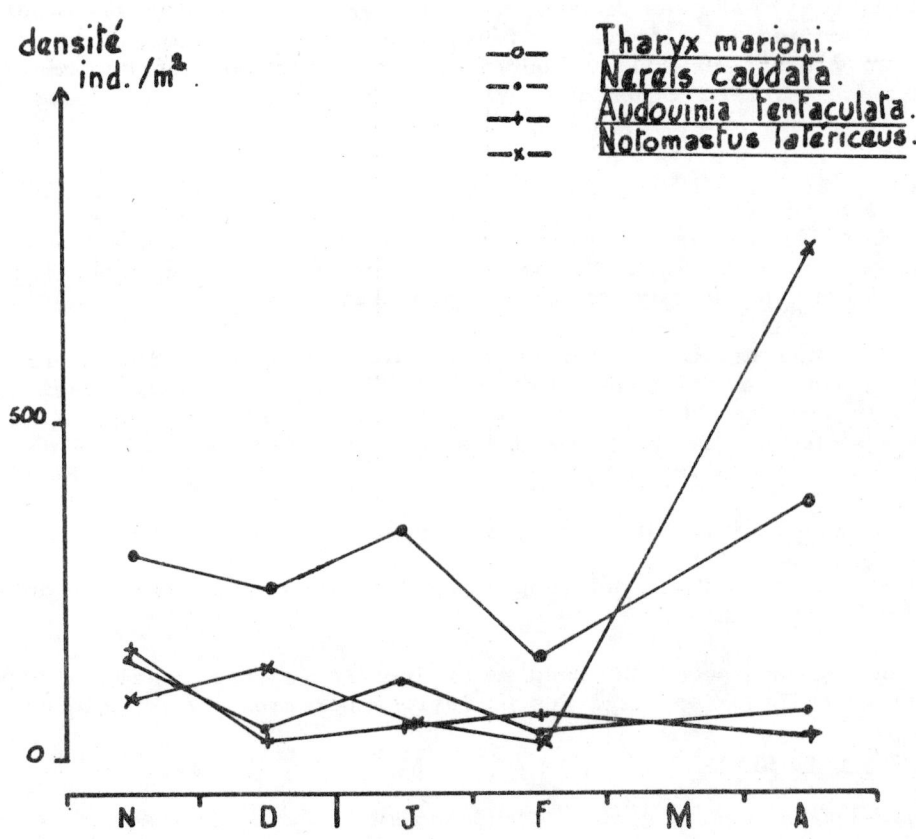

Figure 9. Evolution des densités des principales espèces de Polychètes

Elle présente des densités élevées en novembre et décembre, et une densité maximale (760 individus m^{-2}) en avril. Les valeurs minimales de densité se rencontrent en hiver (janvier et février).

Nereis caudata est considérée par Bellan (1967) comme une espèce indicatrice de pollution caractéristique de la zone polluée, telle qu'il la définie. Cognetti (1974), considère Nereis caudata comme une espèce caractéristique des milieux portuaires polluées. Hily (1983) la classe comme une espèce tolérante ou indifférente à la pollution appartenant à ce qu'il définit comme zone de déséquilibre dans les milieux perturbés.

Les conditions du milieu semblent être favorables au développement de cette espèce, notamment en novembre et janvier où ses densités sont élevées (respectivement 150 et 122 individus m^{-2}).

Audouinia tentaculata est classée par Bellan (1967) comme espèce indicatrice de pollution appartenant à la zone polluée. C'est également une espèce caractéristique des milieux portuaires pollués (Cognetti, 1974).

Hily (1983) considère Andouinia comme une espèce à comportement opportuniste qui appartient à la zone polluée des milieux perturbés.

Cette espèce est abondante dans le milieu en novembre (162 individus m^{-2}) les autres mois son abondance est moindre.

Prionospio malmgreni est une espèce relativement importante du peuplement des Polychètes, en particulier en janvier et février où elle est abondante. Hily (1983) signale son abondance dans les milieux perturbés et il la considère comme une espèce tolérant un excès de matière organique, il en est de même pour l'espèce Polydora antennata.

Les espèces Lumbriconereis latreilli et Heteromastus filiformis sont quantitativement bien représentées dans le peuplement de Polychètes du port d'Alger. Dans le port de Marseille Bellan (1967) signale les fortes abondances de ces espèces dans la zone subnormale. Stora et Romano (1980) considèrent Lumbriconereis latreilli comme une espèce indicatrice de richesse en matière organique.

Eteone picta bien que faiblement représentée sur le plan quantitatif sauf en avril (20 individus m^{-2}) dans le milieu étudié, semble cependant trouver dans ce milieu, riche en matières organiques, des conditions favorables à son développement; Stirn et al., (1975) aussi signalent la présence d'Eteone picta dans les sédiments riches en matières organiques de la baie de Tunis.

Il ressort de cette étude préliminaire des Polychètes:

- c'est en automne et au printemps que les Polychètes ont un développement quantitatif important.

- les principales espèces du peuplement des Polychètes sont des espèces bien adaptées aux milieux soumis à des pollutions par matières organiques.

4.4 Autres groupes zoologiques:

En dehors des espèces des 3 grands groupes zoologiques: Mollusques, Crustacés et Polychètes, un petit nombre d'espèces appartenant à d'autres groupes a été recensé; comme espèce la plus importante nous avons Cerianthus membranaceus (Anthozoaires).

Cerianthus membranaceus, présente dans le milieu toute l'année sauf en juillet et août, a une abondance relativement élevée: notamment en janvier, mai (densité maximale = 116 individus m^{-2}), juin et septembre. Cerianthus semble être une espèce indicatrice de richesse en matière organique car déjà Bellan (1967) signale sa présence dans les zones polluées de la région marseillaise.

Comme autre groupe zoologique nous avons celui des Echinodermes représenté en novembre 1981, juillet et octobre 1982 par quelques individus de Psammechinus microtuberculatus.

Quelques individus des groupes Nemertes et Ascidies ont été récoltés mais n'ont pas été déterminés.

5. DISCUSSION

Cette étude dynamique du peuplement benthique d'une station du port d'Alger nous apporte un certain nombre de précisions en ce qui concerne la macrofaune d'un milieu semi-fermé perturbé.

Bellan (1967) définit une zonation des peuplements des fonds meubles soumis à l'influence de la pollution, il met ainsi en évidence 3 zones:

- la zone I ou zone de pollution maximale, dépourvue de toute vie animale macroscopique.

- la zone II ou zone polluée caractérisée par des espèces indicatrices de pollution en particulier Scolelepis fuliginosa et Capitella capitata. A ces espèces "leaders" s'adjoignent certaines indicatrices de pollution comme Nereis caudata et Audouinia tentaculata. Au sein de cette zone dans le port de Marseille il défine une zone III à faciès à Tapes aureus".

- la zone III ou zone subnormale caractérisée par des espèces à large répartition écologique: Lumbriconereis latreilli, Hyalinoecia bilineata, Heteromastus filiformis et Thyasira flexuosa.

- une zone de transition qui assure le passage avec les peuplements considérés comme non perturbés.

Lors de son étude de la macrofaune de l'ensemble portuaire marseillais Bellan (1967) a mis en évidence l'existence de ces 3 types de zone dans le port de Marseille.

Si nous comparons nos résultats à ceux de Bellan (1967) sur le port de Marseille, nous notons une plus grande richesse qualitative pour le port d'Alger. Il est également à signaler que le peuplement étudié présente des pics printanier, estival et automnal; les densités maximales sont enregistrées au printemps.

En nous basant sur les caractéristiques écologiques des principales espèces du peuplement il est possible de regrouper les espèces ayant des affinités écologiques voisines en 3 groupes:

- 1er groupe: les espèces tolérant la pollution: Abra alba, atteignant de grandes densités, et Corophium acherusicum.

- 2ème groupe: les espèces tolérantes à un excès de matière organique dans le milieu, sont les plus abondantes au sein du peuplement. Ce sont essentiellement des Polychètes: Prionospio malmgreni, Polydora antennata, Eteone picta, Lumbriconereis latreilli, Heteromastus filiformis, auxquelles s'ajoutent 2 Bivalves: Corbula gibba et Venerupis aureus.

- 3ème groupe: espèces tolérantes à une pollution par matières organiques. Ce sont des espèces appartenant aux differents groupes zoologiques: Nereis caudata et Andouinia tentaculata (Polychètes), Cardium exigum (Mollusques), Corophium acutum (Crustacés) et Cerianthus membranaceus (Anthozoaires); elles sont considérées comme des espèces indicatrices de pollution (Bellan 1967; Stirn et al., 1975).

L'essentiel des espèces du 2ème groupe: Lumbriconereis latreilli, Heteromastus filiformis, Corbula gibba et Venerupis aureus sont considérées par Bellan (1967) comme caractéristiques de la zone subnormale. Tandis que celles du 3ème groupe caractérisent la zone polluée.

La station du port d'Alger que nous avons étudiée semble être une station de transition entre la zone polluée et la zone subnormale ou appartenir à une "zone de mélange II/III".

6. BIBLIOGRAPHIE

Bakalem, A., Contribution à l'étude des peuplements benthiques de la baie d'Alger.
1979 Thèse 3ème Cycle. Brest, UBO, 241 p., (mimeo)

_____, Le peuplement des sables fins à Ophiura texturata-Donax semistriatus
1981 de la baie d'Alger: évolution dans le temps. Rapp.P.-V.Réun.CIESM, 27(2):131-2

Bakalem, A. et J.C. Romano, Pollution et peuplements benthiques dans la région
1983 algéroise (baie d'Alger et baie de Bou-Ismaïl). Journ.Etud.Pollut.CIESM, 6(1982):687-97

Bellan, G., Pollution et peuplements benthiques sur substrat meuble dans la région de
1967 Marseille. Deuxième partie. Rev.Int.Océanogr.Méd., 8:51-95

Bellan, G., D. Bellan-Santini et J. Picard, Mise en évidence de modèles éco-biologiques
1980 dans les zones soumises aux perturbations par matières organiques. Oceanol.Acta, 3(3):382-90

Bourcier, M., et al.,. Répartition des biocénoses benthiques en fonction des substrats
1979 sédimentaires de la rade de Toulon (France). Téthys, 9(2):103-12

Cognetti, G., La distribuzione dei Policheti in ambienti litorali inquinati. In
1974 atti della Tavola Rotonda Internazionale "La biol.mar.per la difesa e per la produttività del mare", Livorno, 20-21 May, 1974, edited by G. Cognetti and P. Tongiorgi, pp. 153-62

Ergen, Z. et M. Onen, The effects of pollution in the macrobenthic fauna of the soft
1983 substratum in the Urla Harbour (Izmir-Turkey). Rapp.P.-V.Réun.CIESM, 28(3):207-8

Glemarec, M. et C. Hily, Perturbations apportées à la macrofaune benthique de la baie
1981 de Concarneau par les effluents urbains et portuaires. Acta Oceol. (Oceol.Appl.), 2(2):139-50

Hily, C., Modifications de la structure écologique d'un peuplement de Melinna palmata
1983 (Annelide-Polychète) soumis aux effluents urbains et industriels en rade de Brest. Ann.Inst.Océanogr.,Paris(Nouv.Sér.), 59(1):37-56

Kocatas, A., Evolution cyclique du benthos dans les zones de pollution du Golfe d'Izmir
1981 (Turquie). Journ.Etud.Pollut.CIESM., 5(1980): 643-8

Pearson, T.H., and R. Rosenberg, Macrobenthic succession in relation to organic
1978 enrichment and pollution of the marine environment. Océanogr.Mar.Biol., 16:229-311

Picard, J., Recherches qualitatives sur les biocénoses marines des substrats meubles
1965 de la région marseillaise. Recl.Trav.Stn.Mar.Endoume, 52 (Bull.36):1-160

Reish, D.J., An ecological study of pollution in Los Angeles-Long Beach Harbors,
1959 California. Occas.Publ.Allan Hancock Found., (22):117 p.

Romano, J.C., Etude cyclique des peuplements marins de substrats meubles situés au
1973 débouché en mer du grand collecteur de Cortiou (Marseille). Thèse Doct.spécial.Océanogr.biol.,Univ. Aix-Marseille II, 94 p.

Stirn, J. et al., Selected biological methods for assessment of marine pollution. In
1975 Marine pollution and marine waste disposal, edited by E.A. Pearson and E.D.F. Frangipane. Oxford, Pergamon Press, pp.307-27

Stora, G., et J.C. Romano, Influence de la construction et l'exploitation d'un complexe
   1980   portuaire et industrial sur la distribution et l'évolution des peuplements
          benthiques littoraux d'un golfe mediterranéen (Golfe de Fos).
          Prog.Water Technol., 12(1):137-44

Swartz, R.C., Biological criteria of environmental change in the Chesapeake Bay.
   1972   Chesapeake Sci., 13 Suppl. S17-41

ETUDE COMPARATIVE DES MODIFICATIONS DES ECOSYSTEMES DES EAUX COTIERES
DU GOLFE ET DU PORT DE PATRAS SOUMISES A L'INFLUENCE DES POLLUANTS

par

J. CASTRITSI-CATHARIOS, G. GANIAS et D. PANAGOPOULOS
Université d'Athènes, Laboratoire de Zoologie,
Panepistimiopolis, GR 15771, Athènes, Grèce.

1. INTRODUCTION

Le sujet de cette étude entre dans les objectifs des programmes du MED POL Phase II et particulièrement dans l'activité I "Modifications des ecosystèmes par la pollution".

Bien que le Phylum des Bryozoaires soit le sujet des études très étendues des Instituts Internationaux ainsi que des Universités du monde entier, chez nous, en Méditerranée orientale, et plus précisément dans l'Archipel grec, il n'y a que quelques rapports isolés.

Les Bryozoaires sont des organismes cosmopolites et le développement de quelques espèces est favorisé dans des conditions spéciales du milieu, par exemple, ports (Prenant et Bobin, 1966).

Spécialement la distribution quantitative et qualitative dépend:

- De la zone géographique de la navigation des bateaux, de la durée du voyage et de la période de son mouillage.

- Des facteures saisonniers.

- Des modifications écologiques du milieu, dues à l'influence de l'activité humaine.

Le golfe de Patras présente un intérêt important du point de vue de la recherche, pour une étude approfondie des conséquences de la pollution sur les communautés benthiques locales des Bryozoaires.

2. TOPOGRAPHIE DE LA REGION ETUDIEE

Les prélèvements ont été effectués dans la zone sublittorale du golfe et du port de Patras (Fig. 1). Au point de vue morphologique ce littoral est peu profond. Les rivages sont dentelés, il y a de nombreuses cavités et des zones relativement protégées des vagues, où l'eau est plus calme. Par contre vers la mer Ionienne les vagues déferlent.

Il y a plusieurs milieux écologiques très différents entre eux, par exemple:

Dans le port de Patras où la pollution portuaire est très grande. (Le mazout lui-même, si abondant sur certaines plages, même éloignées du port).

Près de l'embouchure de la rivière Evinos où la salinité diminue, cette multitude de biotopes paraît plus clairement aux analyses: soit granulométriques du substrat soit chimiques de l'eau de mer où ont été faits les prélèvements. Nous verrons plus loin les conséquences de cette multitude sur la faune Bryozoologique.

3. CONDITIONS DE PRELEVEMENT DES ECHANTILLONS

La récolte des spécimens a été faite pendant l'été tout au long des côtes (zone sublittorale) dans un réseau de stations denses (distance entre chacune d'entre elles: 1,5 km environ).

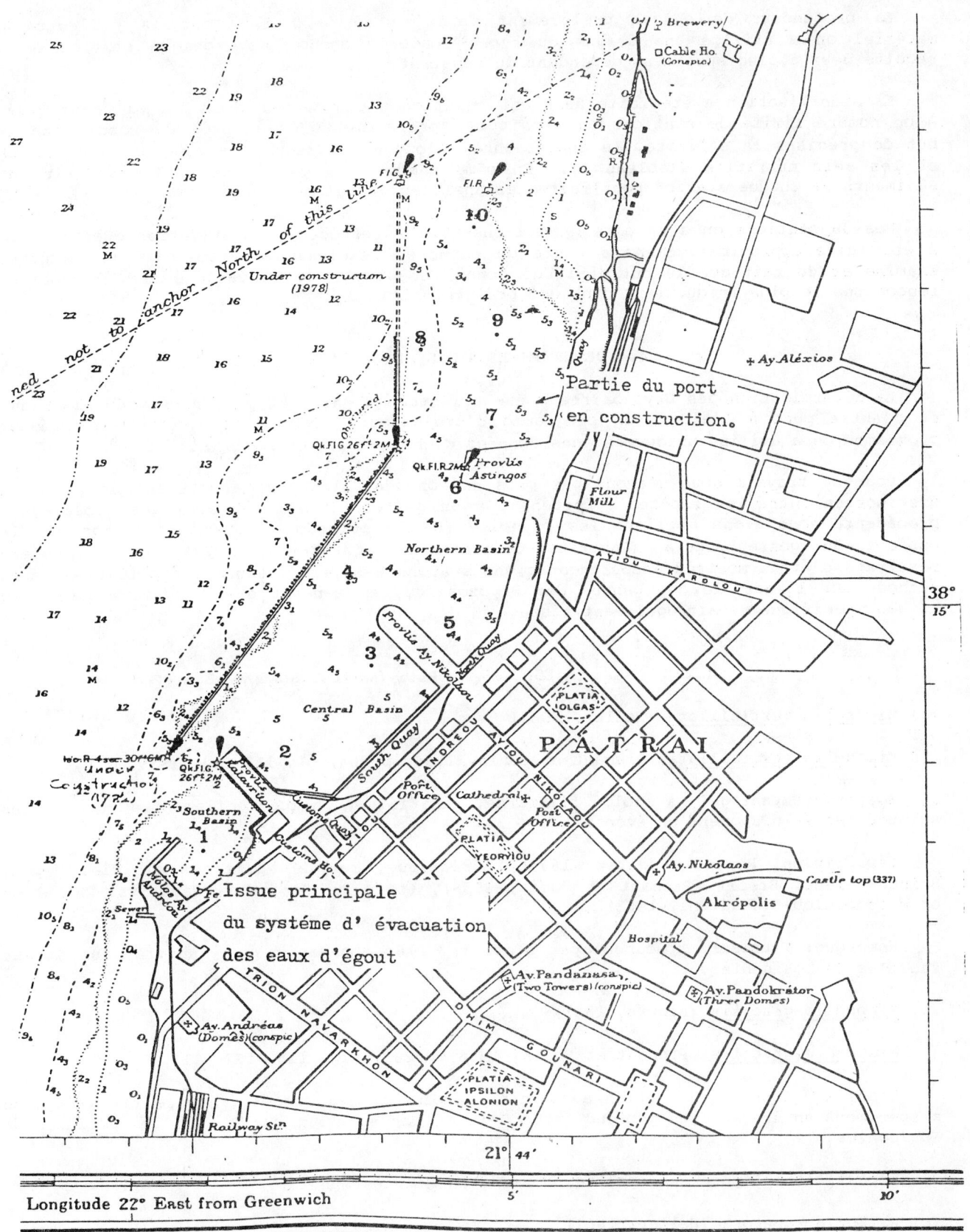

Figure 1. Carte de la région du port de Patras où ont été effectués les échantillonnages (localisation des stations)

La profondeur de chaque prélèvement varie entre 5-20 m. Pour la collecte du matériel on a utilisé une drague du type "Foster's anchor". A chaque station on a récolté deux à cinq spécimens dépendant du substrat.

La stabilisation a été faite dans une solution d'eau de mer avec formol neutralisé. A un nombre limité de stations on a fait des prélèvements d'eau près du fond, dans le but de préciser la salinité, la température, l'oxygène dissous, la conductivité, le pH et les sels nutritifs (Tableau I). En même temps on a gardé une petite quantité de sédiments de chaque station pour faire les analyses granulométriques. (Tableau II).

Des 88 stations on a récolté dans l'ensemble 95 espèces. L'estimation quantitative a été faite approximativement à cause du caractère exclusivement colonial du matériel examiné et du fait que le but était de donner une image de la variété des Bryozoaires rencontrés le plus fréquemment dans nos prélèvements.

4. DETERMINATION DES BRYOZOAIRES

La détermination des Bryozoaires a été effectuée d'une part par la méthode classique et d'autre part à l'aide d'un programme d'ordinateur. Pour l'élaboration du présent programme on a utilisé uniquement des données dichotomiques ou multigrades.

Pour ce travail nous n'avons pas pris en considération des caractères servant à la distinction entre les espèces appartenant à un genre. Toutefois, dans une publication précédente nous avons appliqué les principes de la méthode à la détermination des 18 espèces du genre Bugula (Castritsi-Catharios et Catharios, 1973). Ces premiers résultats s'étant montrés encourageants, nous avons essayé d'étendre l'application de la méthode sur l'ensemble des genres des Bryozoaires, en pensant que la détermination au niveau spécifique serait également aisée.

5. DESCRIPTION DES ESPECES LE PLUS SOUVENT RENCONTREES

(a) <u>Mimosella verticillata</u> (Heller, 1867)

<u>Mimosella verticillata</u>, Prenant et Bobin, 1956, p.264, pl. 117 fig. I-IV.

Zoarium rampant gel, à stolon filiforme, ramifié. A chaque renflement du stolon les autozoécies se disposent en éventail.

Répartition: Roscoff (Joliet, 1877), Adriatique, mer de Marmara, Malaisie, Japon, Sainte-Hélène, Brésil (Prénant et Bobin, 1956), Golfe de Korinthe (Castritsi-Catharios et Marcopoulou-Diakantoni, 1983).

Rémarque: Plusieurs spécimens par station. Plus abondantes sur substrat dur et sur feuilles de posidonies.

(b) <u>Mimosella gracilis</u> (Hincks, 1851)

<u>Mimosella gracilis</u>, Prenant et Bobin, 1956, p.262, fig. 116, 122-VIII.

Zoarium dressé, ramification abondante, autozoécies allongées. Espèce très répandue à spécimens en dragage peu profond entre 5-15 m, sur substrat dur ou sur les feuilles de Posidonies.

Répartition: Espèce cosmopolite, commune en Méditerranée.

(c) <u>Walkeria tuberosa</u> (Heller, 1867)

<u>Walkeria tuberosa</u>, Prenant et Bobin, 1956, p.256, pl. 112, fig. I-III.

Draguées dans la zone sublittorale, trouvées sur feuilles de posidonies. Espèce très répandue et cosmopolite en mers temperées.

Tableau I

Concentrations de différents facteurs physicochimiques pris dans le port de Patras.

Station	Température	Salinité °/oo Surf.	Salinité °/oo Prof.	pH	Conductivité micromhos	Oxygène dissous %	Nitrite µg atoms l^{-1} NO$_2$-	NO$_2$-	Sels carboniques (en somme)°dH
1	17	38	39	7,70	6500	84,8	0,250	0,526	7
2	17	31	36	7,90	6750	94,6	0,033	0,053	7
3	17,4	31	36	7,80	6500	84,8	0,065	0,368	8,4
4	17,4	32	34	7,70	6500	83,7	0,037	0,368	8,4
5	16,9	30	32	7,85	6250	90,2	0,028	0,474	8,4
6	16,9	35	36	7,75	6500	84,2	0,022	0,442	8,4
7	16,6	36	37	7,85	6250	90,2	0,037	0,453	9,8
8	16,0	38	39	7,90	6500	85,9	0,022	0,242	8,4
9	16,0	34	37	7,80	6750	82,6	0,022	0,179	8,4
10	16,0	38	39	7,80	6750	86,9	0,043	0,007	9,8

Tableau II

Substrat du Port de Patras (d'après l'analyse granulométrique)

STATION

1	Sable vaseux, présence de sapropel
2	" " " " "
3	" " " " "
4	" " " " "
5	" " " " "
6	Sable argileux
7	Vase, présence de sapropel
8	Sable vaseux, présence de sapropel
9	Sable, " " "
10	"

(d) Aetea sica (Couch, 1844)

Hippothoa sica (Couch, 1844)

Aetea sica Gautier, 1961, p.28, Prenant et Bobin, 1966, p.83, fig. 19-21 I-III, pl. 2, fig. 1, fig. 3.

(e) Aetea truncata (Landsborough, 1852)

Anguinaria truncata, Landsborough, 1852

Aetea truncuta, Gautier, 1961, p.29, Prenant et Bobin, 1966, p.89, fig. 20, I-V, 21 V, pl. 2, fig. 1.

Le zoarium aux deux espèces est partiellement dressé et se ramifie normalement. A. sica est plus robuste et plus calcifiée que celle de A. truncata. On remarque aussi des rayures sur la partie dressée de l'autozoécie.

Gautier (1954) a décrit des ramifications triangulaires à bords concaves. On l'a constaté aussi, aux colonies d' A. sica provenant du golfe de Korinthe et celles du golfe de Patras.

On les a trouvées fréquemment en abondance aux stations 1, 2, 3, 4, 5, 9, 11, 12, fixées sur feuilles de Posidonies ou sur substrats durs. Espèces cosmopolites dans les zones tempérées et chaudes, communes en Méditerranée où elles viennent de 1m à 150 m.

(f) Electra posidoniae Gautier, 1954

Electra posidoniae, Gautier, Gautier, 1954, p.66, fig. 1-4

Electra posidoniae, Gautier, Prenant et Bobin, 1966, p.147, fig. 41

Electra posidoniae, Gautier, Harmelin, 1973, p.676

Electra posidoniae, Gautier, Hayward, 1975, p.349

D'après Gautier cette espèce n'est connue que dans la Méditerranée où elle est très abondante.

On l'a trouvée uniquement sur feuilles de posidonies. Les zoariums sont blancs, encroûtants. Les zoécies présentent une grande variabilité dans le nombre des pores et des épines.

(g) <u>Turbicellepora</u> <u>armata</u> (Hincks, 1860)

<u>Cellepora</u> <u>armata</u> Hincks, 1860, p.278, pl. 30, fig. 5

<u>Schismopora</u> <u>armata</u> (Hincks), Canu et Bassler, 1930, p.76, pl. II, fig. 1-12

<u>Schismopora</u> <u>armata</u> (Hincks), Gautier, 1961, p.256

<u>Turbecellepora</u> <u>armata</u> (Hincks), Buge et Debourle, 1977, p. II, fig. 3

Plusieurs spécimens non ovicellés sur Algues.

Répartition: Atlantique oriental Méditerranée occidentale, Méditerranée sud, Tripoli (Libye).

(h) <u>Chorizopora</u> <u>brongniarti</u> (Audouin et Savigny, 1826)

<u>Flustra</u> <u>brongniarti</u>, Audouin et Savigny, 1826

<u>Chorizopora</u> <u>bronghiarti</u> Gautier, 1961, p.126

D'après J. Harmelin(1973) c'est une espèce à très large distribution. Abondante aux stations 2, 4 sur feuilles de phanérogames. Colonies fragiles, structure des parois vitreuses, rides à la frontale. Sur toute les zoécies fértiles un aviculaire distal médiane.

(i) <u>Lichenopora</u> <u>radiata</u> (Audouin et Savigny, 1826)

<u>Melobesia</u> <u>radiata</u>, Audouin et Savigny, 1826, p.235, p.6, fig. 3

<u>Desporella</u> <u>radiata</u>, (Audouin et Savigny), Waters, 1879, p.276, pl. 24, fig. 11

<u>Lichenopora</u> <u>radiata</u> (Audouin et Savigny), Calvet, 1902, p.44

<u>Lichenopora</u> <u>radiata</u> (Audouin et Savigny), Gautier, 1958, p.205

On l'a rencontrée en abondance sur substrat dur et feuilles de posidonies. Cette espèce vit dans l'Atlantique oriental mais surtout en Méditerranée à faible profondeur.

6. DISCUSSION

Par nos résultats on a essayé de trouver quelques paramètres qui ont une influence sur la faune Bryozoologique. Cette faune montre un caractère dépendant du biotope (Paramètres chimiques ou physiques).

Etant donné que la salinité joue généralement un rôle décisif sur les organismes marins on a comparé deux régions toutes proches l'une à l'autre. Près de l'embouchure de la rivière d'Evinos (stations 16-20) où la salinité était d'environ 32°/oo et dehors de la saline de Tourlis (stations 21-48) où elle était environ 40°/oo. La pauvreté des spécimens qui provennaient du prémier biotope (quantitative et qualitative) était caracteristique, bien que les dragages ont été fait à de differentes profondeurs.

Par contre aux stations suivantes où la salinité était haute on a trouvé des grandes quantités des familles <u>Mimoselidae</u>, <u>Valkeriidae</u>, <u>Aeteidae</u>, <u>Electridae</u>, <u>Celleporidae</u>, <u>Crisiidae</u> et <u>Lichenoporidae</u>. Les espèces qui appartiennent aux familles précedentes sont caracteristiques de la pollution portuaire.

Au cours de cette étude 56 espèces ont pu être distinguées (Tab. III). La majorité des espèces déterminées étaient des Chilostomes avec prédominance des espèces encroûtantes. Mais, parmi les espèces plusieurs étaient endémiques pour la Méditerranée: <u>Celleporina hassallii</u>, <u>C. globulosa</u>, <u>Turbicellepora armata</u>, <u>Tubulipora</u>

hemiphragmata, Adeonella passasi. Aussi 4 espèces qui sont considérées comme rares en Méditerranée, ont été trouvées: Codonellina sp., Hippaliosina depressa (Busk, 1854), Escharina armata (Hincks, 1862), Palmicellaria aviculifera (Canu et Bassler, 1930).

Tableau III

Espèces distingueés.

DIVISION SYSTEMATIQUE	NOMBRE DES GENRES DETERMINEES		NOMBRE DES ESPECES DETERMINEES	
ORDRE: EURYSTOMATA	46		70	
SOUS-ORDRE: CTENOSTOMATA	9		14	
FAMILLE: FLUSREILLIDAE		1		1
FAMILLE: MIMOSELLIDAE		1		2
FAMILLE: NOLLELIDAE		1		1
FAMILLE: VALKERIIDAE		2		3
FAMILLE: VESICULARIIDAE		3		6
FAMILLE: VICTORELLIDAE		1		1
SOUS-ORDRE: CHEILOSTOMATA	37		56	
SECTION: ANASCA		15		27
FAMILLE: AETEIDAE		1		2
FAMILLE: ALDERINIDAE		1		2
FAMILLE: BICELLARIELLIDAE		2		5
FAMILLE: CELLARIDAE		1		1
FAMILLE: CHLIDONIIDAE		1		1
FAMILLE: CRIBRILLINIDAE		4		5
FAMILLE: ELECTRIDAE		1		2
FAMILLE: MICROPORIDAE		1		1
FAMILLE: ONYCHOCELLIDAE		1		2
FAMILLE: SCRUPOCELLARIIDAE		2		6
SECTION: ASCOPHCRA		22		29
FAMILLE: ADEONIDAE		2		2
FAMILLE: CELLEPORIDAE		1		2
FAMILLE: CHEILOPORINIDAE		3		3
FAMILLE: ESCHARELLIDAE		3		3
FAMILLE: EXOCHELLIDAE		1		1
FAMILLE: HIPPOTHOIDAE		1		1
FAMILLE: MARGARETTIDAE		2		2
FAMILLE: MICROPORELLIDAE		2		2
FAMILLE: SCHIZOPORELLIDAE		5		10
FAMILLE: SERTELLIDAE		1		1
FAMILLE: UMBONULLIDAE		1		2
ORDRE: CYCLOSTOMATA	11		23	
FAMILLE: CRISIIDAE		1		7
FAMILLE: DIAPEROECIIDAE		2		2
FAMILLE: DIASTOPORIDAE		3		3
FAMILLE: FRONDIPORIDAE		1		1
FAMILLE: LICHENOPORIDAE		1		3
FAMILLE: TUBULIPORIDAE		3		7

Les modifications des écosystèmes marins sous l'effet des polluants sont visibles dans la zone côtière près du port de Patras. C'est la raison pour laquelle on les a étudiées séparément. Malgré le grand nombre des stations la pauvreté des spécimens était très caractéristique (Tab. IV). En plus, la plupart des colonies étaient mortes et le polymorphisme absent. Le pH de l'eau était plus bas que le physiologique. Ce paramètre paraît jouer un rôle définitif à la multitude de la faune bryozoologique.

On peut expliquer l'abaissement du pH dans cette région à cause de la décomposition naturelle des substances organiques du fond, créée par les déchets du système d'évacuation des eaux d'égout. (La teneur des eaux en gaz carbonique joue un rôle déterminant dans l'ajustment de leur pH.).

Tableau IV

Espèces des Bryozoaires encroûtants, morts △ et vivants ▲, trouvés sur differents substrats du port de Patras (10 stations)

ESPECES	1	2	3	4	5	6	7	8	9	10
EURYSTOMATA										
CTENOSTOMATA										
CHEILOSTOMATA										
ANASKA										
Tricellaria sp.			1△							
ASCOPHORES										
Cryptosula pallasiana (Moll, 1803)			1▲					1△		
Hippaliosina depressa (Busk, 1854)								2△		
Margaretta cereoides (Ellis et Solander, 1786)						1△		5△	1△	
Schizomavella gardensis			1△							
Schizomavella linearis (Hassal, 1841)			1△							
Schizoporella errata (Waters)								1△		
CYCLOSTOMES										
Frondipora sp.					1△	1△				
Lichennopora radiata (Andouin, 1826)					1△	1△			5△	
Stomatopora major (Johnston)				1△	1△	1△				
Tubulipora liliacea (Pallas, 1766)									1△	

Dans cette région aussi, la mer est fermée par un retranchement artificiel; c'est pourquoi il n'y a pas de renouvellement d'eau.

L'image présentée par les plus proches stations immédiates: 61, 62, 63 est riche en espèces contraires à celles des stations qui sont près de la sortie de l'eau du port, et dans un autre conduit d'évacuation des eaux d'égout, est confirmée la pauvreté des spécimens récoltés. Le petit nombre de bryozoaire qui a été trouvé dans cette region, appartient à la catégorie de ceux qui ont été caractérisés, sous réserve, comme des témoins de pollution portuaire.

A la suite de ce qui a été précité nous concluons que:

- La taxonomie systématique des Bryozoaires est extrêmement difficile (au niveau de l'espèce). Ils ont l'avantage de se trouver sur tous les substrats. (sable argileux, vaseux, rocailleux, près de Posidonie, etc.) et ils sont des indices de pollution portuaire.

- Leur répartition quantitative et qualitative montre que cette répartition peut avoir une relation avec une autre forme de pollution, comme les eaux d'égout.

- Ce programme, pour l'étude du golfe de Patras, nous a aidés:

(a) à connaître la faune bryozoologique de la région susdite, où il n'y a pas d'autres études relatives.

(b) A mentionner le rapport direct de la pollution avec cette faune.

(c) A localiser le problème de la modification des écosystèmes que présente le port de Patras et ses régions environnantes.

Il serait recommandé de faire une étude plus étendue dans le but de confirmer les remarques ci-dessus et cela pourrait se réaliser à la suite d'observations systématiques et continues aux biotopes problématiques ou pas.

Enfin, on veut souligner le besoin imminent de prendre des mesures pour la ville côtière de Patras, où la pollution de la mer a atteint le point critique et qui menace les merveilleux rivages avoisinants.

7. BIBLIOGRAPHIE

Audouin, V., et C. Savigny, Explication sommaire de planches de polypes de l'Egypte
1826   et de la Syrie. In Description de l'Egypte. Hist.Nat., Paris, 1(4):225-44

Busk, G., Catalogue of marine Polyzoa in the collection of the British Museum. London,
1854   British Museum (Natural History), Part 2:55-120

Buge, E. et A. Debourle, Ecologie de la faune de Bryozoaires d'une plage des environs de
1977   Tripoli (Libye). Bull.Cent.Rech.Explor.Prod.Elf-Aquitaine, 1:321-77

Calvet, L., bryozoaires marins des côtes de Corse. Trav.Inst.Zool.Univ.Montpellier (2),
1902   12:52 p.

Canu, F. et R.S. Bassler, Bryozoaires marins de Tunisie. Ann.Inst.Océanogr.Salammbô, 5:
1930   91 p.

Castritsi-Catharios, J. et G. Catharios, Essai d'application d'un programme d'ordinateur
1973   pour la détermination taxinomique d'organismes causant des salissures en Méditerranée. Journ.Etud.Pollut.CIESM, 1(1972):105-6

Castritsi-Catharios, J. et A. Marcopoulou-Diacantoni, Les bryozoaires fossiles et vivants
1983   de la côte N.E. du golfe de Corinthe (Hellas Méridionale) Note préliminaire. Rapp.P.-V.Réun.CIESM, 28(4):225

Couch, J., A Cornish fauna, being a compendium of the natural history of the country.
1844   Truro, Cornwall, Part 3:164 p.

Gautier, Y., Sur l'Electra pilosa des feuilles de Posidonies. Vie Milieu(A Biol.Mar.),
1954   5:66-70

_____, Bryozoaires de la côte ligure. Ann.Mus.Civ.Stor.Nat.Giacomo Doria,
1958   70:193-206

_____, Recherches écologiques sur les Bryozoaires chilostomes en Méditerranée
1961   occidentale. Thèse Faculté des Sciences.Aix-Marseille, 91:434 p.

Harmelin, J.C., Bryozoaires de l'herbier à Posidonies de l'île de Port-Cros.
1973   Rapp.P.-V.Réun.CIESM, 21:675-7

Hayward, P.J., Studies on the chilostome bryozoan fauna of the Aegean island of Chios.
1975   J.Nat.Hist., 8:369-402

Heller, C., Die Bryozoen des Adriatischen Meeres. <u>Verh.Zool.Bot.Ges.Wien</u>, 17:77-136
   1867

Hincks, T., Note on British Zoophytes, with descriptions of some new species,
   1851   <u>Ann.Mag.Nat.Hist.(2)</u>, 8:353-62

_____, Descriptions of new Polyzoa from Ireland. <u>Q.J.Microsc.Sci.</u>, 8:275-80
   1860

_____, A catalogue of the Zoophytes of South Devon and South Cornwall. <u>Ann.Mag.</u>
   1862   <u>Nat.Hist.(3)</u>, 9:22-30, 200-7, 303-10, 467-75.

Joliet, L., Contribution à l'histoire naturelle des Bryozoaires des côtes de France.
   1877   Thèse Paris, <u>Arch. Zool.Exp.Gén.</u> (1) 6:.193-304

Landsborough, D., A popular history of British zoophytes. Polyzoa. London, pp.265-386
   1852

Prenant, M. et G. Bobin, Faune de France Bryozoaires, 1er partie (Entopro-proctes,
   1956   Phylactolèmes, Cténostomes). Paris, Lechevallier, vol.60:398 p.

_____, Faune de France, Bryozoaires. 2ème partie: Chilostomes Anasca, Paris,
   1966   Fédération française des sociétés de sciences naturelles, vol.68:647 p.

Waters, A.W., On the Bryozoa (Polyzoa) of the bay of Naples. <u>Ann.Mag.Nat.Hist.</u>, 3:28-43,
   1879   114-26, 192-202, 267-81

RECOVERY OF A SOFT BOTTOM COMMUNITY AFTER
EXTENSIVE DREDGING. I: MOLLUSCA
by
M.C. CURINI-GALLETTI
Istituto di Biologia Marina
Università degli Studi di Pisa
Pisa, Italy

1. INTRODUCTION

Although dredging is among the most common human-induced disturbances on marine benthos, its impact on benthic communities (as pointed out by Poiner and Kennedy, 1984) has received much less attention than the effects of other pollutants - organic and chemical agents above all. Particularly scanty are prolonged observations covering a considerable span of years, and the detection of patterns of full recovery, if any.

Recently, Bonvicini Pagliai et al., (1985) have examined the impact of extensive dredging in the Gulf of Cagliari (Sardinia, Italy) before, during and six months after the end of the operations. The authors noticed that, albeit the pioneer community evolved very rapidly, the process of recovery could not be considered as fully completed six months after the dredging. A further survey, covering a period of two more years from the last samplings, was then undertaken. The present contribution deals with data concerning the Mollusca collected during the whole survey. Although the information concerns a single taxocenosis, it is worth mentioning that the importance of Molluscs as "efficient descriptors" of benthic communities has been recently stressed (Gambi et al., 1982). Furthermore, molluscs are quickly determinable in general and a good knowledge has been achieved concerning their patterns of larval development and trophic guilds. Consequently, they are useful for an economic monitoring of environmental conditions and, in the specific case of dredging, they can offer a potential for a better understanding of the patterns of recruitment and trophic successions within the community.

2. METHODS

The surveys were performed during the Springs of 1979 (preoperative), 1980 (two months after dredging), 1981 and 1982 in the Gulf of Cagliari (Italy) (see Bonvicini Pagliai et al., 1985). Samples were taken at the center of the dredged area (site A); immediately at the border of the excavations (site B) and in a neighbouring area (site C), far enough from the dredging to be considered unaffected by it, at a depth of about 20 m. (Fig. 1).

From every site, samples of water and sediment were collected. The following parameters were determined on the water: dissolved oxygen, ammonia, nitrites, dissolved phosphates and transparency. From the sediment (collected by means of a grab) particle-size distribution and total organic carbon (TOC) content was evaluated. For further details on analytical techniques see Bonvicini Pagliai et al., (1985).

Benthic macrofaunal samples were collected with a 0.07 m^2 grab with hydraulic closure. Three replicates were collected for each sample. The material was processed through a 0.5 mm mesh sieve, preserved in 4% neutral formaldehyde and then sorted in the laboratory.

The Shannon diversity index (H') (Shannon and Weaver, 1949), its evenness componet (Pielou, 1966) and the Kulczynsky index (Kulczynsky, 1927) were calculated from the community data. Numerical data were analysed with a multivariate ordering and classification technique (Zurlini, 1983). The ordering was performed by factorial analysis of correspondence (Benzecrì, 1973). The significance of results was evaluated from Lebart tables.

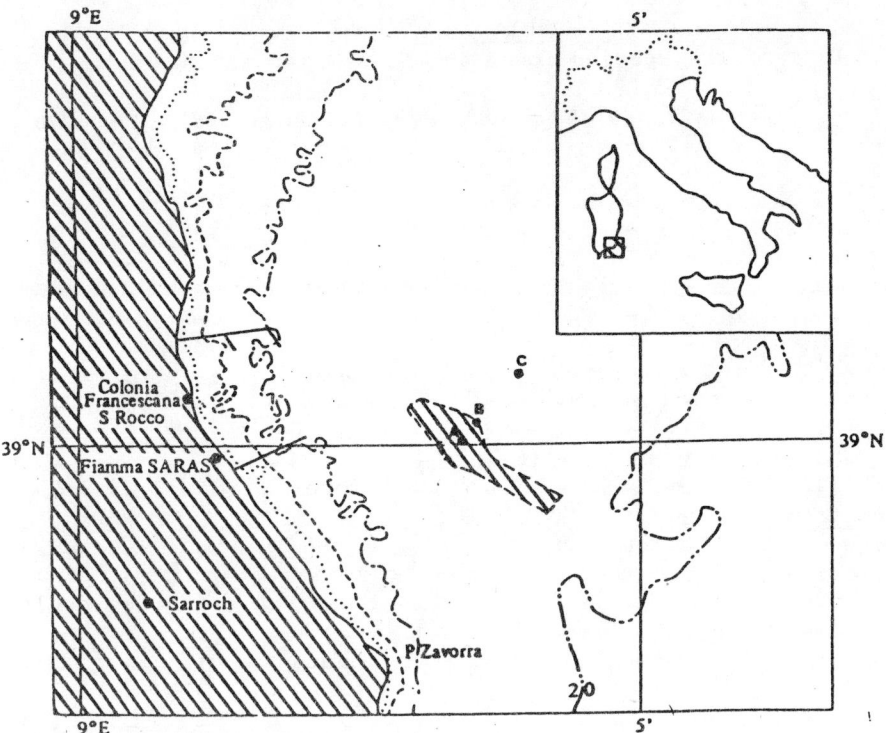

Figure 1. Map of the area subject to dredging, showing the sampling stations (from Bonvicini Pagliai et al. (1985), modified)

3. RESULTS

Abiotic Data

A full report of the data concerning chemical analysis and transparency for the preoperative survey and the immediately postoperative samples is reported by Bonvicini Pagliai et al., (1985). Data relative to the following surveys will be presented by Bonvicini Pagliai et al., (in preparation).

The results of main interest for this study is that only TOC and transparency were significantly affected by dredging. In the 1980 survey, TOC was markedly lower in the dredged area (site A) while it was unaffected in the neighbouring areas (sites B and C). A significant diminution in water transparency was observed in both sites A and B.

Sediments were remarkably mixed, consisting of silt and clay with small amounts of sand and gravel. Only minor variations of the texture were observed both among sites and over the years.

Biotic Data

A total of 1712 specimens, belonging to 54 species (57.4% Bivalves, 40.7% Gastropods, 1.8% Scaphopods), were obtained. The species list and their occurrence in the three sites during the whole survey is presented in Table I.

The predredging community is referable to a facies of "Vase terrigène côtière" (VTC) biocenosis (sensu Peres and Picard, 1964). Species as Mysella bidentata, Nucula nucleus and Corbula gibba were present, together with more mud-tolerant species as Abra alba (see also Febvre-Chevalier, 1969). However, the values of similarity between sites, evaluated on the basis of the Kulczynsky coefficient are low, ranging from 16% to 32%.

Table I
List of the species found in the sampled sites.

	A79	A80	A81	A82	B79	B80	B81	B82	C79	C80	C81	C82
Nucula nucleus(L.)	1	-	-	10	2	-	-	6	-	-	3	3
Nuculana pella (L.)	-	-	-	-	-	-	-	1	-	-	-	-
Glycymeris glycymeris(L.)	-	-	-	-	-	-	-	-	-	1	-	-
Musculus subpictus(Cantr.)	-	-	8	1	1	-	1	13	-	2	3	-
Modiolus adriaticus(LmK)	-	1	1	-	-	-	-	1	-	3	-	-
Mantellum inflatum(Chemn.)	-	-	1	-	-	-	-	6	-	-	3	-
Pecten jacobaeus (L.)	-	-	-	1	-	-	-	-	-	1	-	-
Kellia suborbicularis(Mtg)	-	-	-	-	-	-	-	3	-	-	-	-
Mysella bidentata (Mtg)	23	-	49	22	4	-	2	15	1	-	-	5
Lepton squamosum (Mtg)	-	-	-	-	-	-	-	-	-	1	-	-
Thyasira flexuosa (Mtg)	-	-	1	4	-	3	-	3	-	-	-	1
Anodontia fragilis (Phil.)	4	-	-	-	-	-	-	-	-	-	-	-
Divaricella divaricata(L.)	1	-	-	-	-	-	-	-	-	-	-	-
Acantocardia paucicostata (G.B.Sowerby)	-	3	2	4	-	6	2	2	-	-	-	-
Plagiocardium papillosum (Poli)	-	-	1	-	-	-	-	13	3	1	6	-
Parvicardium exiguum(Gmel)	-	-	-	1	-	-	-	-	-	5	6	-
Gouldia minima (Mtg)	-	-	-	1	-	-	-	1	4	-	-	-
Pitar rudis (poli)	-	-	-	1	1	-	-	2	-	-	-	1
Tapes Juv.	-	-	-	-	2	-	-	-	-	-	-	1
Abra prismatica (Mtg)	-	6	3	48	-	1	-	4	-	1	-	1
Abra alba (Wood)	-	-	3	3	1	7	10	2	-	-	3	7
Tellina nitida Poli	1	-	-	-	-	-	-	-	-	-	-	-
Tellina tenuis Da Costa	-	-	2	-	-	-	-	1	1	-	-	-
Tellinella distorta (Poli)	-	-	-	1	-	-	-	5	-	2	-	-
Spisula subtruncata (Da Cos.)	-	-	-	2	-	3	-	81	-	-	-	-
Pharus legumen (L.)	-	2	-	-	-	-	-	1	-	-	-	-
Corbula gibba (Olivi)	4	379	94	82	11	380	10	80	3	51	6	27
Sphaenia binghami Turton	-	-	1	-	-	-	-	5	-	-	2	-
Hyatella arctica (L.)	-	-	1	-	-	-	-	17	-	7	-	-
Thracia corbuloides Desh.	-	-	1	-	-	-	-	-	-	-	-	-
Lyonsia norvegica (Chemn.)	-	-	-	-	-	1	-	-	-	-	-	-
Hyala vitrea (Mtg.)	-	-	-	-	-	-	-	-	-	-	-	2
Turboella radiata (Phil.)	-	2	1	-	-	15	-	-	-	-	-	-
Turboella inconspicua(Alder)	-	1	-	-	-	1	-	1	-	-	-	3
Rissoa juv.	-	1	-	-	-	-	-	-	-	-	-	-
Acinopsis cancellata(Da Cos.)	-	-	-	-	-	-	-	-	-	-	1	-
Caecum trachea (Mtg)	-	-	1	-	-	-	-	-	-	-	-	-
Balcis curva (Jeffr.)	-	-	-	1	-	-	-	-	-	1	-	-
Strombiformis glabra(Da Cos.)	-	-	1	-	-	-	-	-	-	-	-	-
Aporrhais pespelecani (L.)	-	-	-	2	-	-	-	1	-	-	-	-
Lunatia juv.	-	-	1	-	-	2	-	-	-	-	-	-
Naticarius juv.	-	-	-	-	-	-	-	1	-	-	-	-
Hinia incrassata (Ström.)	-	1	-	1	-	1	-	-	-	-	-	-
Hinia pygmaea (Lmk.)	-	-	3	1	-	-	4	-	2	-	3	-
Bela brachystoma (Phil.)	-	-	-	-	-	-	1	-	-	-	-	-
Odostomia juv.	-	-	1	-	-	-	-	-	-	-	-	-
Ringicula conformis Mtrs.	-	-	1	-	-	-	-	-	-	-	-	-
Cylichnina subcylindrica (Brown)	-	-	1	-	-	-	-	-	-	-	-	-
Haminoea navicula (Da Cos.)	-	-	-	-	-	-	1	-	-	-	-	-
Retusa truncatula (Brug.)	-	-	6	-	-	-	-	-	-	-	-	-
Retusa obtusa (Mtg.)	-	-	1	2	-	-	-	-	-	-	-	-
Philine aperta (L.)	-	-	7	10	-	-	-	-	-	-	-	-
Akera bullata Müller	-	-	2	-	-	-	-	-	-	-	-	-
Dentalium dentalis L.	-	-	2	-	-	-	-	-	-	-	-	3

This community, on closer observation, can be referred to what was formerly considered to be a distinct biocenosis ("Fonds meubles instables" (F.M.I.), according to Peres and Picard, 1964). Successively, Picard (1965) reconsidered it as a facies of other communities (mainly VTC or DC) centered in areas such as bays or gulfs, possibly surrounded by lower depths and close enough to the land to receive a continuous input of sediment. This closely fits with the location of the sampling sites in the Gulf of Cagliari. Those areas practically act as traps for fine sediments and larvae from neighbouring environments. Therefore, the occurrence in the samplings of species like Spisula subtruncata, Hinia pygmaea (characterizing SFBC communities) or Plagiocardium papillosum, Pitar rudis (characterizing DC communities) is easily explained on this basis.

The impact of dredging on such a community mostly affected the relative abundance of species. The relative frequency of Corbula gibba, for example, rose, in the 1980 survey, from 12% to 95.7% in site A, and from 50% to 90.3% in site B. A more limited rise in abundance of this species was detectable also in site C (from 20% to 67.1%).

In the 1981 survey, the relative frequency of Corbula gibba decreased (A:47.9%; B: 32.3%; C: 18.2%), while in the area affected by dredging, and particularly in A, a substantial enhancement of Gastropods both in number of species and individuals has been detected since the end of the operations.

In 1982, the final year of the survey, the frequency of C.gibba was still lower in A (41.6%) and B (30.9%), while it was high in C (50%). Sites A and B still maintained a high enhancement of their community, both in number of species (A: 19; B: 24; C: 11) and specimens (A: 197; B: 259; C:54) with respect to the unaffected site.

Fig. 2.1 and 2.2 summarize the total number of species per site and total number of individuals per site respectively. In both, site C shows a good constancy in time, while the site from within the dredged area (A) shows a noticeable increase in numbers of species-taxa and individuals immediately after dredging. The pattern is similar in site B, although 1981 samples were rather poor in species and individuals.

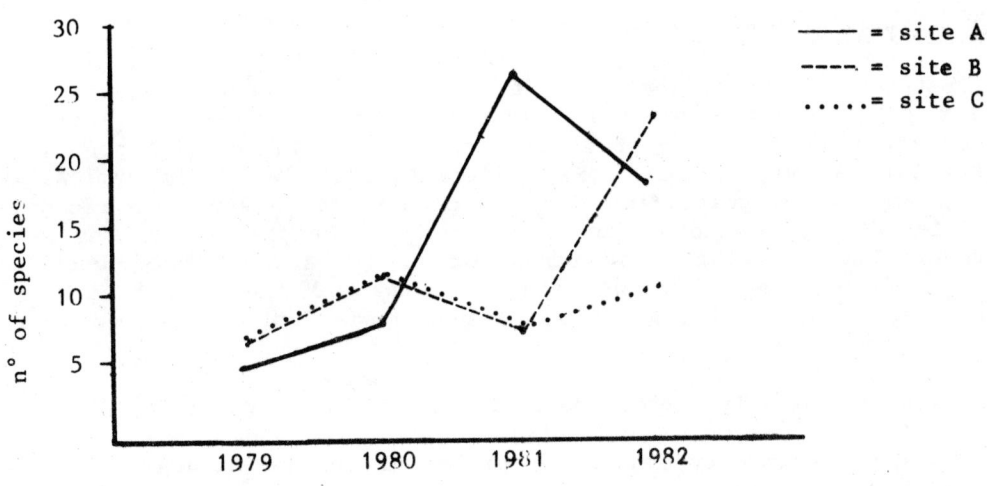

Figure 2.1. Number of species per sites over the years.

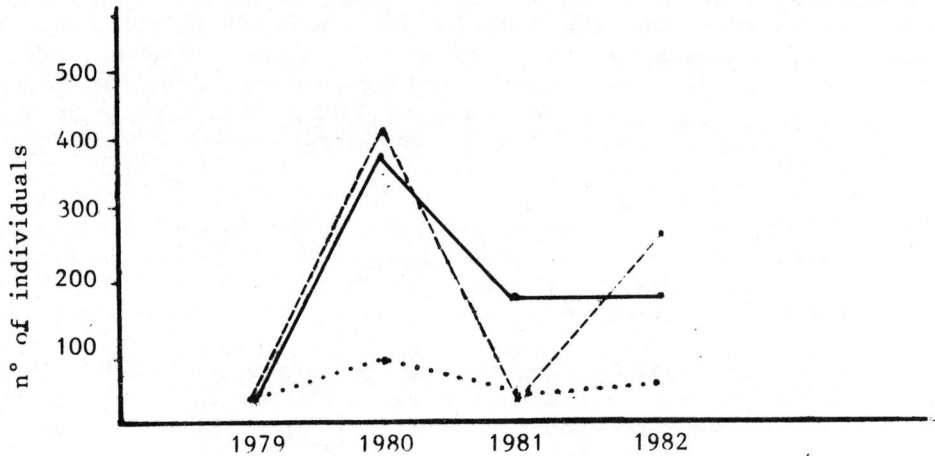

Figure 2.2. Number of individuals per site

Diversity Indices

Figure 3.1. presents the Shannon diversities (H'). It shows a reasonable constancy in time in the unaffected area (C) while it drops in sites A and B immediately after dredging. A rise to normal values is however already present in 1981, and in 1982 both sites A and B have H' values slightly higher than C. The values of equitability (Figure 3.2.) somewhat recall the pattern of Shannon diversities. However, in 1980 a more marked decrease in C is also present.

Factorial Analysis of Correspondence

The first three terms of the factorial analysis of correspondence are significant at a level of $p < 0.05$ and account for 54.05% of total variance. The ordering model obtained in the plane formed by the first two axes is shown in Figure 4. Most of the variability lies along the first axis, a conformation which is typical of situations where a unitary strong polarizing factor exists. This factor is recognizable as the greater or lesser degree of dominance of C. gibba. The stations at the lefthand side of the graph all have a relative abundance of C. gibba of 90% or more, while C '79, situated at the extreme righthand side has one of the lowest values (20%). The scattered location of station A '79 is presumably due to its high content of 'endemic' species.

The results generally support most of the statements presented above:

(i) the original community is widely dispersed in the diagram, thus showing a low degree of similarity
(ii) the unaffected site C also shows a wide displacement along the first axis, thus revealing the existence of natural fluctuations in the composition of the original community, due to the redundance of C. gibba itself.
(iii) sites A and B, immediately after dredging have an isolated position on the lefthand side of the diagram. Site C in the same year occupies a close position.
(iv) after dredging both sites A and B mostly come back to a sort of 'average' localization to whom also C 1982 refers.

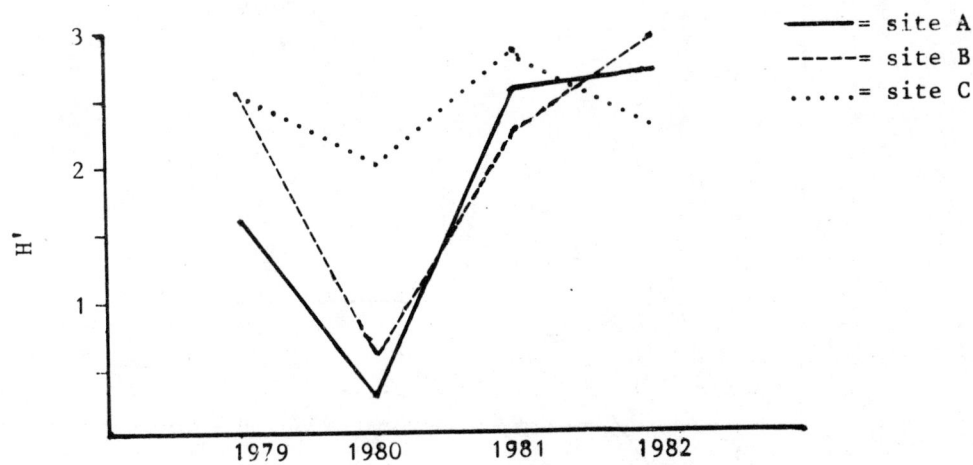

Figure 3.1. Trends of Shannon diversity index during the survey.

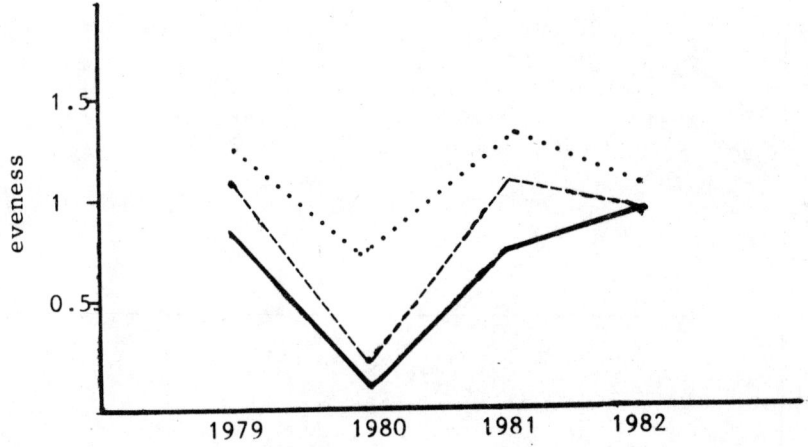

Figure 3.2. Trends of the eveness.

4. DISCUSSION

As a rule, during the recovery of benthic communities after dredging, two main classes of phenomena arise : (i) those connected with the progressive recolonization of a (supposedly) barren bottom; (ii) those connected with interactions among the successful colonizers.

(i) Two main factors can be expected: larval recruitment and adult vagility. The latter seems to be limited to a few agile bivalves (such as Abra) or Gastropods (as Hinia and Naticarius). However, in the specific case of this report this factor seems to have been of little or no importance.

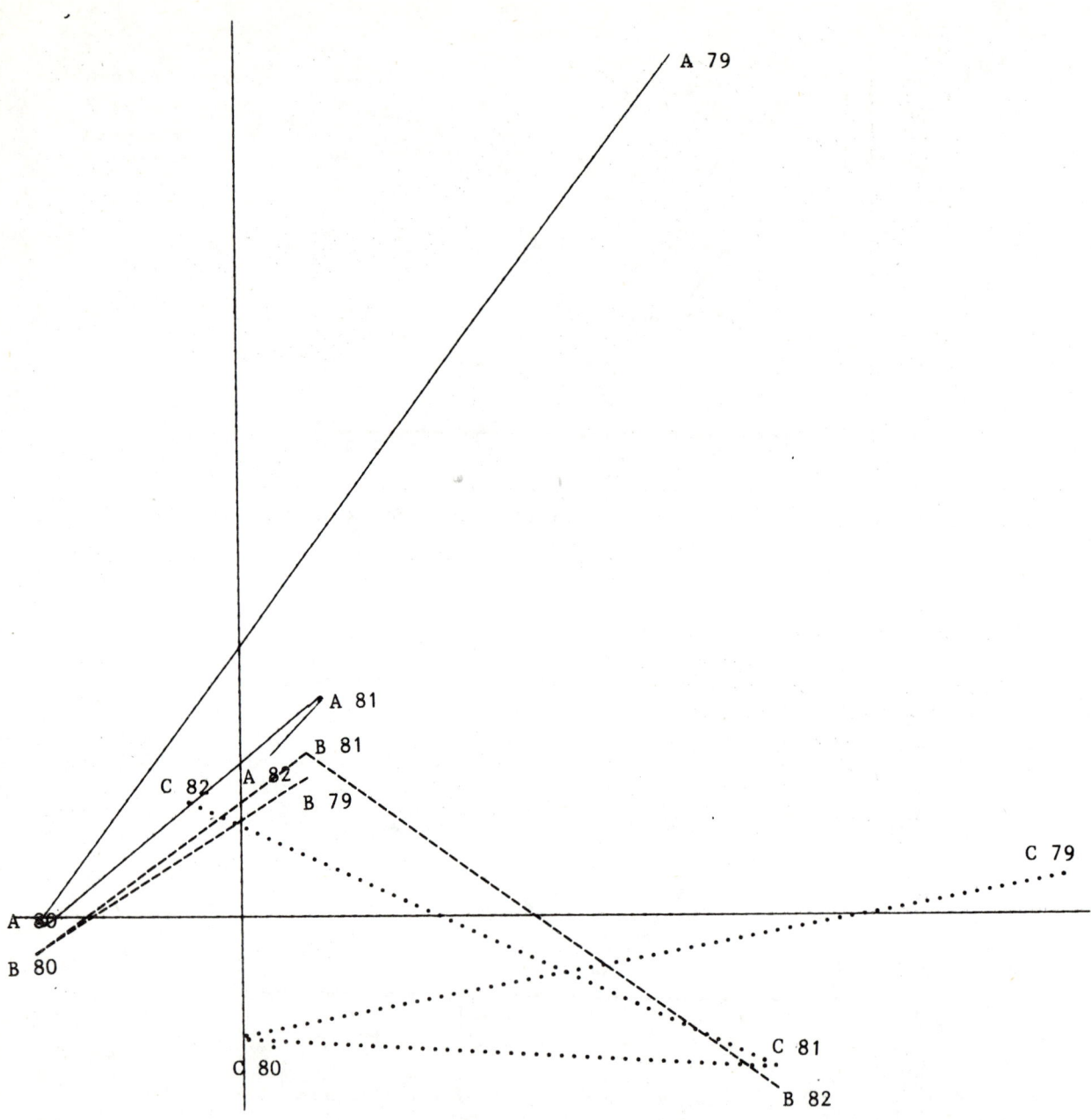

Figure 4. Ordering model of the Factorial analysis of Correspondence.

Much greater was the incidence of larval recruitment. It is well known that species with a long planctonic stage tend to present sharply differing good and bad years for settlement, while species with short or no planctonic stage usually present much more uniform populations (Thorson, 1946). An example of species with long planctonic life is Corbula gibba. Although it is considered typical of environments with unpredictable sedimentary rates (Picard, 1965) and a 'pioneer' species (Salen-Picard, 1981) and therefore its occurrence in sites A and B appears well justifiable, its relative abundance in the unaffected site C as well allows the supposition that in 1980 (and, to a lesser extent, in 1982) a large swarm of that species settled in the entire area.

A similar fluctuation of good or bad years, with the addition of a much greater patchiness, is shown by Abra alba. Generally represented by few specimens, it was among the dominant molluscs in the 1982 samples of site A. Also Spisula subtruncata - whose patchy distribution is well known (Cattaneo et al., 1983) - was practically absent from over the examined area, but became the dominant mollusc in the 1981 survey of site B. However, in both cases, only post-settlement individuals were concerned.

Among the other bivalves, it is worth mentioning that the only species attaining a considerable size ($>$30 mm) found during the survey (Acanthocardia paucicostata) is markedly more common in sites subject to dredging (A: n= 9; B: n= 10; C: n= 0). This species has been already reported for dredged areas (Russo, 1982). It is interesting that A. paucicostata shows, when compared with other Mediterranean Cardiidae, certain characters of the shell (i.e. smaller size, brittler shell and a general 'immature' appearance, which closely resembles the juveniles of both A. echinata and A. aculeata) which can be interpreted as the result of a process of paedomorphosis by progenesis. It is to be underlined that progenesis is considered linked to r-strategies (Gould, 1980) and that A. paucicostata is the only species of its genus liable to quickly colonize disturbed environments.

As a whole, the occurrence of gastropods is among the most striking dissimilarities among sites. Site C only has a very reduced number, both of species and specimens, and they are mostly represented by scavengers (Hinia) and detritus feeders (Hyala vitrea). In the other sites, and more specifically in A, after the dredging a marked enhancement of the group, with a conspicuous number of species, was found. Predatory Opistobranchs are mainly concerned, but parasitic Proso- and Opistobranchs (Balcis, Odostomia, Strombiformis), detritus feeders (Aporrhais) and predatory Prosobranchs (Lunatia, Naticarius) are present.

A marked general enhancement of community in dredged areas in the years following the dredging has already been pointed out by Poiner and Kennedy (1984). These authors suggest that this effect may be determined by the mobilization of resources by the dredging operations and their transport to neighbouring areas by the sediment plume. In our case, at least in site A, the effect seems to be particularly marked on predatory gastropods. Possibly, it depends upon the large number of young preys (mostly bivalves) in the initial steps of the recolonization. A similar model of patterns of abundance and species diversity has been recently proposed by Hughes (1984) on a theoretical ground.

(ii) The predredging community has a filter/deposit feeders ratio dominated by the former, with the exception of site A, where the local abundance of Mysella bidentata (species associated with Echinoderms, and therefore patchy in its distribution) shifts the ratio toward deposit feeders.

In the 1980 survey the almost complete totality of the species was represented by filter feeders, as is to be expected from a sediment with very low organic matter content. The relative frequency of filter feeders (depending mostly on the single species C. gibba) is however also high in site C. Nonetheless it must be underlined that C. gibba occupies a rather peculiar trophic niche: it is basically a filter-feeder, but is also liable to exploit the supply of diatoms, bacteria and organic debris - when present (Yonge, 1946).

During the following years, with the rise of TOC to normal values in all the stations, a slow recovery in number of deposit feeders occurred, with a readjustment of ratio between the main trophic guilds which approximates the values found in the predredging survey, even if it does not exactly reproduce them.

5. CONCLUSIONS

According to the analysis of the physiognomy of pre-and post-dredging communities, a number of considerations, some of which supposedly of general value, can be made:

a) the natural occurrence of marked fluctuations in species composition in nearshore soft-bottom communities is shown even by a relatively (4 years) short-term survey. McCall (1976) displays a large number of possible explanations of the phenomenon. In our case, the marked fluctuations found in the unaffected site C seem mostly to rely upon the more or less successful settlement of the single species C. gibba. Most of the specimens found appear to be juvenile, and the whole community seems to be in a perennial 'pioneer stage', possibly according to the uneven rate of sedimentation. As the actual recruitment into established communities may depend mostly on the availability of space (as pointed out by Hughes, 1984), it is therefore liable to be less successful than in barren areas. Dredging seemingly leads eventually to a magnification of the natural fluctuations of this kind of soft bottom.

b) the areas subject to dredging have shown a significant enhancement of the whole community, especially for certain groups, both in the number of species and specimens. However, in our case young specimens are mostly concerned, and the survival of the species in the following years is very scanty. Diversity however remained high, and postdredging communities are rich in species and individuals. This supposedly is the reason why the 1982 swarm of C. gibba affected site C more directly than sites A and B, where a smaller part of the substratum was available for settlement.

c) the dominant species in the first steps of recolonization (C. gibba) seems to be constantly present in areas subject to dredging (see Russo, 1982). C. gibba shows certain characters (high reproductive potential, way of feeding, possibility of anchoring itself to mixed substrata via byssus threads or in muddy substrata according to the thickness of the shell) which allow it to be considered an r-species, particularly adapted, anyway, to environments with unpredictable sedimentary rates. The possibility that A. paucicostata is also an r-oriented species has already been mentioned. The occurrence of r-species in the initial steps of recolonization is well known in ecology (Odum, 1969). It has been recently confirmed for the marine benthic biota by artificial experiments of recolonization (Arntz and Rumohr, 1982). However, data are too scanty to determine if this community has greatly evolved towards k on the r-k continuum. It may be presumed that, due to the peculiarity of this environment, it remains continuously close to a perennial r-stage.

d) two years after dredging, the community structure still does not correspond perfectly to the predredging situation. It seems rather utopian to evaluate the definitive recovery rate in an environment subject to strong natural fluctuations. Apparently, apart from the redundance of C. gibba, a basic stock of species was already present two months after dredging. The particular community found in the Gulf of Cagliari seems therefore -to a certain extent- well adapted to react to even apparently catastrophic perturbations of its environment.

6. ACKNOWLEDGEMENTS

I gratefully acknowledge the indispensable help of Dr. Carlo Nike Bianchi (Centro E.N.E.A. C.R.E.A., S. Teresa, SP) for his invaluable contribution in the statistical elaboration of data and his useful suggestions.

7. REFERENCES

Arntz, W. and H. Rumohr, An experimental study of macrobenthic colonization and
    1982   succession, and the importance of seasonal variation in temperate latitudes.
    J.Exp.Mar.Biol.Ecol., 64:17-45

Benzercrì, J.P., L'analyse des données. Vol. 2. L'analyse des correspondances. Paris,
1973 Dumond, 619 p.

Bonvicini Pagliai, A.M., et al., Environmental impact of extensive dredging in a coastal
1985 marine area. Mar.Pollut.Bull., 16:483-8

Cattaneo, M., et al., Macrorépartition des post-larves de Spisula subtruncata (da Costa)
1983 lors du recrutement sur le fond. Rapp.P.-V.Réun.CIESM, 28(3):241-3

Febvre-Chevalier, C., Etude bionomiques des substrats meubles dragables du Golfe de
1969 Fos. Téthys, 1(2):421-76

Gambi, M.C., E. Fresi and A. Giangrande, Descrittori efficaci di comunità bentoniche.
1982 Nat.Sicil., 6(Suppl.): 489-97

Gould, S.J., Ontogeny and phylogeny. Cambridge, Mass., Harvard University Press,
1980 501 p. 4th ed.

Hughes, R.G., A model of the structure and dynamics of benthic marine invertebrate
1984 communities. Mar.Ecol.(Prog.Ser.), 15:1-11

Kulczynsky S., Die Pflanzenassoziationen des peinenen. Bull.Int.Acad.Pol.Sci.
1927 Math.Nat.(B Sci.Nat.), Suppl. 2: 57-203

McCall, P.L., Community patterns and adaptive strategies of the infaunal benthos of Long
1976 Island Sound. J.Mar.Res., 35(2):221-47

Odum, E.P., The strategy of ecosystem development. Science, Wash., 164: 262-70
1969

Peres, J.M. and J. Picard., Nouveau manuel de bionomie benthique de la mer Méditerranée.
1964 Recl.Trav.Stn.Mar.Endoume, 31(47):1-137

Picard, J., Recherches qualitatives sur les biocenoses marines des substrats meubles
1965 dragables de la région marseillaise. Recl.Trav.Stn.Mar.Endoume, 52(36): 1-160

Pielou, E.C., The measurement of diversity in different types of biological collections.
1966 J.Theor.Biol., 13: 131-44

Poiner, I.R. and R. Kennedy, Complex patterns of change in the macrobenthos of a large
1984 sand-bank following dredging. I. Community analysis. Mar.Biol., 78:335-52

Russo, G.F., Distribuzione della malacofauna di fondo mobile nella rada di Augusta.
1982 Boll.Mus.Ist.Biol.Univ.Genova, 50(Suppl.):325-31

Salen-Picard, C., Evolution d'un peuplement de vase terrigène côtière soumis à des
1981 rejets de dragages dans le Golfe de Fos. Téthys, 10(1):83-8

Shannon, C.E. and W. Weaver, The mathematical theory of Communication. Urbana, Illinois,
1949 University of Illinois Press, 117 p.

Thorson P., Reproduction and larval development of Danish marine bottom invertebrates.
1946 Medd.Komm.Dan.Fisk.Havunders.(D Plankton), (4):1-523 p.

Yonge C.M., On the habits and adaptation of Aloidis (Corbula) gibba.
1946 J.Mar.Biol.Assoc.U.K., 26(3): 358-76

Zurlini, G., Applicazioni di metodi multivariati all'analisi ecologica. In Atti del
1983 Convegno E.N.E.A., Rome, June 1983 "Un esempio di analisi ecologica del
sistema marino-costiero da Capo Circeo all'Isola d'Ischia", edited by G.
Zurlini and V. Damiani, ENEA, Serie Simposi pp.13-34

LES EFFETS DE LA POLLUTION SUR LES PEUPLEMENTS BENTHIQUES DE
SUBSTRATS ROCHEUX DU PORT D'AUGUSTA (SICILE ORIENTALE)

par

S. I. DI GERONIMO
Istituto di Scienze della Terra dell'Università
Corso Italia, 55 Catania, 95129
Italie

1. INTRODUCTION

Le Port d'Augusta, depuis plusieurs dizaines d'années, est le siège d'une importante base de la Marine Militaire Italienne, il abrite également une petite flotte de bateaux de pêche. Ce n'est qu'après la Seconde Guerre Mondiale qu'il est devenu un des plus grands ports industriels d'Italie et d'Europe. De nombreux complexes industriels s'y sont établis, rejetant leurs eaux usées à l'intérieur du Port.

Les rejets en mer des eaux usées de la ville ont considérablement augmentés, le taux de pollution urbaine ayant pratiquement doublé depuis l'après-guerre. De plus, l'installation, dans la proche baie de S. Panagia d'un terminal pour les pétroliers de la zone industrielle ISAB, a fait augmenter le trafic maritime.

Tant ceci, a provoqué une accumulation imposante et continue de polluants à l'intérieur du port (Sciacca et Fallico, 1978) amenant un bouleversement et parfois la disparition des biocénoses benthiques qui proliféraient dans ce secteur avant l'industrialisation.

Ces deux dernières années (1983-1984) un programme de contrôle biologique continu sur les substrats rocheux méso- et infralittoraux supérieurs, financé par la FAO, a été conduit dans la zone du Port d'Augusta (MED POL Phase II) afin de mettre en évidence les rapports entre la pollution globale et la composition et structure des peuplements benthiques.

La pollution marine, en fait, provoque une grande variété d'altérations tant dans l'eau que sur les sédiments du fond, ces dernières viennent inévitablement transformées en altérations plus ou moins complexes de nature biologique. La pollution, par conséquent, peut être considérée comme un phénomène de désordre écologique, induit par l'activité humaine, qui affecte tant les individus que les biocénoses.

Pour connaître le niveau de dégradation du milieu il convient d'étudier la réponse des organismes individuels (méthode autoécologique) ou celle des peuplements (méthode synécologique) en rapport avec la nature et la puissance des sources de pollution, et ceci, non seulement pour les effets immédiats mais également pour ceux à longue durée.

Le "Manuel des méthodes de recherche sur l'environnement aquatique-Huitième partie" (Stirn, 1981) présente un vaste panorama des méthodologies d'étude de la pollution marine et des critères pour l'estimation des effets, dont ceux biologiques.

Les peuplements d'organismes benthiques (Algues, Mollusques, Polychètes) considérés lors de l'étude sur la pollution du Port d'Augusta sont très représentatifs, chacun d'entre eux étant suffisant pour établir l'appartenance à la biocénose originaire. Les changements advenus dans la composition et la structure de ces peuplements benthiques, en fait, temoignent assez fidèlement des transformations écologiques sur le milieu, pour autant que les peuplements se comportent comme des enregistreurs de facteurs écologiques (Giaccone, comm.pers.).

2. PRESENTATION DU BIOTOPE

Le biotope étudié (Fig. 1) se situe dans la Baie d'Augusta (Sicile orientale) laquelle est divisée en trois parties du Nord au Sud: le Port Xifonio, une anse naturelle entre la Pointe Izzo et l'île sur laquelle se dresse Augusta. Ce port communique par un canal étroit avec le Port Magarese qui est délimité au Sud par une digue artificielle qui le sépare du Seno du Priolo et qui se termine au Sud contre la péninsule de Magnisi.

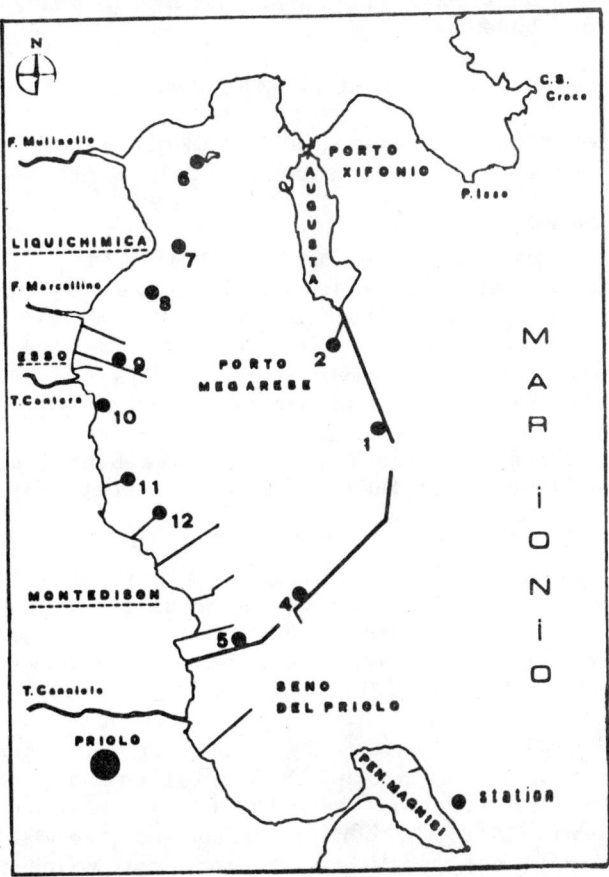

Figure 1. Baie d'Augusta: localisation des stations mesolittorales et infralittorales échantillonnées en 1983 et 1984.

Dans le Port Megarese débouchent les torrents Mulinello, Marcellino, Cantera et dans le Seno du Priolo le torrent Canniolo. Dans le Port Megarese comme dans le Seno du Priolo sont rejetées les eaux usées de nombreux établissements industriels dont les plus importants sont: Montedison-établissement de Priolo; Esso - Raffinerie d'Augusta; Enichimica - Augusta S.p.A. (ex Liquichimica). Dans la partie septentrionale du Port Megarese il y a, de plus, l'Arsenal militaire d'Augusta et les rejets à la mer des égouts de la ville.

Le dépurateur du groupement industriel, inauguré l'année dernière, aurait dû empêcher les décharges de ces industries à l'extérieur du Seno del Priolo. Cette installation a mal fonctioné et à régime réduit.

La zone portuaire proprement dite (Port Megarese) communique avec les eaux externes de la baie grace à trois entrées: une naturelle au Nord, dont la profondeur ne dépasse pas le mètre, pour une largeur d'une trentaine de mètres; deux artificielles, beaucoup plus profondes, au Sud et Sud-Est du port. Une telle disposition des accès au port facilite une circulation générale des eaux à l'intérieur de la zone portuaire dirigées du Nord vers le Sud selon des circuits anti-horaires discontinus. Les vents d'Est et du Sud-Est (Levante et Scirocco) très fréquents pendant toute l'année provoquent des phénomènes d'accumulation et stagnation des eaux dans la partie nord du port.

La côte occidentale est rocheuse (calcarénites pleistocènes) ou artificielle (jetées en ciment armé); la côte orientale est également rocheuse (calcarénites pleistocènes) dans sa partie septentrionale (zone d'Augusta) alors que le secteur central et méridional est artificiel, édifié de gros blocs de ciment armé. La côte septentrionale est basse et vaseuse, elle est entourée de salines aujourd'hui abandonnées.

Le fond du port (Di Geronimo, 1983) est couvert, dans la partie septentrionale, par des sables fins vaseux alors que la partie restante est recouverte d'une couche épaisse, dépassant parfois 20 cm, de vase argileuse très fine, fluide, de couleur grise ou noire, irisée, riche en hydrocarbures.

3. ETUDES PRECEDENTES

La construction des digues externes du Port Megarese est probablement à l'origine du processus d'érosion auquel ont été soumises les limites externes de l'herbier de Posidonies du Port Xifonio (Molinier et Picard, 1958). L'influence des eaux du port, polluées par les décharges industrielles, sur la zone externe de la baie a été clairement mise en évidence dans le Seno du Priolo à partir de la teneur des sels nutritifs et la charge des bactéries hétérotrophes nettement supérieure à celles rencontrées au large, mais aussi par la dégradation de l'herbier de Posidonie en relation avec l'apport de la fraction fine provenant des décharges (De Domenico, 1972; Genovese et De Domenico, 1975; De Domenico et al., 1978). Vers le large la situation, bien que moins critique, est toujours soumise à la pollution de la zone côtière.

La majeure partie des études sur les peuplements benthiques de la zone portuaire et de la baie ont été realisée par l'industrie ou à la suite de procédures judiciaires, de ce fait elles ne sont pas utilisables.

En 1975 une campagne d'étude a été réalisée à l'intérieur du Port Megarese. Elle a porté sur l'étude des métaux lourds et des hydrocarbures dans les sédiments ainsi que sur les peuplements benthiques. Les concentrations élevées des hydrocarbures et des métaux lourds ont été corrélées avec les cycles de productions des établissements industriels (Sciacca et Fallico, 1978).

L'étude des Polychètes a permis de délimiter 6 zones à pollution croissante également en relation avec les décharges industrielles (Cantone et Pilato, 1981), tandis que l'étude des Mollusques (Di Geronimo, 1983), en suivant la terminologie de Bellan (1967) et Bellan-Santini (1968), a permis de reconnaitre dans la baie une "zone d'eau pure" avec des peuplements non perturbés et une "zone moyenne soumise à la pollution"; cette dernière est divisée en deux zones: zone subnormale et zone polluée, dans laquelle l'influence de la pollution s'accentue conduisant à la disparition des Mollusques dans le secteur sud-ouest du port. Di Geronimo (1983) conclut qu'il y a un rapport étroit entre la dégradation des peuplements à Mollusques, les zones polluées et la position des industries, mais il est difficile d'attribuer à chaque type de pollution (urbaine, industrielle) la part qui lui revient dans les modifications relatives apportées aux peuplements. De même, l'abondance du Mollusque immigré de la Mer Rouge Brachidontes variabilis a été attribuée à la pollution des hydrocarbures (Di Geronimo, 1971). En outre de hautes concentrations d'hydrocarbures chlorurés ont été rencontrées dans des algues marines ainsi que dans les Mytilus galloprovincialis récoltés près de Priolo (D'Amico et al., 1979).

Récemment deux notes publiées sur les Polychètes de la baie d'Augusta (Giangrande et Gambi, 1983) et sur les Mollusques (Russo, 1983) ont expliqué l'altération de la structure et de la composition de tels peuplements par un "notable et diffus degré de déséquilibre sédimentologique".

Enfin, l'application de la méthode synécologique (Cormaci et al., 1985; Di Geronimo et al., 1985) a seulement récemment été finalisée au contrôle écologique continu du Port Megarese, pendant que la partie externe du port a fait l'objet de recherches de type naturaliste sur la végétation (Cormaci et al., 1983).

4. METHODES

Douze stations de l'Infralittoral supérieur ont été séléctionnées à l'intérieur du Port d'Augusta (Fig. 1) sur lesquelles ont été effectuées deux prélèvements par an: en mai et en octobre 1983 et 1984. Les stations sont toutes sur substrat dur rocheux naturel (stat. A_2, A_6, A_{10}), ou artificiel stat. (A_1, A_4, A_5), sur des pilons (A_8, A_{11}, A_{12}) ou sur des bouées (A_3, A_7, A_9).

Pour chaque station ont été prelévés deux échantillons de 400 cm^2: un pour l'étude qualitative et l'autre pour l'étude quantitative.

Les méthodes utilisées sont celles employées par Pérès et Picard (1964), Bellan-Santini(1969), Giaccone(1977) et Stirn(1982).

L'étude bionomique qualitative et quantitative a été menée sur les Algues, les Mollusques et les Polychètes. Dans l'étude des échantillons une évaluation a été effectuée pour la couverture en pourcentage totale et pour chaque espèce. L'identification taxonomique a été accompagnée de notes phénologiques et bionomiques propres à chacune des espèces. Les résultats obtenus ont été élaborés pour chaque échantillon et pour chaque peuplement algale et animal. L'indice de diversité de Shannon a été calculé; l'indice R/P, qui exprime le rapport entre Rodophicées et Phéophicées a également été étudié.

L'indice de diversité est fréquemment utilisé par les écologistes pour décrire l'état de santé d'un peuplement ou d'une biocenose. En effet, si cet indice donne bien un panorama synthètique de la structure d'un peuplement à partir du nombre des espèces et des exemplaires d'un échantillon, il doit être utilisé avec grande prudence car, entre autre, sa valeur dépend de la qualité de l'échantillon et de l'expérience du chercheur. De plus, dans la même station la valeur de cet indice varie selon le groupe systématique étudié, mais également, avec les saisons car le nombre des espèces et des individus change selon les cycles biologiques et saisonniers. Pour cette raison, les indices ont été calculés séparément pour les Algues, les Mollusques et les Polychètes. Ces trois groupes d'organismes sont considérés comme caractéristiques des biocénoses et suffisant à eux seuls pour permettre non seulement l'individualisation d'une biocénose mais aussi pour juger de la structure. Chacun des groupes dans les diverses biocénoses est structuré de manière significative autonome ou parfois, de façon à calquer la structure des biocénoses d'appartenance. De même, l'indice de diversité a été calculé pour une même station comme valeur moyenne globale pour une année de façon à réduire la différence. Il est ainsi possible de corréler cette valeur avec les conditions d'équilibre écologique.

Pour les Algues Belsher (1977) a trouvé, dans le Golfe de Marseille, des valeurs qui varient de 0.9 à 2.2 en zone polluée, et 3.7 dans un biotope en équilibre. Giaccone et al. (sous presse) ont trouvé des valeurs semblables à Palermo, avec des variations de 2.0 à 4.0 d'une zone perturbée vers une zone en équilibre.

Pour des peuplements de Mollusques de la Biocénose SFBC du Golfe de Catania des valeurs du même ordre ont été trouvées (données inédites) avec un indice de diversité qui s'échelonne de 2.13 à 4.24 pour des eaux pures à peu polluées.

Une même tendance de valeurs basses en zones polluées à des valeurs plus élevées de l'indice de diversité en zones moins polluées a été relevée par Desrosiers et al. (1982) dans le Golfe de Fos. Ce dernier présente une valeur moyenne de H, pour une période d'échantillonage légèrement supérieure à une année, qui varie de: 1.13 dans les peuplements à Ulva cf.rigida les plus polluées, à 2.43 dans les peuplements à Mytilus galloprovincialis, à 2.32-2.83 dans les peuplements à Corallina cf.mediterranea et à 2.72-3.28 dans les peuplements à Cystoseira stricta d'eaux toujours moins polluées.

Giaccone et al. (sous presse) retiennent que l'indice R/P, pour évaluer l'impact de la pollution sur l'environnement à partir des Algues, est plus significatif que l'indice de diversité, même s'il dépend également de la qualité de l'échantillonage et de l'expérience du chercheur. Feldman (1937), le considerait, au contraire, comme un indicateur biogéographique. L'indice R/P décroît avec une augmentation de latitude et de profondeur et est sensible à l'exposition et à la pente du substrat. Il varie également dans les différents niveaux de végétation et à l'intérieur de ceux-ci en rapport avec l'équilibre du milieu (Belsher, 1977; Giaccone, 1977). Cet indice se calcule en considérant les listes complètes de la végétation d'une zone déterminée ou, comme cela a été fait dans ce travail, en se référant aux valeurs moyennes de la couverture des espèces calculées pour les différentes stations dans un tableau. Les stations pour lesquelles l'indice ne peut être calculé à cause du manque des Algues brunes ne peuvent être prises en considération, un pourcentage de l'indice doit alors être évalué sur l'ensemble des stations.

L'indice R/P doit également être corrigé pour pallier aux variations brutales de sa valeur pour chaque prélèvement. Dans ce but ont été calculées les valeurs moyennes bisannuelles (1983-1984) obtenues en additionnant les R/P de chaque relevé et divisant par le nombre des relevés avec un R/P différent de zéro.

De telles valeurs moyennes des indices, calculées pour l'infralittoral, donnent ces chiffres: 3.1 à Ustica (Giaccone et al., sous presse), valeur issue de 16 relevés en zone peu ou pas polluée; 4.3 à Palermo (Giaccone et al., sous presse), valeur provenant de 26 relevés en zone polluée; 7.3 à Marseille (Belsher, 1977), valeur issue de 22 relevés en zone polluée; 5.0 dans le Port d'Augusta (Cormaci et al., sous presse), valeur provenant de 44 relevés en zone polluée.

Les rapports entre biomasse et pollution d'un secteur sont peu connus. Il semble, cependant, qu'il ait une relation entre ces deux paramètres à travers la présence de la composante algale et celle des Mytilus et Cirripèdes. Dans la zone plus polluée la composante algale diminue ou tend à disparaître et à certains stades de pollution peu marquée la composante animale augmente notablement en poids par la prolifération des faciès à Mytilus et à Cirripèdes.

Pour chaque prélèvement réalisé dans le Port d'Augusta ont été mesurés: poids humides, poids décalcifiés, poids secs, poids de cendre et matière organique. Tous ces paramètres semblent se comporter de manière homogène.

5. RESULTATS ET DISCUSSION

5.1 Algues

Le stock végétal récolté en deux ans est composé de 187 espèces et 11 taxa inférieurs à l'espèce. Ce nombre est plutôt élevé pour un milieu portuaire, mais pas excessif si on considère que dans le Port d'Augusta il y a plusieurs stations sur substrat rocheux naturel.

La richesse végétale des stations est très différente pour les prélèvements effectués en deux ans, mais le secteur Est du Port (stat. A_1 - A_6) possède un nombre moyen d'espèces par échantillon (29) supérieur au double du nombre (13) du secteur Ouest. La même différence est mise en évidence par les couvertures moyennes en pourcentage qui donnent 83% à l'Est et 45% à l'Ouest; par les indices de diversité avec des moyennes de 2.01 à l'Est et de 1.34 à l'Ouest; par les indices R/P qui, au contraire, sont plus faibles (6.2) à l'Est et plus forts (9.8) à l'Ouest.

La valeur globale moyenne de R/P pour 1983 et 1984 est, comme cela a déjà été dit, de 5.0. Cette valeur est moins élevée que celle trouvée à Marseille par Belsher (1977), où il existe de forts taux de pollution; elle est éloignée des valeurs de 2.5-4 qui indiquent en Méditerranée une zone en équilibre. Il faut considérer aussi que dans le 20% et le 54% des stations à l'Est et à l'Ouest du port R/P n'est pas évaluable pour la manque des Phéophicées.

Les espèces recoltées ne sont jamais structurées en associations climax et elles ne sont pas hiérarchisables. Le tableau phytosociologique, synthétisé graphiquement sur le diagramme circulaire de la végétation (Fig. 2), montre un peuplement destructuré dans lequel on enregistre une forte remonté des espèces sciaphiles (Rhodimenietalia 13.5%) avec un stock significatif d'espèces indicatrices de pollution (Pterocladio-Ulvetum 9.5%) et un cortège excessif d'espèces ubiquistes (54%). Les espèces photophiles (Cystoseiretalia 14.13%) et les espèces caractéristiques des peuplements méso-littoraux (Archrochaetetalia 8%) sont également très réduites confrontées à une zone en équilibre.

Les différences entre les données sur deux années sont faibles: on note une petite remontée des espèces ubiquistes et des Achrochaetetalia ainsi qu'une diminution, toujours très legère, des autres stocks, mais significative pour les Rhodimenietalia (de 16% à 10.9%).

Le déséquilibre du peuplement algale est prouvé aussi par les indices de diversité moyens (Tab. I) qui varient de 1.47 à 2.41 dans le secteur Est et de 0.68 à 1.61 dans le secteur Ouest du Port, avec une moyenne globale de 1.67. Les peuplements algales au cours de la période choisie conservent une même structure, même si quelques variations physionomiques sont observées, notamment dans le secteur oriental, où s'effectue un renouvellement des espèces dominantes.

La couverture varie également de façon significative entre les deux secteurs du Port: la moyenne des données de printemps est 89.16% dans le secteur oriental et 44.23% dans le secteur occidental; en automne elle est respectivement de 85% et 37.05%.

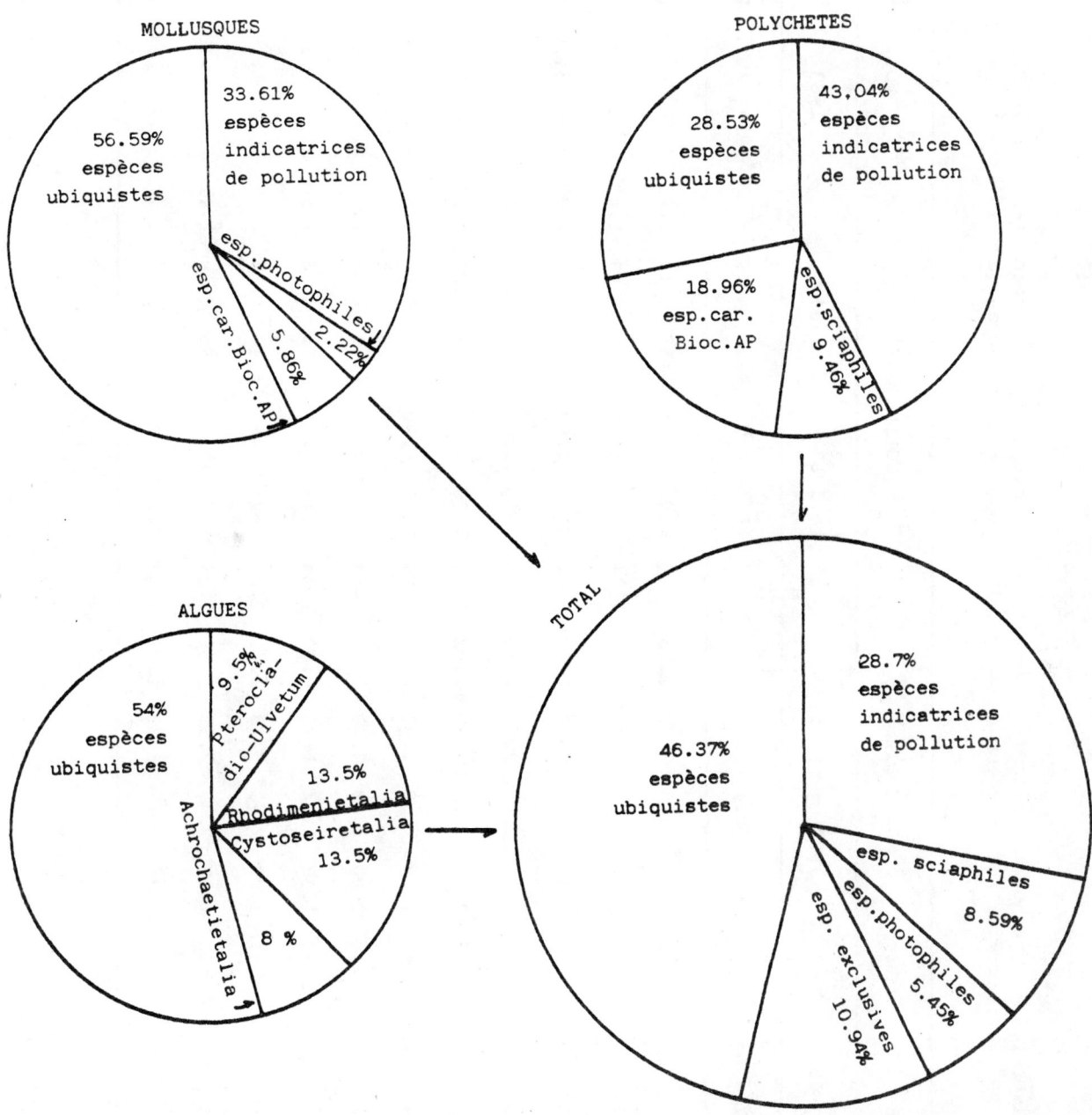

Figure 2. Diagrammes circulaires montrants les rapports quantitatifs entre les différents stocks biocénotiques des prélèvements du Port d'Augusta. L'angle des secteurs circulaires est proportionel au nombre d'individus de chaque stock. Les pourcentages se réfèrent à la moyenne globale des valeurs de printemps et d'automne de 1983 et 1984.

Tableau I

Indices de diversité de Shannon (H) calculés pour les différents groupes d'organismes dans le Port d'Augusta
PR. = Printemps; AUT. = Automne; M. = Valeur moyenne rélative aux années 1983-1984

	ALGUES					POLYCHETES					MOLLUSQUES					POL. + MOLL.				
	1983		1984		M.	1983		1984		M.	1983		1984		M.	1983		1984		M.
	PR.	AUT.	PR.	AUT.		PR.	AUT.	PR.	AUT.		PR.	AUT.	PR.	AUT.		PR.	AUT.	PR.	AUT.	
1	2.10	1.33	3	3.2	2.41	4.07	2.98	3.32	2.55	3.23	2.97	1.94	1.53	2.10	2.13	4.74	3.39	3.54	3.19	3.71
2	1.38	0.79	2.5	3.1	1.94	3.40	1.69	1.60	1.17	1.96	2.28	1.15	—	0.87	1.43	3.38	2.43	1.77	1.98	2.51
4	1.87	1.60	2.6	2.2	2.07	4.10	2.88	2.99	2.87	3.21	2.16	1.93	1.72	1.85	1.91	4.28	3.00	3.12	2.61	3.25
5	0.65	0.53	2.5	2.2	1.47	2.91	2.40	2.36	2.44	2.53	1.75	1.03	1.53	1.75	1.51	3.38	2.31	2.83	2.55	2.77
6	1.44	1.66	2.9	2.6	2.15	2.70	2.15	2.89	1.02	2.19	1.88	2.38	2.47	3.26	2.50	3.13	3.17	3.63	3.25	3.29
7	2.41	0.96	0.9	1.6	1.47	2.19	1.50	2.44	0.14	1.57	0.75	0.90	0.42	0.63	0.69	2.63	2.22	1.40	1.34	1.90
8	1.10	0.49	2.2	2.1	1.47	0.00	1.49	2.79	1.89	1.54	0.75	2.27	1.19	1.16	1.34	0.74	2.90	3.10	2.40	2.28
9	1.44	0.77	1.7	1	1.23	1.53	1.34	1.74	0.80	1.35	1.48	2.19	1.75	0.86	1.57	2.15	2.82	2.75	1.79	2.38
10	0.64	1.32	2.5	2	1.61	3.30	2.75	3.29	2.00	2.83	1.26	0.87	1.44	1.19	1.19	1.58	1.18	2.09	1.45	1.57
11	0.23	0.00	1.5	1	0.68	1.12	2.42	2.58	1.53	1.91	0.73	2.42	1.61	0.88	1.41	1.35	3.13	3.18	1.84	2.37
12	2.22	1.92	2	0.3	1.61	2.38	1.17	3.23	1.59	2.09	1.81	2.02	0.65	1.39	1.47	3.05	2.44	3.29	2.48	2.81

5.2 Polychètes

Les Polychètes sont représentées par 84 espèces dont 13 apparaissent dans les prélèvement de 1984. Il n'y a aucune relation entre le nombre des espèces et le nombre des individus en fonction de la localisation des stations, mais le total des individus recueillis au printemps est supérieur à celui de l'automne.

L'espèce dominante est Hydroides elegans caractéristique des ports plus ou moins pollués, suivie par l'espèce cavitaire Ceratonereis costae. En outre, dans les prélèvements printaniers du secteur oriental Platynereis dumerilii est bien representé.

Les peuplements des stations orientales sont mieux structurés que ceux des stations occidentales, dans lesquelles les indices de diversité de Shannon sont nettement plus bas (Tab. I). L'exception la plus évidente est fournie par les valeurs de la station A_{10} représentée par un peuplement à Corallina elongata sur substrat rocheux naturel.

Les indices de diversité des prélèvements de printemps sont presque toujours plus élevés que ceux d'automne.

Le diagramme circulaire de la Fig. 2 montre que les espèces indicatrices de pollution (43.04%) forment le stock le plus important suivies par les espèces ubiquistes, les espèces caractéristiques de la Biocénose des Algues Photophiles et, enfin, les espèces sciaphiles. En 1984 on enregistre une petite diminution des espèces ubiquistes (30.79% à 26.28%) et indicatrices de pollution (46.23% à 39.86%) et, d'autre part, une augmentation des espèces caractéristiques de la Biocénose AP (15.65% à 22.28%) et sciaphiles (7.39% à 11.59%).

5.3 Mollusques

Les Mollusques sont représentés par 44 espèces: 21 Bivalves, 20 Gastropodes, 3 Polyplacophores. Les Bivalves sont les plus abondants; les espèces dominantes sont les mytilides: Mytilaster minimus, Mytilus galloprovincialis et Brachidontes variabilis, suivis par le Bivalve endobionte Petricola lithophaga.

L'indice de diversité (Tab. I) des stations du secteur occidental est inférieur à celui des stations du secteur oriental, comme pour les Algues et les Polychètes. Il faut remarquer que les indices de diversité d'automne 1983 dans les stations du secteur Ouest dépassent tant les valeurs printanières du même secteur que celle de toute l'année dans le secteur oriental du Port.

L'analyse bionomique dont les résultats sont synthétisés dans le diagramme circulaire de Fig.2, montre que le stock dominant est celui des espèces ubiquistes (56.59%) suivi par le stock des espèces indicatrices de pollution (33.61%) et par le stock des espèces caractéristiques exclusives de Biocenoses infralittorales (5.86%). Les espèces définies comme photophiles (2.22%) sont insignifiantes. Il faut remarquer qu'en 1984, contrairement à ce qui a été enregistré pour les Algues et les Polychètes, le stock des espèces indicatrices de pollution a doublé sa valeur (22.25% à 44.95%). Les espèces exclusives (3.35% à 8.37%) et les photophiles (0.38% à 4.06%) ont considérablement augmenté, tandis que les espèces ubiquistes ont diminué fortement (60.61% à 42.58%). La destructuration du peuplement à Mollusques est également souligné par une structure trophique dominé par les filtreurs (85%), tandis que les détritivores, les herbivores et les carnivores sont insignifiants.

5.4 Biomasse

Les valeurs de la biomasse (g pour $1/25\ m^2$) rencontrées (Tab.2) sont parfaitement identiques à celles des peuplements de même faciès en milieu portuaire ou pollué (Bellan-Santini, 1969).

La quantité importante de carbonates présents dans les échantillons est dûe principalement aux coquilles de Mytilus galloprovincialis. Pour toutes les stations du secteur occidental, également au Cirripède Balanus amphitrite et, en proportion moindre, aux tubes de Serpulidés.

Tableau II

Port d'Augusta: valeurs de la biomasse calculées à partir des moyennes saisonnières des années 1983-1984. PR. = Printemps; AUT. = Automne; VEG. = Végétal; AN. = Animal

STAT.		POIDS HUMIDES		P.DECALCIFIES		POIDS SECS		P.DE CENDRES		MAT.ORGANIQUE	
		VEG.	AN.	VEG.	AN.	VEG.	AN.	VEG.	AN.	VEG.	AN.
1	PR.	87.38	331.40	41.41	72.12	6.23	7.59	1.43	7.24	4.8	6.94
	AUT.	173.82	116.44	76.25	14.19	10.62	3.93	6.07	1.74	4.56	2.04
2	PR.	179.12	916.78	82.17	133.87	21.59	42.23	4.38	22.29	17.22	19.93
	AUT.	53.76	485.88	27.4	21.26	5.46	23.05	1.06	17.42	4.4	5.64
4	PR.	68.42	378.92	32.94	49.92	4.4	15.66	0.93	8.03	3.46	7.63
	AUT.	64.15	164.94	22.36	24.01	4.16	4.67	1.37	3.58	2.79	1.08
5	PR.	113.17	358.43	78.54	84.53	8.09	15.38	2.57	22.31	5.52	6.62
	AUT.	31.54	298.9	9.95	72.41	1.7	25.95	0.25	21.07	1.44	4.88
6	PR.	161.23	301.32	71.6	59.01	4.07	20.03	1.92	7.62	8.83	12.41
	AUT.	89.11	290.47	45.03	64.53	5.13	24.07	1.02	14.92	4.10	9.15
7	PR.	27.98	646.1	11.14	109.94	1.73	24.53	0.49	5.94	1.25	18.59
	AUT.	26.21	332.18	3.79	71.31	1.22	13.9	0.46	3.22	0.76	10.68
8	PR.	18.05	878.07	9.57	148.18	0.82	33.97	0.21	4.1	0.87	29.86
	AUT.	12.65	628.56	1.99	109.24	0.5	23.06	0.14	3.23	0.35	19.82
9	PR.	35.92	879.52	9.05	171.5	1.23	46.96	0.16	15.82	1.07	31.14
	AUT.	1.69	548.13	0.31	76.74	0.07	19.01	0.01	6.4	0.06	12.61
10	PR.	128.64	199.58	55.26	23.11	7.85	9.76	0.8	5.07	7.04	4.69
	AUT.	104.03	233.73	39.85	50.16	3.96	12.9	0.34	8.6	3.62	4.3
11	PR.	14.05	606.04	7.06	91.61	0.59	18.31	0.1	5.59	0.49	12.71
	AUT.	-	561.14	-	78.85	-	9.44	-	1.69	-	7.75
12	PR.	5.99	365.08	2.5	82.33	0.31	15.12	0.04	2.86	0.27	12.26
	AUT.	-	402.45	-	39.32	-	7.04	-	1.32	-	5.72

La biomasse végétale des stations du secteur occidental est plus basse que celle du secteur oriental; la biomasse animale des prélèvements de printemps est toujours plus élevée que celles des prélèvements d'automne. Les stations plus riches en <u>Mytilus</u> ou en <u>Cirripèdes</u> sont les plus pauvres en biomasse végétale.

6. CONCLUSION

L'étude qualitative et quantitative des peuplements benthiques de substrats durs de l'infralittoral supérieur a mis en évidence l'état de dégradation du Port d'Augusta, causée par une pollution complexe industrielle, urbaine et portuaire. Il est très difficile, voire impossible, d'évidencier analytiquement les divers produits industriels déversés dans le Port puis dilués dans l'eau marine. Il est aussi délicat d'apprécier quantitativement leur contribution à la pollution, mais la méthode synécologique nous a permis de reconnaître la présence globale des polluants industriels en analysant les effets sur les peuplements benthiques. Ces effets sont accentués quand dans une boue liquide industrielle manquent les composants eutrophisants, qui, dans le Port d'Augusta, sont sûrement recueillis en totalité par la centrale d'épuration (Rizzi-Longo et al., 1982).

Le peuplement benthique algale et animal est déstructuré par rapport à la composition et structure de la biocénose originale des algues photophiles présentes sur les côtes avant la construction des digues du Port et l'implantation des industries. L'analyse biocoenotique se réfère aux données de 1983 et 1984, mises en évidence dans le cyclogramme de la Fig.2. Elle montre un stock important d'espèces indicatrices de pollution (28.7%) accompagné par une grande quantité d'espèces ubiquistes (46.37%). La présence du 10.94% d'espèces exclusives de la Biocenose AP s'explique par l'influence du stock des Polychètes (18.96%) et, probablement, par le fonctionnement, même réduit, de la centrale d'épuration fournissant des eaux plus claires à l'intérieur du Port.

Il y a une grande différence entre les stations du secteur oriental du Port et celles du secteur occidental. Les premières sont soumises aux apports urbains principalement dans la partie septentrionale, avec cependent des échanges d'eaux à travers la digue externe, de la partie centrale et méridionale, qui est formée de gros blocs. Les secondes subissent l'effet des décharges industrielles et des turbulences sur le fond causées par les navires le long des pontons d'amarrage.

Cette différence est bien reconnaîssable grace aux indices de diversité de Shannon (Tab.I, Fig.3) plus élevés dans les stations orientales que dans le secteur occidental.

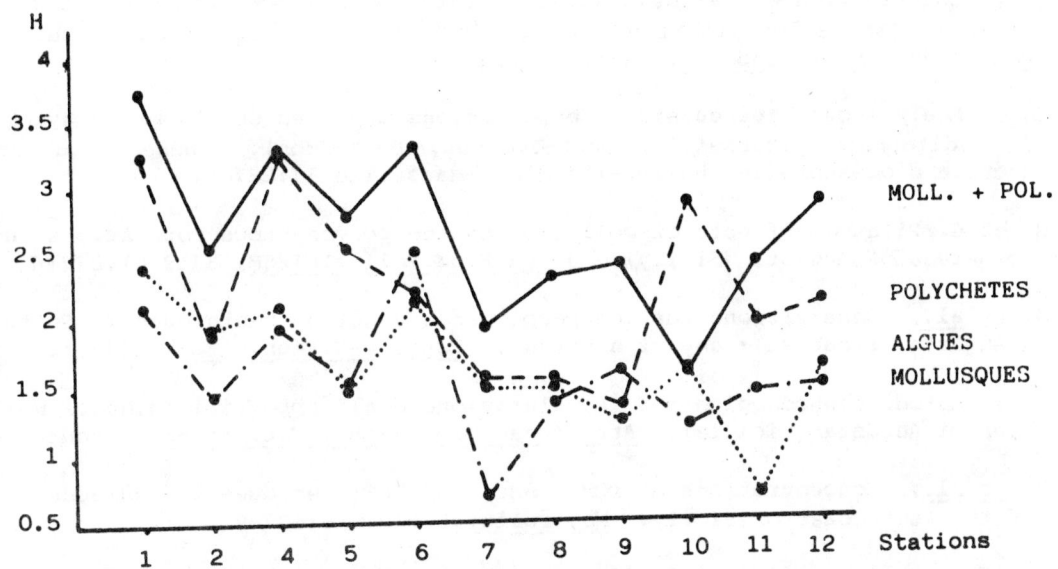

Figure 3. Variations de l'indice de diversité de Shannon (H) globale pour les années 1983 et 1984, dans le Port d'Augusta.

Cet indice présente des valeurs différentes selon les groupes d'organismes considerés: basses pour les Mollusques (Tabl. I) et plus élevés pour les Algues et les Polychètes. L'indice calculé avec les deux groupes de Mollusques et des Polychètes est toujours plus élevé que l'indice de chaque groupe composant.

Il est plus important de remarquer que, quelque soit le groupe d'organisme, l'indice de diversité présente toujours la même allure.

La différence entre les deux secteurs du Port est traduite aussi par l'indice R/P (Rhodophycées/Phéophycées) qui caractérise l'état de dégradation du milieu. La valeur globale de R/P pour le Port d'Augusta (5.0) est moins importante de celle du Port de Marseille (7.3) mais dénonce également la pollution très poussée du port sicilien.

Les données de la biomasse, enfin, concordent avec toutes les autres données et soulignent seulement la faiblesse en poids ou même l'absence des Algues dans le secteur occidental pollué du port.

En conclusion, il ressort de cette étude que l'analyse bionomique, les indices de diversité et l'indice R/P sont de bons temoins de l'état des biocenoses et que leur emploi dans l'étude des peuplements de substrat dur du Port d'Augusta à donné des résultats satisfaisants qui ont permis de contrôler le degré et la marche de la pollution. Les récherches conduites en 1983 et 1984 permettent de déduire qu'en 1984 la situation générale c'est très légèrement améliorée mais le Port d'Augusta demeure toujours dans un état de déséquilibre très poussé, assez comparable à celui du Port de Marseille.

7. REMERCIEMENTS

Je remercie Mme G. Cantone pour la détermination des Polychètes, Ms G. Giaccone pour la détermination des algues et Mme M. Leonardi pour les données sur la biomasse. Un remerciement particulier à Mlle A. Rosso et Ms S. Giacobbe dont l'étroite collaboration m'a aidé à l'éxécution de ce travail.

8. BIBLIOGRAPHIE

Bellan, G., Pollution et peuplements sur substrats meubles dans la région de Marseille
1967 Première partie Le secteur de Cortiou. Rev.Int.Océanogr.Méd., 6-7:53-87

Bellan-Santini, D., Influence de la pollution sur les peuplements benthiques.
1968 Rev.Int.Océanogr.Méd., 10:27-53

_____, Contribution à l'étude des peuplements infralittoraux sur substrat
1969 rocheux (Etude qualitative et quantitative de la frange supérieure). Recl.Trav.Stn.Mar.Endoume, 63(47):1-294

Belsher, Th., Analyse des repercussions de pollutions urbaines sur le macrophytobenthos
1977 de Méditerranée (Marseille, Port-Vendres, Port-Cros). Thèse de doctorat de 3e cycle d'océanologie, Université d'Aix-Marseille II:287 p.

Cantone, G. et G. Pilato, Effects of pollution on the polychaetous populations in the
1981 roadstead of Augusta (Sicily). Journ.Etud.Pollut.CIESM, 5(1980):857-62

Cormaci, M. et al., Observations sur les peuplements végètaux benthiques du Golfe
1983 d'Augusta (Syracuse): aspect hivernal. Rapp. P-V.Réun.CIESM, 28(3):219-20

_____, Metodo sinecologico per la valutazione degli apporti inquinanti nella Rada di Augusta (Sicilia). Atti Accad.Gioenia Sci.Nat.Catania (sous presse)

D'Amico, V. et al., Concentrations of PCB_1, BHC_1 and DDT_1 residues in seaweeds
1979 of the East Coast of Sicily. Mar.Pollut.Bull., 10:177-9

De Domenico, E., Sulle condizioni fisico-chimiche e sulla produzione primaria delle
1972 acque esterne alla Rada di Augusta. Atti Soc.Peloritana Sci.Fis.Mat.Nat., 18:177-212

De Domenico, E. et al., Nuove ricerche idrobiologiche nella zona esterna alla Rada di
1978 Augusta (Luglio 1974-Aprile 1975). Atti Soc.Peloritana Sci.Fis.Mat.Nat., 24:93-179

Desrosiers, G. et al., Evolution spatio-temporelle des peuplements de substrats rocheux
1982 superficiels dans un golfe soumis à de multiple pollutions (Golfe de Fos, France). Tethys, 10(3):245-53

Di Geronimo, I., Prima segnalazione sulle coste italiane di Brachydontes variabilis
1971 (KRAUSS). Boll.Sedute Accad.Gioenia Sci.Nat.Catania, 10(6):847-52

_____, Influence de la pollution sur les peuplements à Mollusques de la baie
1983 d'Augusta (Sicile). Journ.Etud.Pollut.CIESM, 6(1982):715-19

Di Geronimo, I., et al., Relation entre pollution et biocenoses en Baie d'Augusta (Sicile
1985   orientale). Journ.Etud.Pollut.CIESM, 7(1984):791-7

Feldman, J., Recherches sur la végétation marine de la Méditerranée: la côte
1937   des Albéres. Rev.Algol., 10:1-339

Genovese, S., et E. De Domenico, Sulle condizioni microbiologiche delle acque esterne
1975   alla Rada di Augusta. Atti Soc.Peloritana Sci.Fis.Mat.Nat., 21:123-33

Giaccone, G., Effetti sul fitobentos dello smaltimento a mare delle acque di fogna.
1977   Ing.Ambientale, 6(1):37-43

Giaccone, G., et al., Evoluzione e distribuzione della vegetazione marina nei tre golfi
dell'area metropolitana di Palermo (Sicilia). Atti Accad.Gioenia
Sci.Nat.Catania (sous presse)

_____, Flora e vegetazione marina dell'Isola di Ustica. Atti Acc.Gioenia Sc.Nat.
Catania, (sous presse)

Giangrande, A., et M.C. Gambi, Distribuzione dei Policheti nei fondi mobili della Rada
1983   di Augusta (Sicilia). Boll.Mus.Ist.Biol.Univ.Genova, 50 (Suppl.):218-22

Molinier, R., et J. Picard, Notes biologiques à propos d'un voyage d'étude sur les
1958   côtes de la Sicile. Ann.Inst.Océanogr., Monaco, 28(4):163-87

Pérès, J.M. et J. Picard, Nouveau manuel de bionomie benthique de la Mer Méditerranée.
1964   Recl.Trav.Stn.Mar.Endoume, 31(47):1-137

Rizzi-Longo, L. et al., Variazioni dell'attività metabolica di alghe marine bentoniche
1982   in coltura in presenza di liquami industriali. Nat.Sicil., 6(Suppl.1):61-9

Russo, G.F., Distribuzione della malacofauna di fondo mobile nella Rada di Augusta
1983   (Sicilia). Boll.Mus.Ist.Biol.Univ.Genova, 50 (Suppl.):325-31

Sciacca, S., et R. Fallico, Presenza e concentrazione di sostanze inquinanti di origine
1978   industriale nei fanghi della Rada di Augusta (Siracusa). Inquinamento,
20(6):1-4

Stirn, J., Manuel des méthodes de recherche sur l'environnement aquatique. Huitième
1982   partie. Evaluation des modifications des écosystèmes marins dues à la
pollution (Directives destinées au projet commun coordonné FAO(CGPM)/PNUE sur
la pollution en Méditerranée). FAO Doc.Tech.Pêches, (209):75 p. Publié aussi
en anglais

OCCURRENCE OF TINTINNIDS IN TWO POLLUTED AREAS
OF ALEXANDRIA COAST

by

M.M. DORGHAM
*Oceanography Department, Faculty of Science,
Alexandria University, Alexandria, Egypt

1. INTRODUCTION

Tintinnids are permanent components of the plankton in the Egyptian Mediterranean waters. They are characterized by a significant diversity of species and low production (Dowidar, 1965). Little attention has been drawn to the study of the tintinnids in Alexandria coast. Dowidar (1965) and Dowidar and El-Maghraby (1970) referred to the tintinnid population in Alexandria region, mainly in the offshore waters, where they recorded 99 species. El-Maghraby and Halim (1965) sampled two stations only in front of the Eastern Harbour of Alexandria and recorded 22 species over all the year. No study was made on the effect of pollution of the plankton communities in the two areas concerned in this work.

In this work, the species composition and population density of tintinnids in relation to the effect of pollution was stressed. The study was concentrated on the inshore water proper which is affected by the land-based effluents.

The site

The investigated areas are about 10 km from each other (Fig. 1). The first area (El-Mex) is off the western Port of Alexandria city. It extends from the shore seaward to 15 m depth. This area receives through the Umoum drain about 6 million m^3 day^{-1} of land drainage water from cultivated land nearby. In addition, 40000 m^3 day^{-1} of industrial wastes are disposed into the area by chemical industries, petroleum refineries, 16 tanneries and a slaughterhouse. The most significant pollutants in these wastes are mercury (5 tons $year^{-1}$), free chlorine (up to 70 mg l^{-1}), ammonia, sludge and others (Salem, pers.comm.).

The second area is located close to Fort Kayet Bay at El-Anfoushi, where about 150000 m^3 day^{-1} of domestic waste water are discharged with high ammonia content (up to 93 ug at l^{-1}) Hanafy, pers.comm.).

The range of salinity variations was significantly wide in the two areas; El-Mex: 18.02 - 38.44 $^o/oo$ and El-Anfoushi: 35.24 - 39.44 $^o/oo$ (Hanafy, pers.comm.).

These conditions give the two areas particular characteristics, which in turn affect the qualitative and quantitative structures of the marine biota inhabiting them.

2. MATERIALS AND METHODS

Samples were collected at 7 stations representing the different water qualities in the two areas (Fig. 1). The first area (El-Mex) was represented by stations 1,2 and 3, while the second area (El-Anfoushi) was sampled at stations 4,5 and 6. Station R at El-Agami, 13 km west of Alexandria, was chosen as reference. Collection of samples was carried out bimonthly during the period from April 1982 - August 1983. Qualitative samples were collected by a coarse plankton net. Quantitative samples were obtained by filtering 50 liters of sea water through the coarse net.

Present address: Department of Marine Sciences, Faculty of Science, Qatar University, Doha, Qatar

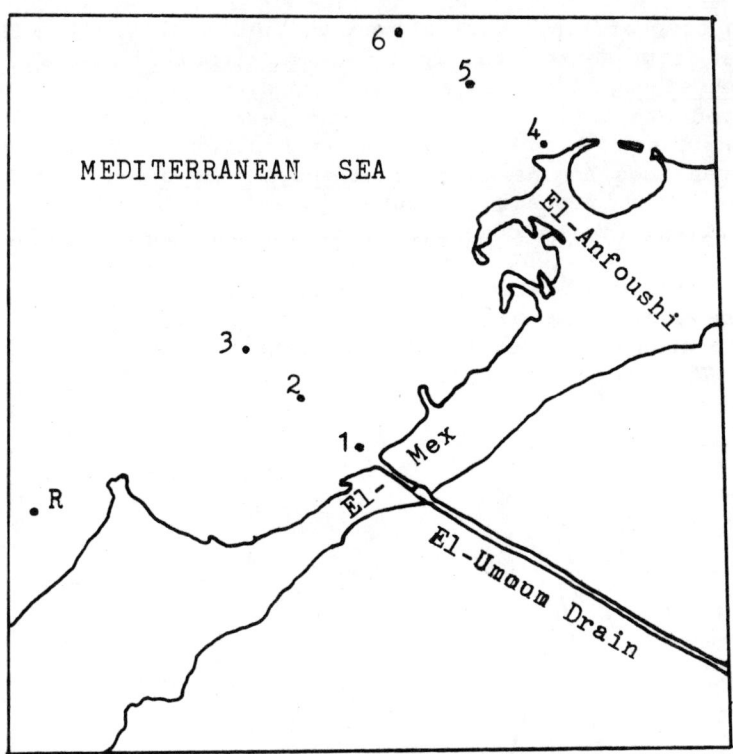

Figure 1. Area of investigation and the sampling stations

3. RESULTS AND DISCUSSION

Forty-six tintinnid species belonging to 20 genera were recorded during the period of investigation. The majority of them are neritic and temperate. Some are oceanic and neritic, while few forms are referred to as oceanic. The latter may be transferred to the inshore area by the water circulation or wave action. The tintinnid population in the investigated areas showed rather high species diversity but it was characterized by low production. This phenomenon was also observed by Dowidar (1965) in Alexandria water and by Jorgensen (1924) for the whole Mediterranean Sea.

The number of the recorded species showed significant seasonal variations. Winter and spring samples contained the same number (26-27), while 20 and 14 species were observed in summer and autumn respectively. The most important genera were Favella (8 sp.), Tintinnopsis (6 sp.) and Eutintinnus (5 sp.).

Helicostomella subulata, Tintinnopsis beroidea, Eutintinnus fraknoi, E. lusus-undae and Favella ehrenbergii were observed over the year as perennial species. Others dominated the tintinnid population in different seasons. Winter population was dominated by Tintinnopsis campanula, but in spring the dominant species were Favella markuzowskii, F. taraekaensis, F. adriatica and F. campanula, particularly at station 3. In Summer, Crateralla urceolata and Favella markuzowskii co-dominated with the previously mentioned perennial species. Only two perennial species (Favella ehrenbergii and Helicostomella subulata) dominated in autumn.

Quantitative samples contained only 3-4 species all the year round, mostly those of small size like Helicostomella subulata and Tintinnopsis beroidea.

The most important species observed by Dowidar and El-Maghraby (1970) in the neighbouring area were more or less different. These were; Favella markuzowskii, F. franciscana, F. azorica, Tintinnopsis beroidea, T. campanula, Helicostomella subulata, Codonellopsis morchella and Coxliella annulata.

In the present work, 6 species are observed as new records to the Egyptian Mediterranean water; namely, Amphorella minor, Dictyocysta obtusa, Helicostomella edentata, Undella attenuata, Undellopsis sp. and Poroecus apiculatus.

The average standing crop of tintinnids showed two peaks (Fig. 2). The summer peak (4200 organisms m^{-3}) was represented mainly by Helicostomella subulata, Tintinnopsis beroidea and Favella ehrenbergii. The maximum standing crop (9000 organisms m^{-3}) during summer was observed at station 2. The second peak of the average standing crop (2770 organisms m^{-3}) was observed in spring with a maximum of 7200 organisms m^{-3} at station 3. The spring peak was dominated by Tintinnopsis beroidea. It is to be noted that, another high count of this species (5600 organisms m^{-3}) in spring was recorded at station 5, near the effect of the domestic wastes. This may indicate the tolerance of Tintinnopsis beroidea to some extent to the pollution effect. On the other hand, in addition to its summer maximum at station 2, Helicostomella subulata showed another maximum in autumn. These two maxima of the latter species at the same station (2) may point to its greater tolerance to temperature variations and pollution effect than Tintinnopsis beroidea. The same can be claimed for Favella taraikaensis, which was abundant in spring at station 2.

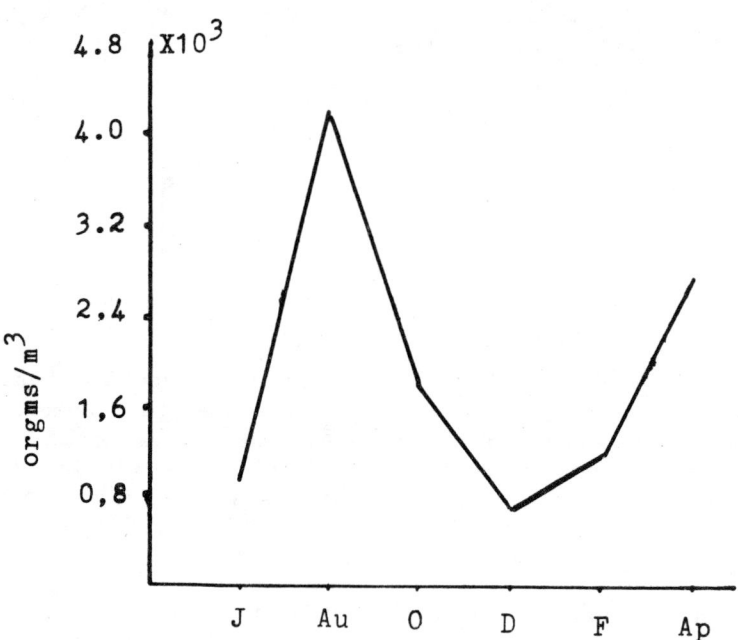

Figure 2. Average standing crop of Tintinnids in the area of investigation all the year round.

In the net samples, several species were recorded in significant numbers at different salinities and temperatures. Stenosemella nivalis, Tintinnopsis campanula, Amphorella quadrilineata, Eutintinnus latus and Favella fistulicauda were found during both winter and summer with a difference in temperature of about $10^{o}C$. Other species including the perennials were observed at different salinities. This phenomenon allows us to suppose that several tintinnids showed no obvious affinities with certain temperature or salinity in their maximum growth. Dowidar (1965) indicated that, tintinnids showed no temperature preference to build up a large population and their appearance was spasmodic and apparently not controlled by particular environmental factors. Davis (1955) pointed to the susceptibility of tintinnids to relatively slight environmental changes and the wide fluctuation of their number.

During most of the year the number of tintinnid species at the less polluted stations (3 and 5) was larger than that observed at the polluted stations. Eight species were restricted mainly to station 4, which is directly affected by the domestic wastes, but they were represented by small numbers of assemblages. These species were; Crateralla urceolata, Favella fistulicauda, Metacylis mediterranea, Proplectella angustior, Rhabdonella elegans, Stenosemella nivalis, S. ventricosa and Undellopsis sp.

It must be noted that tintinnids were not recorded at station 1, located just in front of Umoum drain. This can be attributed to the low water salinity, which was lower by almost 5 °/oo.

As mentioned before, the two areas are mostly 10 km far from each other, but the pollutants discharged into them are different. That at El-Anfoushi is mainly domestic waters, while at El-Mex, agricultural, industrial and petroleum wastes are disposed. Comparing the two areas qualitatively as regards to the tintinnid population, it is clear that, 37 species occurred at El-Anfoushi against only 22 at El-Mex. This had a noticeable effect on tintinnids as several species were absent from this area. On the other hand, the sewage outflow in El-Anfoushi seems to be less effective than the industrial wastes at El-Mex. The species recorded in both areas all the year round and those recorded only at El-Anfoushi are listed in Table I.

The Tintinnid population exhibited significant annual variations in species composition and standing crop. Some species, which were recorded in significant numbers during 1982, disappeared from samples at the same stations in 1983 while others appeared. At station 3, Favella adriatica, F. campanula and Metacylis mediterranea were observed in significant numbers in samples of April 1982, and completely disappeared in April 1983. This figure was repeated at station 4 for Favella composita and F. ehrenbergii. The annual replacement of species was observed at several stations (Table II). This is attributed to the continuous changeable inflow of the land drainage loaded with variable quantities of different pollutants.

Table I
Tintinnid species recorded in both investigated areas and at El-Anfoushi only

In both areas	At El-Anfoushi
Coxliella ampla	Amphorella minor
Crateralla urceolata	" quadrilineata
Eutintinnus elongatus	Codonella sp.
" fraknoi	Dictyocysta lepida
" latus	" obtusa
" lusus-undae	Epiplocylis undella
Favella adriatica	Eutintinnus tubulosus
" composita	Poroecus apiculatus
" ehrenbergii	Proplectella angustior
" fistulicauda	Rhabdonella conica
" markuzowskii	" elegans
" taraikaensis	Salpingella attenuata
Helicostomella edentata	Steenstrupiella steenstrupii
" subulata	Stenosemella nivalis
Metacylis mediterranea	" ventricosa
Rhabdonella spiralis	Tintinnopsis buetschlii
Tintinnopsis beroidea	" radix
" campanula	Undella attenuata
	Undellopsis sp.

4. CONCLUSIONS

The Tintinnid population in the investigated areas was characterized by high species diversity and low production. The high production of some species was not restricted to a certain temperature or salinity. Some species showed slightly high production in the areas affected by the wastes disposal. On the other hand, a low number of species was observed at the polluted stations particularly in El-Mex area affected by the industrial wastes.

Table II

Tintinnid species recorded in El-Mex and El-Anfoushi areas
from April 1982 to August 1983

Species / Stations	April '82			June '82					August '82				
	3	4	5	2	3	4	5	6	2	3	4	5	R
Amphorella minor													
" quadrilineata												R	
Codonella sp.												R	
Codonellopsis morchella													
" orthoceros													R
Coxliella ampla												R	
" annulata	R												
" decipiens	R												
Crateralla urceolata													
" sp.													
Dictyocysta lepida													
" obtusa													
Epiplocylis undella													
Eutintinnus elongatus													
" fraknoi													R
" latus	R												
" lusus-undae				R	R		R	R					
" tubulosus													
Favella adriatica	A	R		R						R		R	
" campanula	A			R									
" composita			F							R		R	R
" ehrenbergii	A	F		R		F	R		R			F	R
" fistulicauda									R				
" markuzowskii	A	R				F	R		R			R	R
" serrata													
" taraikaensis										R		R	
Helicostomella edentata				R									
" subulata										F		R	
Metacylis mediterranea	F	A								R			
Poroecus apiculatus		R											
Proplectella angustior													
Rhabdonella conica													
" elegans													
" spiralis													
Salpingella attenuata													
Steenstrupiella steenstrupii													
Stenosemella nivalis													
" ventricosa													
Tintinnopsis beroidea	R			F						R	R		
" buetschlii													
" campanula													
" lobiancoi													
" mortensenii													
" radix		R											
Undella attenuata													
Undellopsis sp.		R											

Table II (cont.)

Species / Stations	October '82						December '82						February '82					
	2	3	4	5	6	R	2	3	4	5	6	R	2	3	4	5	6	R
Amphorella minor															R	R	R	
" quadrilineata															R	R		
Codonella sp.																		
Codonellopsis morchella										R								
" orthoceros																		
Coxliella ampla	R																	
" annulata																		
" decipiens																		
Crateralla urceolata																		
" sp.																		
Dictyocysta lepida															R	R		
" obtusa										R								
Epiplocylis undella																		
Eutintinnus elognatus																		R
" fraknoi		R		R					R						R			
" latus			R		R											R	R	
" lusus-undae		R	R	R											R			
" tubulosus															R			
Favella adriatica			R															
" campanula																		
" composita																		
" ehrenbergii		R		R	F			R									R	
" fistulicuada										R								
" markuzowskii		R																
" serrata		R																
" taraikaensis									R						R		R	
Helicostomella edentata			R	R	F									R	R			
" subulata	R		R	R	F		R		R	R			F	A	R	R	R	R
Metacylis mediterranea		R	R	R	R													
Poroecus apiculatus																		
Proplectella angustior															R	R	R	
Rhabdonella conica																		
" elegans										R								
" spiralis																		
Salpingella attenuata																	R	R
Steenstrupiella steenstrupii										R							R	R
Stenosemella nivalis																R		
" ventricosa																R	R	
Tintinnopsis beroidea			R				R	R	R	R			R	R	R	R	R	
" buetschlii										R	R							
" campanula							R	R		R	R							
" lobiancoi																		
" mortensenii			R															
" radix								R	R									
Undella attenuata																	R	
Undellopsis sp.																		

Table II (cont.)

Species / Stations	April '83					June '83						August '83				
	2	3	4	6	R	2	3	4	5	6	R	2	3	4	5	R
Amphorella minor																
" quadrilineata											R					
Codonella sp.																
Codonellopsis morchella																
" orthoceros																
Coxliella ampla	R															
" annulata												R				
" decipiens																
Crateralla urceolata												R	F	F		
" sp.												R				
Dictyocysta lepida																
" obtusa																
Epiplocylis undella									R							
Eutintinnus elongatus							R	R	R	R						
" fraknoi				R									R	R	R	R
" latus									R							
" lusus-undae													F			
" tubulosus																
Favella adriatica							R									
" campanula																
" composita																
" ehrenbergii		A		F	R											
" fistulicauda																
" markuzowskii	F	A		F									R	R		
" serrata	A															
" taraikaensis	A		R	F	F											
Helicostomella edentata								R								
" subulata		R		R		F	R		R	R						
Metacylis mediterranea		R														
Poroecus apiculatus																
Proplectella angustior																
Rhabdonella conica							R	R								
" elegans																
" spiralis							R		R	R						
Salpingella attenuata																
Steenstrupiella steenstrupii								R								
Stenosemella nivalis																R
" ventricosa																
Tintinnopsis beroidea	F	F	A	R			R	R		R	R	R				
" buetschlii																
" campanula													R			
" lobiancoi																
" mortensenii																
" radix																
Undella attenuata																
Undellopsis sp.																

A: Abundant, F: Frequent, R: Rare

5. ACKNOWLEDGEMENT

The author wishes to thank Prof. Dr. Y. Halim, Oceanography Department, Faculty of Science, Alexandria University, for his help during preparation of this work. Thanks are also due to Prof. Dr. A.M. El-Maghraby, Department of Marine Sciences, Qatar University, for reading the manuscript. Many thanks to Dr. M. El-Samra and Mrs. S. Hanafy, Institute of Oceanography and Fisheries for their help in sample collections and some hydrographic data.

6. REFERENCES

Davis, C.C., The marine and fresh-water plankton. East Lansing, Michigan, Michigan
1955   State University Press, 541 p.

Dowidar, N.M., Distribution and ecology of marine plankton in the region of Alexandria,
1965   Egypt. Ph.D. Thesis, Alexandria University, Egypt

Dowidar, N.M. and A.M. El-Maghraby, The neritic zooplankton of the south eastern
1970   Mediterranean at Alexandria. 1. Distribution and ecology of the zooplankton organisms with special reference to Copepoda. Bull.Inst.Oceanogr.Fish. Cairo, 1:225-73

El-Maghraby, A.M. and Y. Halim, A quantitative and qualitative study of the plankton of
1965   Alexandria waters. Hydrobiologia, 25(1/2):221-38

Jörgensen, E., Mediterranean Tintinnidae. Rep.Dan.Oceanogr.Exped.Mediterr., 3 vol.2,
1924   Biol. 110 p.

EFFECTS OF POLLUTANTS ON MARINE COMMUNITIES AND ECOSYSTEMS
IN LIMASSOL BAY

by

M. HADJICHRISTOPHOROU and A. DEMETROPOULOS
Ministry of Agriculture and Natural Resources,
Department of Fisheries,
Nicosia, Cyprus

1. INTRODUCTION

The Department of Fisheries was involved in pollution work prior to its involvement in MED POL, mainly in the area of projects MED IV and V, and a similar study (on less sophisticated lines) was undertaken in Morphou Bay.

2. STUDY AREAS

The areas studied and the location of the sampling stations are shown in Fig.1. Two series of sub-stations have been set up, one in Limassol Bay and the other in Episkopi Bay. Each series consists of seven sub-stations at 5, 10, 20, 30, 40, 60 and 100 metres for benthos investigations. At the same time nine stations (profiles) were set up for environmental data collection.

Limassol Bay (Station II or Area II)

The sea bed at this station was as follows:

Beach- sand with some shingle
  5m - posidonia meadows/mud
 10m - posidonia meadows/mud
 20m - mud with caulerpa
 30m - sandy mud with caulerpa
 40m - sandy mud with caulerpa
 60m - mud with caulerpa
100m - mud, aphytal

Limassol Bay faces south and is bounded on the west side by the Akrotiri peninsula and on the north by the mainland. The main town in the area is Limassol, which has a population of about 100,000 and two commercial ports and considerable tourism infrastructure. There are no permanent rivers, only streams which flow during the rainy period. The increasing construction of dams also limits the quantity of water reaching the sea.

Surface sea-water temperatures (at 08.00 hours) reach a maximum of approximately $27.0^\circ C$ and minimum of around $14.5^\circ C$. During the summer months, the thermocline is formed at about 25-30 metres depth. Salinity varies from about $38.825^\circ/oo$ to $39.425^\circ/oo$. The dissolved oxygen concentration during winter is approximately 5.5ml l^{-1} to a depth of 5-75 metres. During summer this value is approximately 5m l^{-1} reaching the same maximum at the same depth. Analyses for suspended solids range between 0-400 ppm. The area of the continental shelf for Limassol Bay is approximately 42 square miles (down to 100 fathoms or approximately 200m) and has a gentle slope, becoming steeper near Cape Gata.

Episkopi Bay (Station I or Area I)

The sea bed at this station was as follows:

Beach- sandy beach with low rocks
  5m - sand with scattered posidonia outcrops
 10m - sand with scattered posidonia outcrops

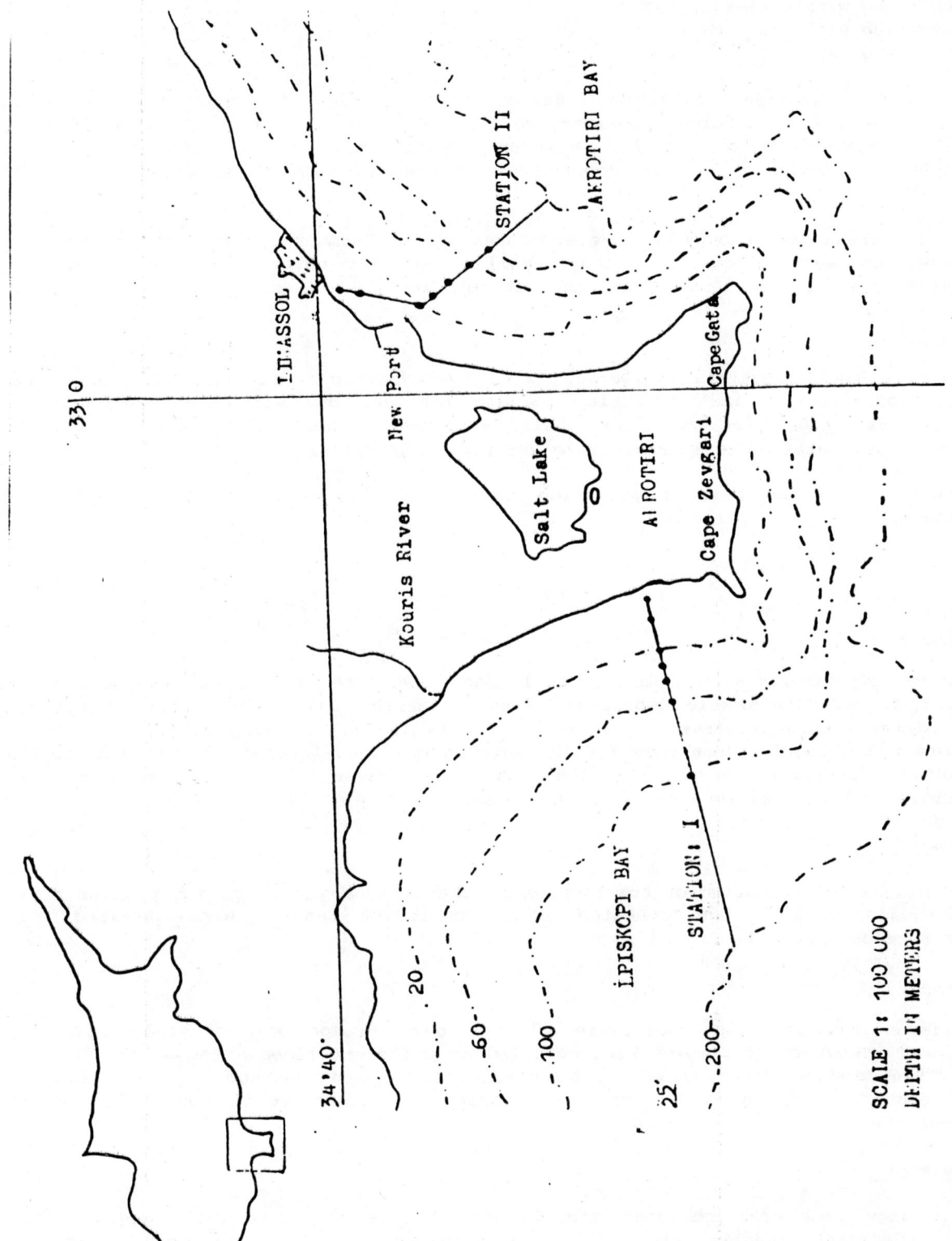

Figure 1. Map showing sampling stations.

```
   20m  - sandy mud with caulerpa
   30m  - sandy mud with caulerpa
   40m  - mud with caulerpa
   60m  - mud with caulerpa
  100m  - mud, aphytal
```

Episkopi Bay lies west of Limassol Bay on the other side of the Akrotiri peninsula. There is no habitation in close proximity to the sea. The area of the continental shelf of Episkopi Bay is 67 square miles (down to 100 fathoms, or 200m) and is similar in topography to Limassol Bay. The hydrographic characteristics are similar to those of Limassol Bay.

The two areas differ only in wave exposure. In the summer in Episkopi Bay there is more consistent wave action due to the prevailing westerly winds. These waves, however, are generally small and do not affect the bottom below 7-8 metres.

Pollutants:

(a) Limassol Bay. Industrial pollution, with few exceptions, is restricted to organic pollution. Local wine industries pollute Limassol Bay with about 100 m.tons of oxygen (as BOD_5) per year for their breakdown. These 100 m.tons of suspended solids are contained in approximately 200,000 m^3 of effluent.

(b) Episkopi Bay. No serious pollution exists in this area and it is therefore considered to be "unpolluted".

3. METHODOLOGY

(a) Infauna:

Infauna was sampled with an orange peel grab. Two sets of samples were obtained at different times. These were labelled Series B (March, 1977) and Series C (October, 1977). Samples were preserved in 4 per cent formalin after being screened through a 1.5mm mesh. 100 Samples were used for the data given, i.e. for each depth (sub-station) 20 grab samples were collected, 5 l each. This was decided on after a cumulative curve showed this to be the minimum required (Holme and McIntyre, 1971).

(b) Biomass:

Dry weight determination of the biomass of the infauna was carried out after drying for 24 hours at 100°C and repeated until consistent results were obtained. The samples were decalcified.

(c) Epifauna:

Epifauna sampling was undertaken with a beam dredge and a sledge of equal dimensions. Two sets of samples were obtained at different times. These were Series A (June, 1976) and B (March, 1977). The results given are the sum of three hauls, 5 minutes each for each series of sampling. Samples were preserved in 4 per cent formalin in sea-water.

(d) Sediments:

An orange peel grab was used and 5 samples were taken at each depth in each station. Sediment gradation was carried out through hydrometer analysis and wet sieve analysis according to B.S. 1377 of 1975.

(e) Environmental data

Standard oceanographic methods were used.

(f) Treatment of the data:

Principal Component Analysis was carried out using the SPSS Programme (Nie et al., 1979).

4. RESULTS AND DISCUSSION

Fines of the sediment

Figure 2 shows the percentage of fines (particles smaller than 0.076 mm) at each depth (sub-station). At the Area I the fines increase from the shallower to the deeper sub-stations. In the polluted Area II there is an increased concentration of fines at the shallower sub-stations which decreases in the middle sub-stations and increases again in the deepest.

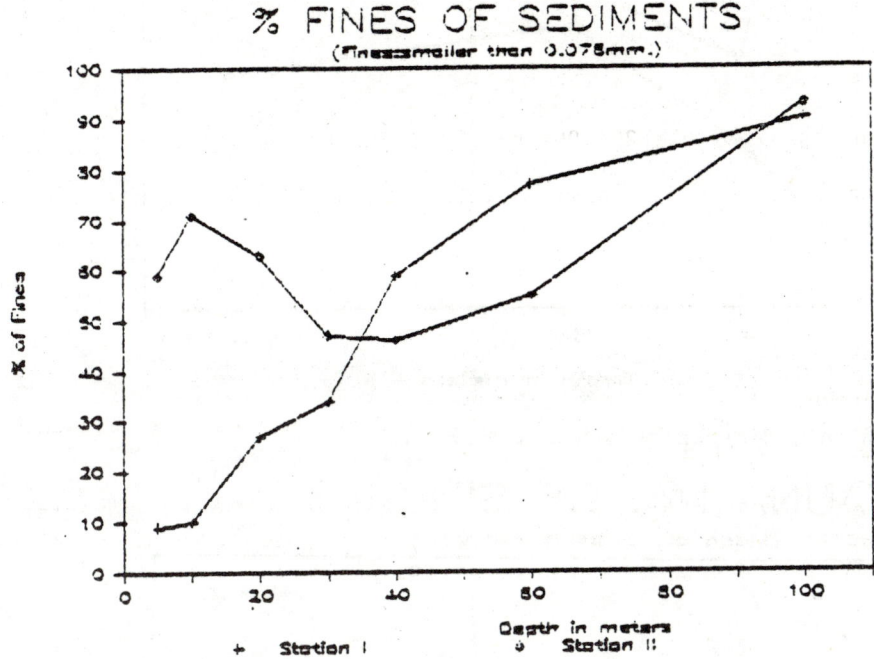

Figure 2. % fines of sediments (<0.076 mm) at station I and II

Organic matter in the sediment

The organic matter in the sediment of Area I range from 0.11% to 0.79% in Series B and from 0.12% to 0.75% in Series C. In Area II the organic matter ranged between 0.42% and 1.20% and between 0.63% and 1.81% in Series B and C respectively. Fig. 3 shows the percentage of organic matter at different depths. The values are the means of the two Series. The higher amount of organic matter in Area II associated with high percentage of fines may be related to the introduction of fine particles of organic matter from the wine industries.

Diversity (Number of species)

The number of species in Area I increases with depth up to 40 metres where it reaches a maximum and then drops (Fig. 4). On the contrary in Area II the maximum of species occurs in the shallower sub-stations. It is obvious that Area II is richer in species than Area I in the shallow sub-stations. Increase in the number of species under moderate organic input has been mentioned in the literature by McIntyre(1977) and Dauer and Conner(1980). The increased number of species may be partly due to the sediment composition. In the shallow stations of Limassol Bay the sediment is mixed containing a high percentage of fine particles, while in Episkopi it is coarser. It has been found (Craing and Jones, 1966) that muddy sand has a greater number of species than pure sand or mud and in general heterogeneous sediments have higher diversity than homogeneous sediment (Gray, 1974).

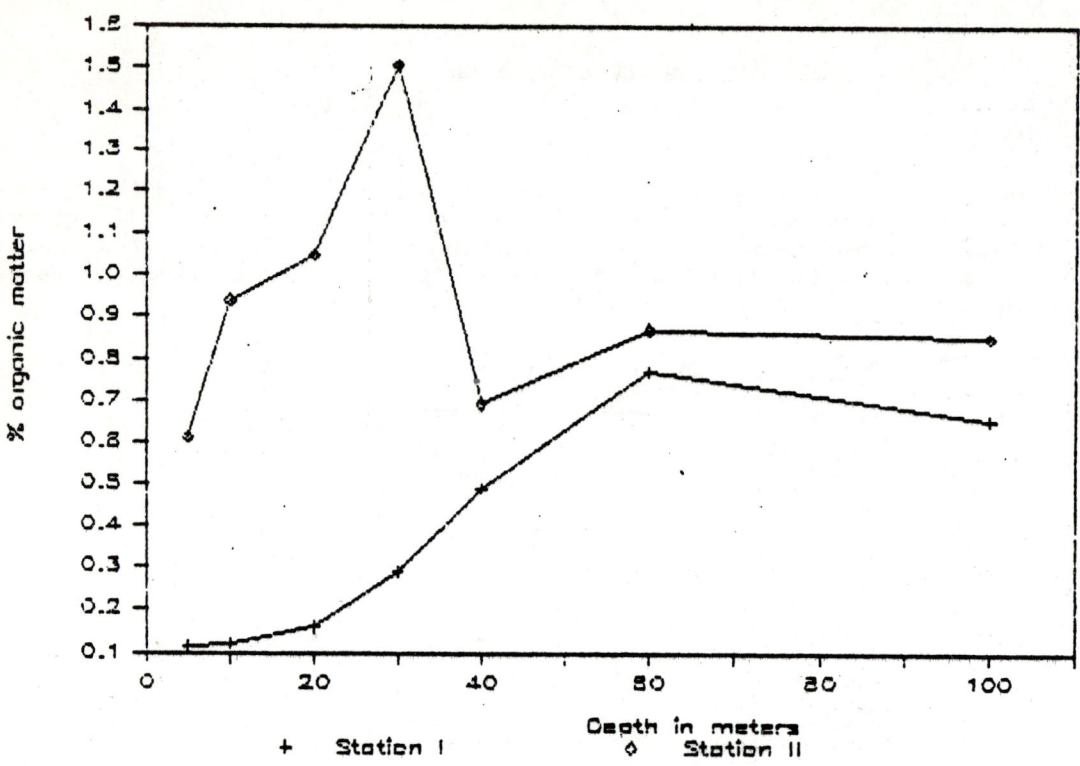

Figure 3. Organic carbon in sediments at station I and II.

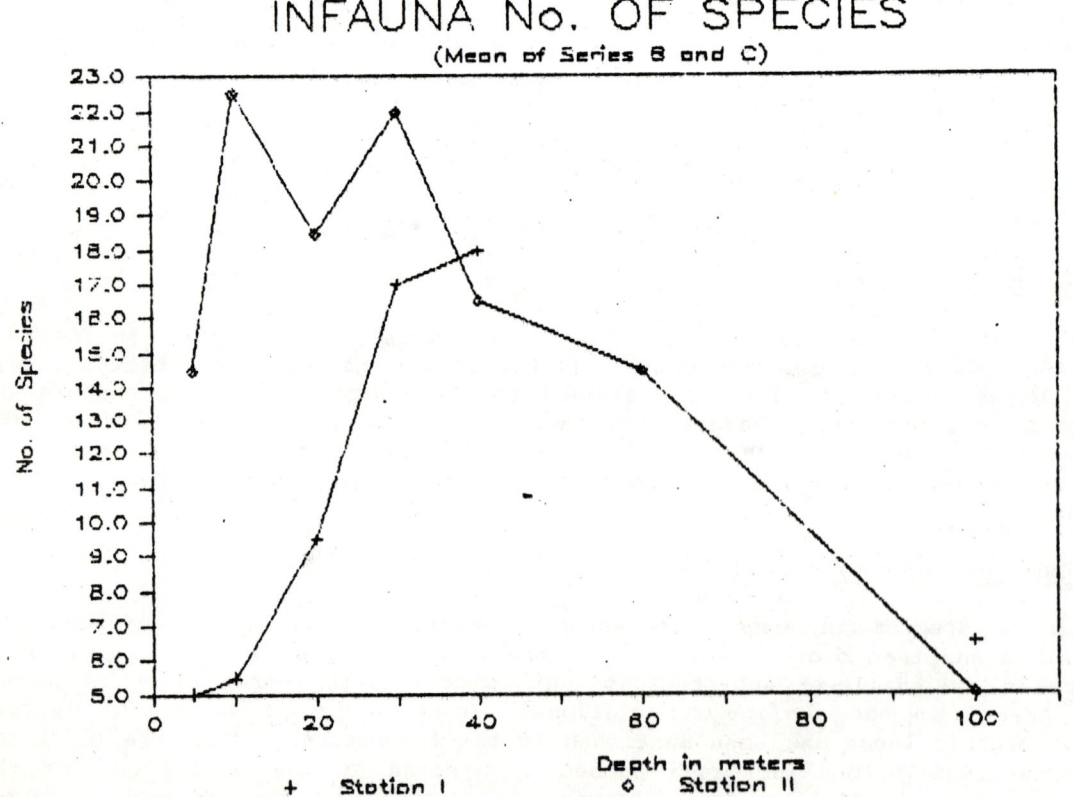

Figure 4. Infauna (No. of species) at stations I and II.

Number of specimens

Figure 5 shows the number of specimens found at each sub-station. The values are the mean of the two series B and C like the values of number of species. The number of specimens like the number of species in the shallow sub-stations is greater in Area II, probably due to the organic enrichment (Pearson and Rosenberg, 1978).

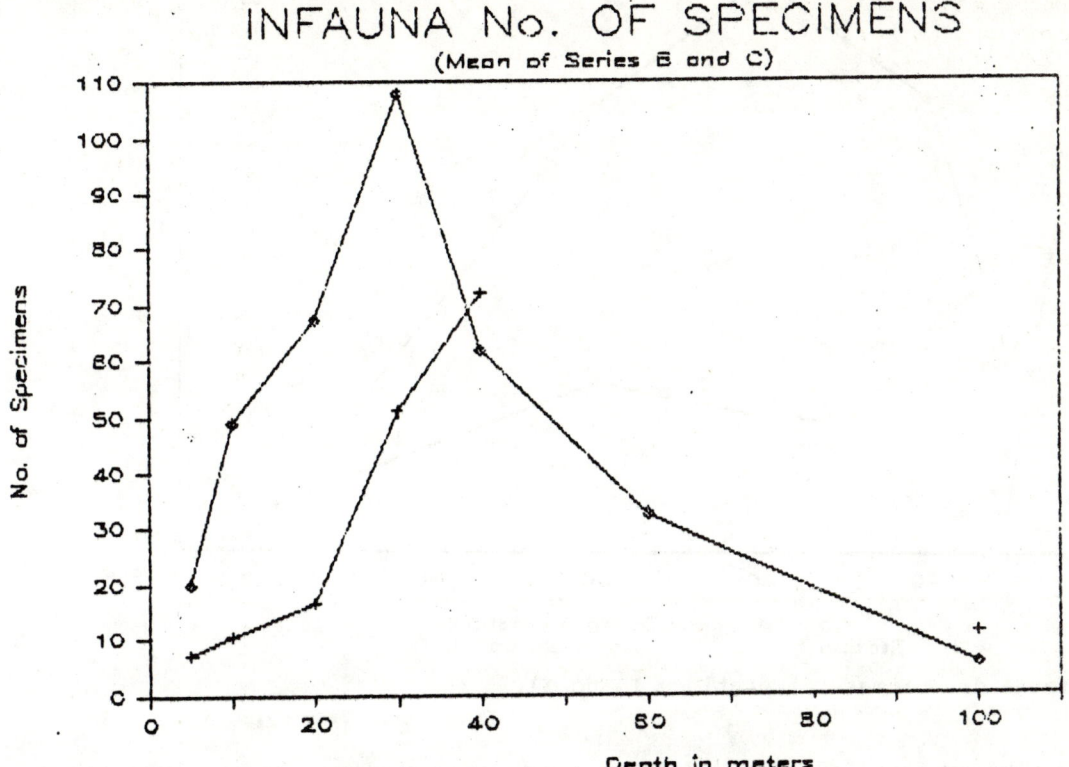

Figure 5. Infauna (No. of specimens) at stations I and II.

Biomass

Figure 6 shows that biomass in Area II is considerably larger than in Area I and that the higher values of it occur in the shallower sub-station. The increase in biomass is a lot greater than could be explained by the increase in numbers of specimens. It must be assumed that there is also increase in the size of individuals, a phenomenon which has been documented in the literature at least for polychaetes (Dauer and Connor, 1980).

Principal components analysis

The effect of organic enrichment was also shown by the Principal Components Analysis. Figure 7 shows Axis 1 (Horizontal Factor 1) plotted against Axis 2 (Vertical Factor 2). Sub-stations with low organic content appear at the lower values of Axis 1 such as IB_5 and IB_{10} (numbers 1 and 2 in Fig. 7) with 0.11% organic matter and IC_{20} (10) with 0.17%. Sub-stations with high percentage of organic matter are found at high scores of Axis 1. For example sub-station IIB_{60} (19) and IIC_{30} (23). Thus, sub-stations of Area II tend to concentrate at the high values of Axis 1, due to the higher concentration of organic matter in the polluted area, while the sub-stations of the clean Area I tend to concentrate at the low values of Axis 1.

Epifauna like infauna has a larger number of species and specimens in Area II than in Area I (Fig. 8 and 9). The effect of pollution shows more clearly in the Echinodermata than in any other group of Epifauna. For this reason, Principal Components Analysis has been done also for the Echinodermata of the Epifauna, although the data are semi-quantitative (Fig. 10 and 11, Table I).

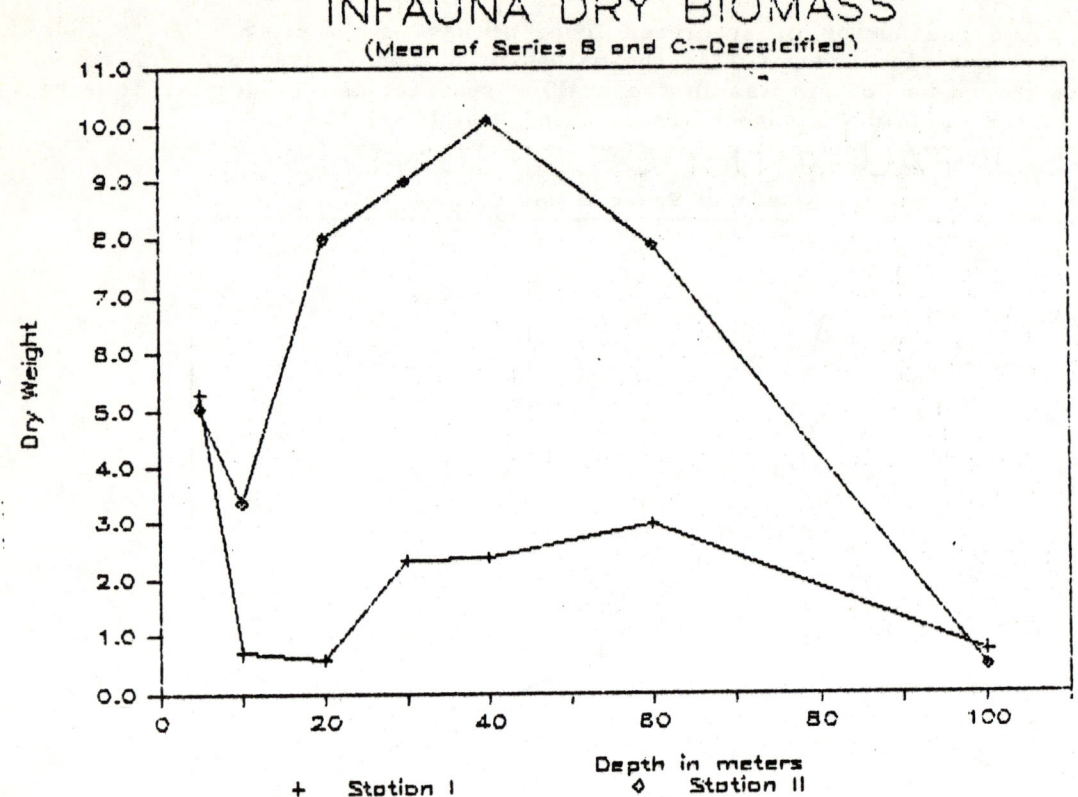

Figure 6. Infauna dry biomass at stations I and II.

A group of Echinodermata found exclusively or in large numbers in the polluted area, formed a group with large scores on Axis 1 (Fig. 10). For example Holothuria tubulosa (15) (refer to Table I for species number) was absent from Area I but was found in very large numbers in Area II, Ophyomyxa pentagona (11) was present with one or two specimens in Area I and almost in all depths and in large numbers in Area II. The species Echinocyamus pusillus (21), Sphaerechinus granulari (2) and Holothuria impatiens (18) were absent from Area I and were present only in Area II in small numbers. At the other end of the Axis I were found species present only in the clean Area I such as Astropecten jonstoni (31), Astropecten bispinosus (33), Echinocardium cordatum (24) and Echinocardium mediterraneum (22). Axis 2 probably represents depth because some species for example Cidaris cidaris (25), Ophiura lacertosa (8) and Ophiura filiformis (9) had high scores on this Axis. The rest of the species did not have a marked depth preference and are found at the same scores approximately, on Axis 2.

It is believed that Axis 1 represents the factor of pollution and it is suggested that the species H. tubulosa and Ophy. pentagona are considered as indicators of organic pollution.

In Figure 11, Axis 3 appears to be related to the amount of fines in the sediment. The species Brissopsis lyrifera (28) was found at depths with 70% to 90% fines in the sediment, Astropecten irregularis (32) and Sphaeriodiscus placenta (2) were only found in sediments with 90% fines. On the contrary Astropecten bispinosus (33) and Astropecten aurantiacus (34 were found at sediments with 27% fines.

The depth distribution and sediment preferences of the Echinodermata agree with that found in the literature (Tortonese, 1965; Demetropoulos and Hadjichristophorou, 1976).

In conclusion, the organic solids introduced by the wine industries cause an increase in fine particles and organic matter in the sediment of Limassol Bay. This, in turn, results in the increase of numbers of species and specimens and of biomass.

Apart from its interest in detecting organic enrichment, the present study is important because several dozens of species were recorded for the first time in Cyprus. The detailed species lists, which will be published elsewhere (Hadjichristophorou and Demetropoulos, in preparation), will provide good baseline material for future reference.

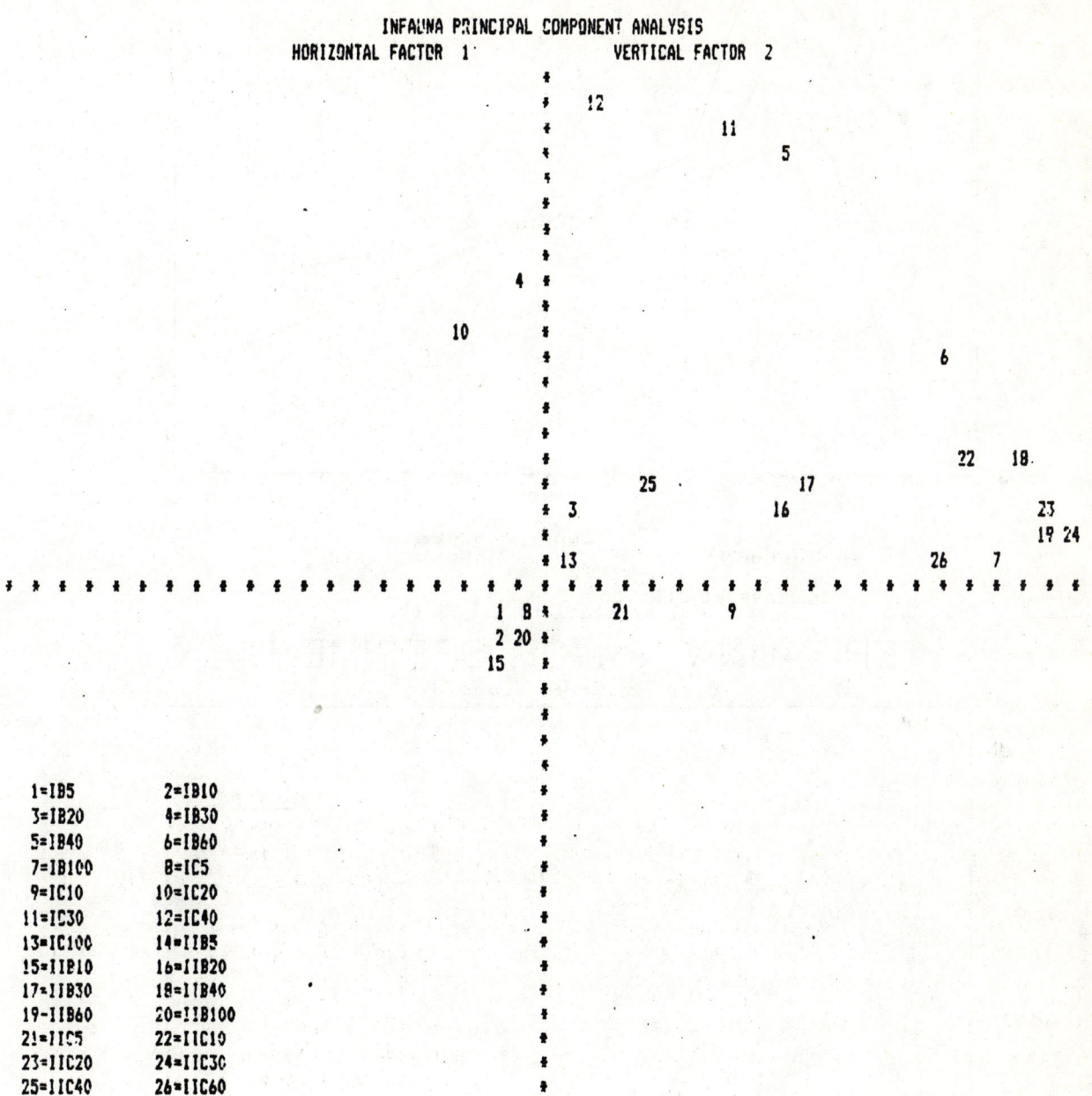

Figure 7. Infauna principal component analysis.

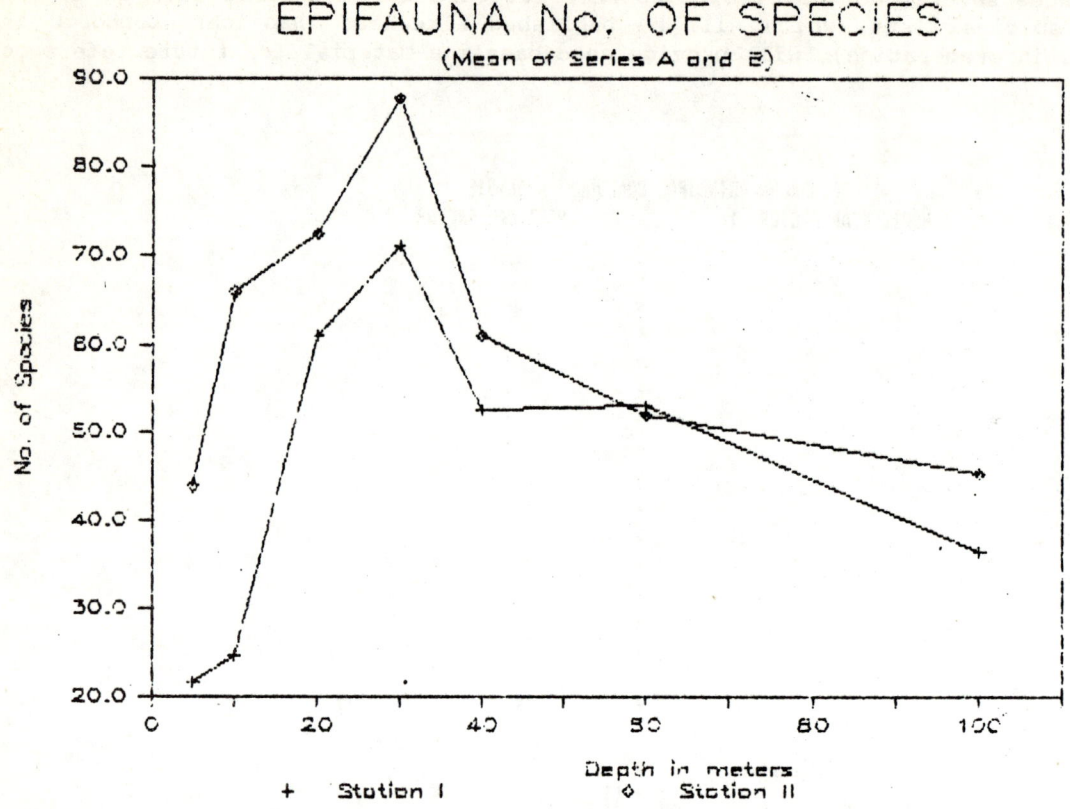

Figure 8. Epifauna (No. of species) at stations I and II.

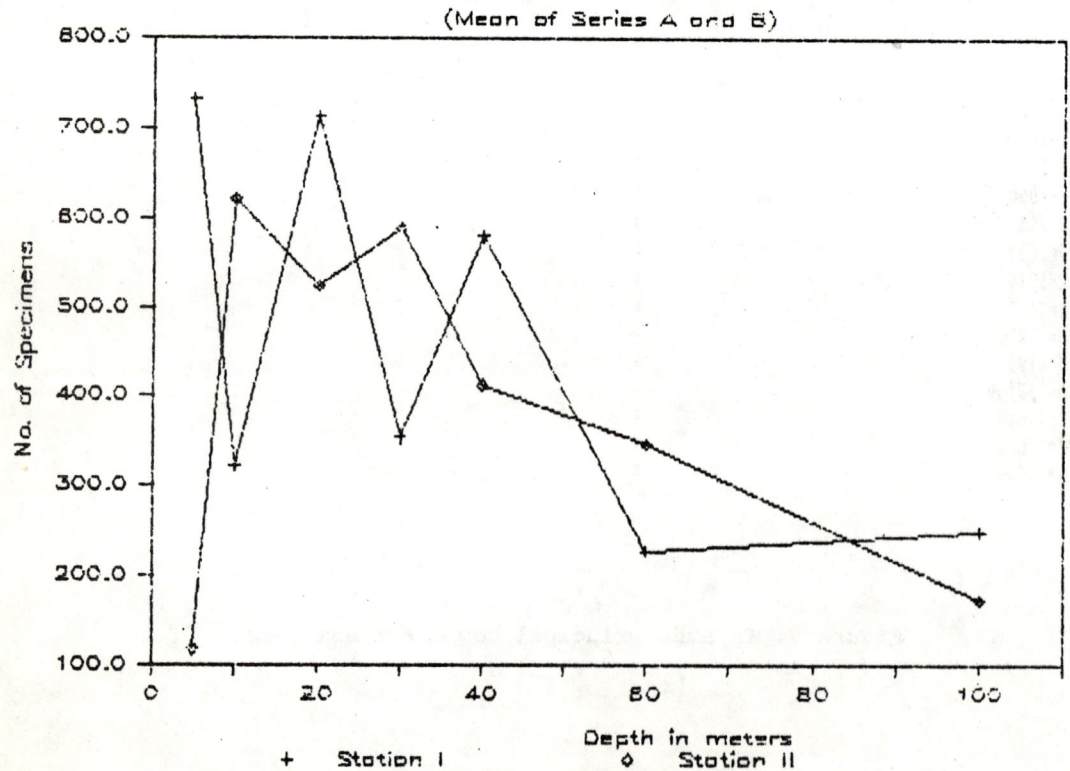

Figure 9. Epifauna (No. of specimens) at stations I and II

Table I

Epifauna Series A and B

S./No.	SPECIES
	ECHINODERMATA
37	Amphiura chiajei
36	Anceropoda placenta
35	Antedon mediterranea
34	Astropecten aurantiacus
33	Astropecten bispinosus
32	Astropecten irregularis
31	Astropecten jonstoni
30	Astropecten platyacanthus
29	Astropecten spinulosus
28	Brissopsis lyrifera
27	Centrostephanus longispinus
26	Chaetaster longipes
25	Cidaris cidaris
24	Echinocardium cordatum
23	Echinocardium corrugatum
22	Echinocardium mediterraneum
21	Echinocyamus pusillus
20	Hacelia attenuata
19	Holothuria helleri
19	Holothuria impatiens
17	Holothuria sanctori
16	Holothuria sp.
15	Holothuria tubulosa
14	Luidia ciliaris
13	Marthasterias glacialis
12	Ophioderma longicauda
11	Ophiomyxa pentagona
10	Ophiura albida
9	Ophiura filiformis
8	Ophiura grubei
7	Ophiura lacertosa
6	Paracentrotus lividus
5	Psammechinus microtuberculatus
4	Schizaster canaliferus
3	Sphaerechinus granularis
2	Sphaeriodiscus placenta
1	Stylocidaris affinis

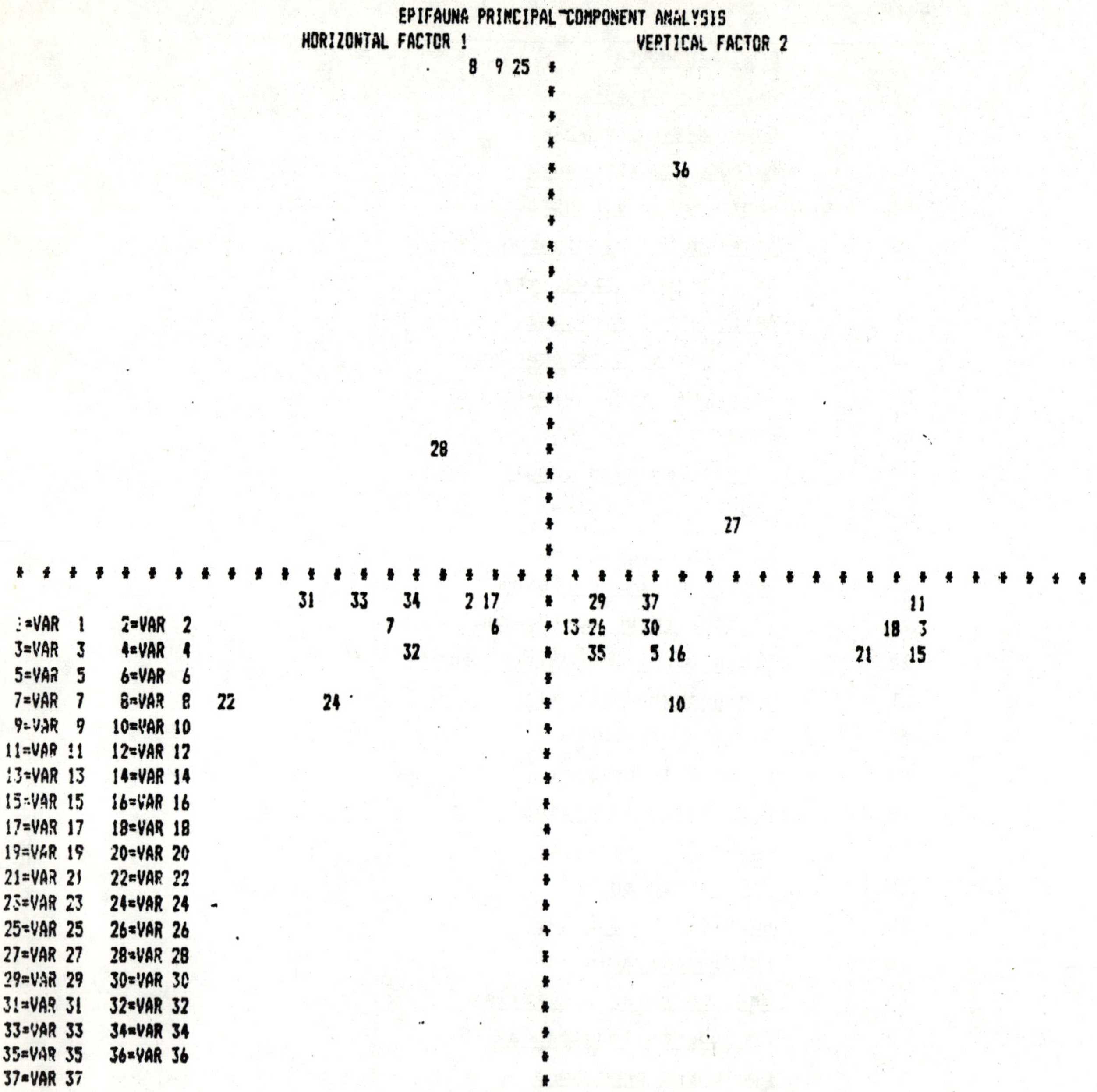

Figure 10. Epifauna principal component analysis.

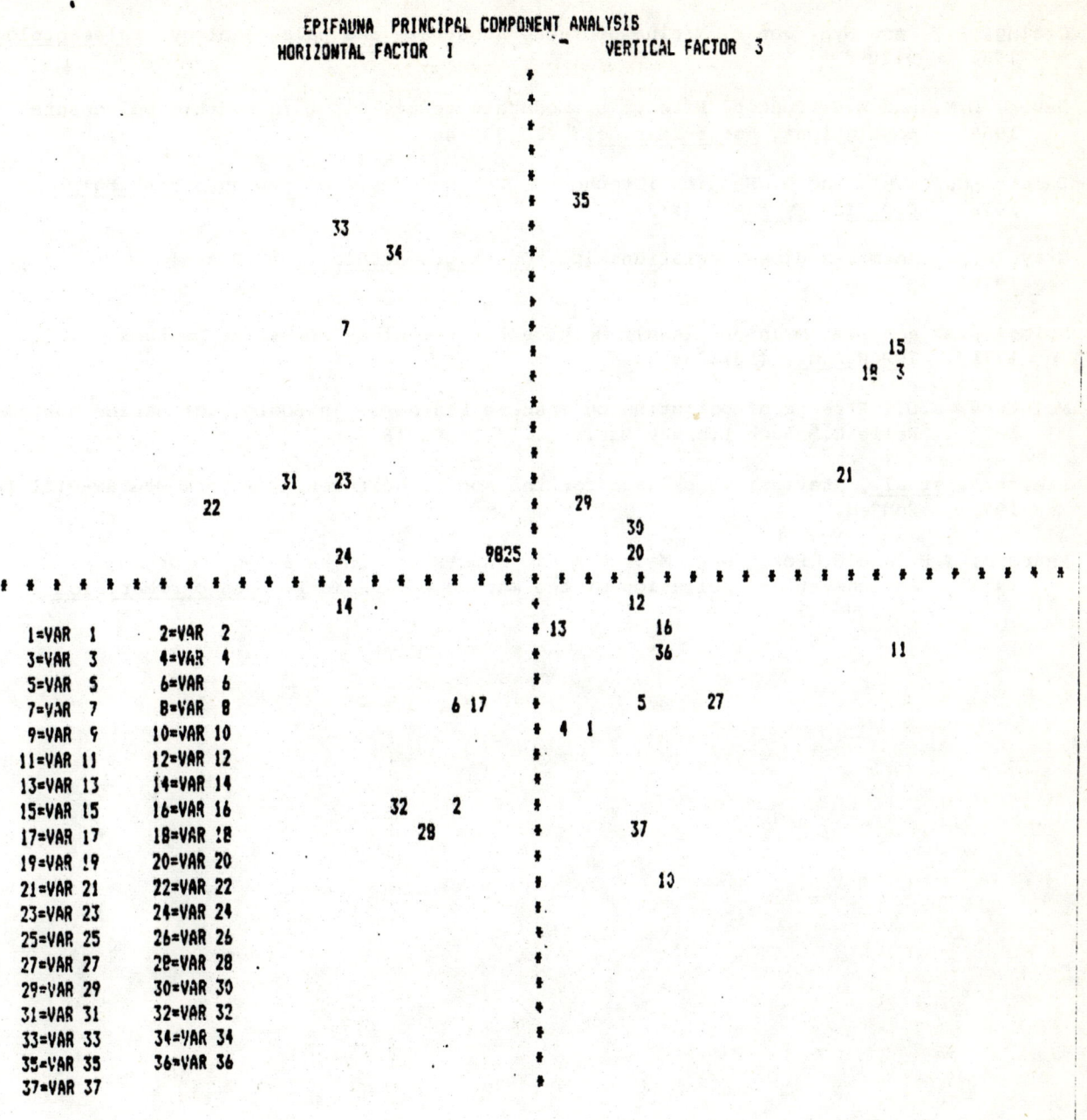

Figure 11. Epifauna principal component analysis.

5. REFERENCES

Craing, G.Y. and N.S. Jones, Marine benthos, substrate and palaeoecology. <u>Palaeontology</u>,
1966   9:30-8

Dauer, D.M. and W.G. Connor, Effects of moderate sewage input in benthic polychaete
1980   populations. <u>Estuar.Mar.Sci.</u>, 10:335-46

Demetropoulos, A. and M. Hadjichristophorou, Echinodermata of Cyprus. <u>Fish.Bull.</u>
1976   <u>Dep.Fish.Cyprus</u>, (4):7-72

Gray, J.S., Animal-sediment relationships. <u>Oceanogr.Mar.Biol.</u>, 12:223-61
1974

Holme, N.A. and D.A. McIntyre (eds), Methods for the study of marine benthos.
1971   <u>IBP Handb.</u>, (16):334 p.

McIntyre, A.D., Effects of pollution on inshore benthos. <u>In</u> Ecology of marine benthos.
1977   Belle C.Baruch Library Mar.Sci., (6):301-18

Nie, N.H. <u>et al.</u>, Statistical package for the social sciences. New York McGraw-Hill Inc.
1979   2nd ed.

Pearson, T.H. and R. Rosenberg, Macrobenthic succession in relation to organic
1978   enrichment and pollution of the marine environment. <u>Oceanogr.Mar.Biol.</u>,
16:229-311

FISH DISEASES, AN INDEX OF WATER POLLUTION : A REVIEW

by

Y. HALIM*, M. FAISAL** and I. AHMED**

*UNARC and Faculty of Science, Alexandria University
Alexandria, Egypt

**Faculty of Veterinary Medicine, Alexandria University
Alexandria, Egypt

1. INTRODUCTION

Recent investigations of marine fish diseases have brought to light, among other things, the high incidence of diseases in fish living in coastal areas receiving water discharges (Möller, 1985; Faisal and Halim, unpub.). Fish diseases, however, do not occur as isolated events but are the end result of interactions between the etiological agents (viruses, bacteria, fungi or parasites), the fish and its environment (Snieszko, 1973) (Fig. 1). The environment is the most important and unstable of these three factors and its importance is intensified by the presence of potential pathogens that commonly coexist in the water.

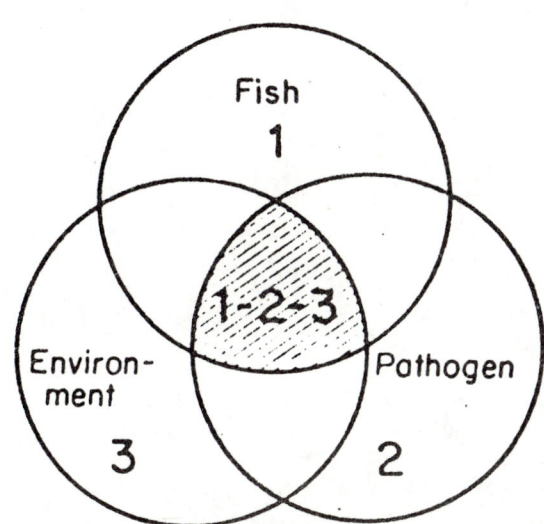

Figure 1. Frequently a fish (1) must interact with a pathogen (2) in a stressful environment (3) for an epizootic (1-2-3) (Snieszko, 1973)

The aquatic environment is the resultant of many independent and interdependent equilibria which are easily, and sometimes abruptly disturbed. Fish dwelling in coastal areas are continuously subjected to environmental fluctuations of physico-chemical factors, as well as to pollution-induced water quality alterations. Such alterations may be directly lethal to fish, but sublethal changes may stress the fish organisms sufficiently to predispose them to infectious diseases (Fig. 2). Stressed organisms are more vulnerable to infections (Wedemeyer and Wood, 1974).

Although facultative fish pathogens (such as Aeromonas, Pseudomonas, Vibrio and Myxobacteria spp.) are continuously present in most coastal aquatic environments, mortality seldom occurs, unless environmental quality and the resistance of the fish also deteriorate (Wedemeyer, 1970). On the other hand, organic pollution, as caused by domestic sewage, enhances the food-chain creating a rich but unhealthy feeding ground, to which herbivorous as well as carnivorous fish are attracted.

Some of the abnormalities occasionally found in fish dwelling in such environments, such as fin erosion, skeletal malformations and tumors, have attracted the attention of scientific and non-scientific observers. The underlying pathological mechanisms, however, have received too little or no attention from Mediterranean scientists.

<u>Chemical</u>

- water chemistry
- pollution

- Diel composition

- nitrogenous & other metabolic wastes

<u>Biological</u>

- population density
- other fish, lateral swimming space requirement
- microorganisms (pathogenic & nonpathogenic)
- macroorganisms (ecto & endo-parasites)

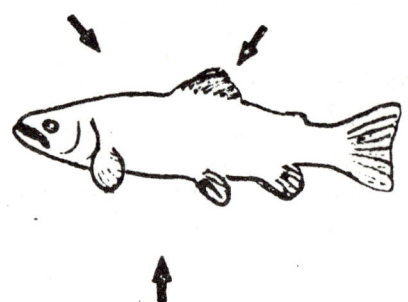

<u>Physical</u>

- temperature
- light
- sounds
- dissolved gas

Figure 2. Biological, chemical and physical environmental factors which can adversely affect fish in intensive culture or coastal areas and cause stress (modified after Wedemeyer <u>et al.</u>, 1976)

The spread of fish diseases in more or less confined coastal zones has far-reaching effects both on the fishery resources and on human health. Parasitic and other infective fish diseases are often transmissible to man and constitute therefore, in many cases, a hazard to human consumers. On the other hand, diseased or stressed fish become more vulnerable to predation, as their escape and defence mechanisms are weakened. Their growth will also be retarded and the development of their gonads impaired. As a result, the quality of the fish is reduced and the survival of the population threatened. It is strongly recommended therefore that concerted efforts be focussed on the study of Mediterranean fish diseases within the frame of the Mediterranean Action Plan and its participants.

A concise review of the major concepts and mechanisms involved is given in this paper.

2. MECHANISMS OF STRESS

Brett (1958) defined the stress as "a state produced by environmental or other factors which extends the adaptive response of an animal beyond the normal range or which disturbs the normal functioning to such an extent that in either case, the chances of survival are significantly reduced".

Within the body of a stressed fish, many morphological, biochemical and physiological changes take place. This activation of the physiological apparatus (also called <u>General Adaptation Syndrome</u>, "GAS") enables the fish body to be adapted to the new unsuitable environmental factors.

Selye (1974) (Fig. 3) differentiated three phases of the "GAS", namely:

(a) <u>The alarm phase</u>, characterized mostly by reversible morphological and physiological reactions. In case of strong rapid stressors, this phase terminates usually with death.

(b) <u>The resistance phase</u>, during which adaptation to achieve homeostasis under the changed circumstances is taking place.

(c) <u>The exhaustion phase</u>, during which adaptation ceases to be adequate and homeostasis cannot be maintained. Most of these changes are irreversible and end with:
(i) immunsuppression, (ii) retardation in growth, (iii) impairment of gonads function, or even death.

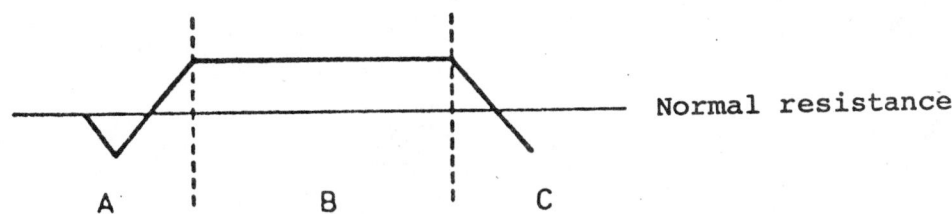

Figure 3. Phases of GAS. A: Alarm; B: Resistance; C: Exhaustion; (modified after Selye, 1974)

Such changes within the stressed fish body are similar in many ways to those recognized in higher vertebrates (Hoar, 1957; Gronow, 1974). The above-mentioned phases with their associated metabolic changes are neither species-specific nor stressor-specific.

The sequence of physiological and biochemical alterations in a stressed teleost, as it attempts to maintain homeostasis in the face of an exogenous or endogenous stressor are mediated by two main reactions, a hormonal and a nervous sequence of events as outlined by Mazeaud et al., (1977) (Fig. 4).

2.1 Hypothalamus-Hypophysis-Interrenal Sequence

Under the effect of a stressor (exogenous or endogenous), the hypothalamus stimulates the anterior lobe of the pituitary gland to secrete Adrenocorticotrophic hormone (ACTH). ACTH in turn stimulates the interrenal tissue of the fish (homologous to the mammalian suprarenal gland cortex) to secrete cortisone (corticosterone, epinephrine) which is also called stress hormone (Fig. 5). Cortisone acts on protein and carbohydrate metabolism as well as on the lymphatic system.

Under the influence of corticosteroids, the ionic equilibrium in the body fluids is disturbed. There is Na^+ and Cl^- retention while K^+ ions are excreted. There is an increase in the levels of blood glucose, lactate and fatty acids with depletion of liver glycogen and muscular proteins (negative nitrogen balance) leading to loss of weight. The increased thyroxin output from the stimulated thyroids <u>accelerates the breakdown of proteins</u>. In addition, there is a marked lymphopenia impairing the defence mechanism.

2.2 Hypothalamus-Sympathetic-Suprarenal Sequence

Activation of the sympathetic nervous system (mediated through the hypothalamus) provokes an increased production of catecholamines (Adrenaline and Noradrenaline) from the chromaffin cells (homologous to the mammalian suprarenal medulla). The synthesis of these hormones consumes ascorbic acid, and that is why blood ascorbic acid decreases during stress. Ascorbic acid (vitamin C) is responsible for the health of the epithelial cells of the fish body, such as epidermis, gill lining and mucosa of the gastrointestinal tract.

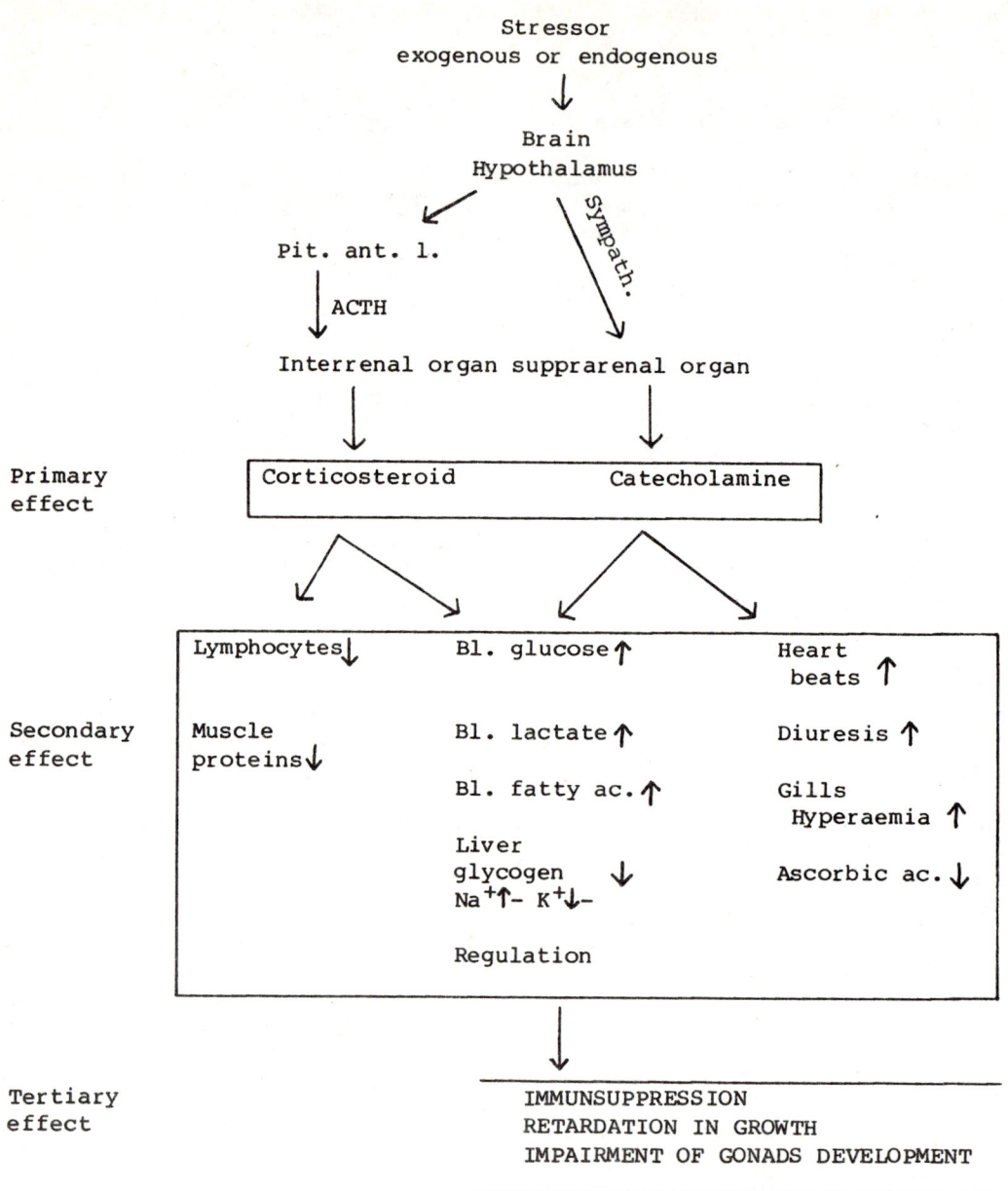

Figure 4. Primary, secondary and tertiary effects of stress (modified after Mazeaud et al., 1977)

Catecholamines induce increased heart beats, diuresis and hyperaemia of the branchial blood vessels disturbing the osmoregulatory mechanism in both fresh and marine water fish.

The stress-induced hormonal changes are primary effects and the consequent changes of the intermediary metabolism are secondary effects of stress (Mazeaud et al., 1977). The macro- and microscopical morphological changes are tertiary effects of stress (Wedemeyer et al., 1976). Tertiary effects of stress are usually observed in the late resistance and exhaustion phase and are, in most cases, irreversible (Faisal, 1985).

In the next chapter the most important stress-induced changes and their relation to disease process are outlined.

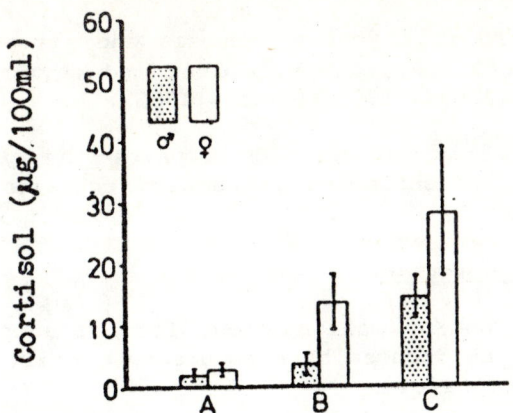

Figure 5. Cortisol concentration in the plasma of <u>Onchorhynchus nerka</u>: A = Resting stage; B = Fishing stress 15 min.; C = Fishing stress 30 min.; (After Fagerlund, 1967)

3. DISEASE PROCESSES IN RELATION TO POLLUTION-INDUCED STRESSES

3.1 <u>Infection pressure</u>

Increased infection pressure from facultative pathogens occurs by pollution with organic substances, particularly with sewage pollution. These organic substances constitute a suitable nutrient medium to <u>Vibrio</u> and <u>Aeromonas</u> spp. which may cause ulcers and septicaemia in fish.

3.2 <u>Latent infections</u>

These are activated into potency, including certain viral infections that may occur more frequently as a result of pollution stress, as in the case of lymphocystis and papillomatosis (Figure 6).

3.3 <u>Deterioration of the epithelial defence barrier</u>.

The epithelial lining of the gills and internal organs, as well as the skin, constitute a natural barrier against the entrance of microorganisms (Munro, 1982). Stress causes hyperaemia (increased blood flow) of the gill blood vessels resulting in increased blood flow in the secondary gill lamellae (Wedemeyer <u>et al</u>., 1976). The secondary gill lamellae are covered with a very thin membrane (1-2 cell in thickness) which easily exfoliates, creating a very short pathway for microorganisms to invade the stressed fish (Hibiya, 1982).

Similar desquamation of the mucosal layer was also recorded in the gastrointestinal tract of stressed eels (Peters, 1984) and African catfish (Ahmed, 1985).

The organic matter overload in the aquatic environment with its associated changes in water quality irritates the skin of the fish with consequent erosion of the soft rays of the fins and gills. The organic matter overload on the other hand provides a suitable substrate for the multiplication of monogenetic trematodes and ectoparasitic crustaceans, which through their bites, induce ulceration of skin and gills creating new pathways for infective organisms (Reichenbach-Klinke, 1980; Faisal <u>et al</u>., 1985).

3.4 **Reduction of effectiveness of internal defence mechanism.**

Long term exposure to pollutants was found to cause necrosis and atrophy of the haematopoietic organs of the fish (anterior kidney, liver, spleen and thymus) causing suppression of the immune defence mechanism (Ahmed, 1985; Peters, 1985).

In addition, increased blood level of corticosteroids impairs the process of phagocytosis (Wedemeyer et al., 1976) and catecholamine production consumes ascorbic acid (vitamin C) from the body.

3.5 **Genetic and cytotoxic effects on eggs and larvae.**

They result in higher mortality rates and/or malformation, for instance skeletal anomaly. These changes could, however, also be induced by exposure to certain chemicals (organochlorine compounds) at a later stage of life.

The relationship between specific adverse environmental factors (stressors) and fish diseases remains, until now, better known for freshwater fish and fish from fish farms than for marine fish (Wedemeyer et al., 1976).

4. THE USE OF FISH DISEASES IN MONITORING MARINE POLLUTION

The relationship between the frequency of a specific fish disease and the presence of a certain contaminant is still unclear, because the effects of various contaminants are included in a complex of ecosystem components and because the effect is often indirect (Christensen, 1981).

In addition, most of the different studies done in this field lack standardized procedures.

To secure uniform and comparable records (in time, and between different localities) of the nature and frequency of the individual diseases, the collection and field examination of fish should be made according to standardized methods and in parallel with contextual information from the area investigated.

In this perspective, guiding notes on field determinations of the commonest fish diseases with a list of contextual information needed to evaluate the results have been prepared in detail after a long term study by the Danish National Agency for Environmental Protection in cooperation with other Danish agencies (European Association of Fish Pathologists, 1982).

The study decided in the initial stages to use a number of marine fish diseases that are visible by a superficial inspection of the fish such as:

- skin ulceration (including Cod Ulcus Syndrome)
- fin erosion
- gill erosion
- skeletal anomalies
- tumors
- lymphocystis
- papillomatosis

These diseases have also been pointed out in similar studies in the Federal Republic of Germany (Möller and Andres, 1983), in USA, France and the United Kingdom (Sindermann, 1970; Christensen, 1981).

Some of these diseases, however, are specific only for fish species inhabiting American and European Seas but not Mediterranean Fish. In addition, skin ulcerations, fin erosion, gill erosion and skeletal anomalies are signs of diseases and not diseases in themselves, and can be caused by a variety of agents.

It is strongly recommended that Mediterranean fish diseases be thoroughly investigated in relation to pollution stress, particularly wastewater discharges. The study should aim at clarifying the exact etiology of the diseases in question. Training courses on the quick diagnosis of fish diseases should also be planned in the context of a large scale survey.

Figure 6. Lymphocystic disease in Sparus aurata caught from Kayet-Bay, Alexandria, Egypt. (Faisal, unpub.)

5. REFERENCES

Ahmed, M.A.S., Study of the histopathological effects of some molluscicides on Clarias
1985 lazera. M.Sc. Thesis, Faculty of Science, University of Alexandria, Egypt

Brett, J.R., Implications and assessments of environmental stress. In The investigation
1958 of fish power problems, edited by H.R. MacMillan. Vancouver, University of British Columbia, pp.69-97

Christensen, N.O., The use of diseases in marine pollution effects monitoring
1981 programmes. Bull.Europ.Assoc. Fish Pathol., (1):7-9

European Association of Fish Pathologists, Field investigations of fish disease.
1982 Bull.Europ.Assoc.Fish Pathol., 2 (supp.1):1-10

Fagerlund, U.H.M., Plasma cortisol in relation to stress in adult Sockeye Salmon during
1967 the freshwater stage of their life cycle. Gen.Comp.Endocrinol., 8:197-207

Faisal, M., Some selections on fish diseases. Alexandria, Egypt, University of
1985 Alexandria Press, pp.180-2

Faisal, M., K.I. Ashmawy and M.H. Rizk, A contribution on dactylogyrases among some
1985 freshwater fishes in Egypt. J.Egypt.Vet.Med.Assoc., 45:95-105

Gronow, G., Nukleinsäure-und Substratgehalte in der dorsalen Rumpfmuskulatur von
1974 Teleosteern während eines "biologischen Stress". Mar.Biol., 24:313-27

Hibiya, T., An atlas of fish histology. Kodansha, Tokyo and Gustav Fischer Verlag
1982   Stuttgart and New York, pp.54-8

Hoar, W.S., Endocrine organs. In The physiology of fishes, edited by M.E. Brown.
1957   New York, Academic Press, vol.1:245-85

Mazeaud, M.M., F. Mazeaud and E.M. Donaldson, Primary and secondary effects of stress
1977   in fish: some new data with a general review. Trans.Am.Fish.Soc., 106:201-12

Möller, H., A critical review on the role of pollution as a cause of fish diseases.
1985   In Fish and shellfish pathology, edited by A. Ellis. London, Academic Press, pp.103-7

Möller K., and K. Andres, Krankheiten und Parasiten der Meeresfische. Stuttgart, Gustav
1983   Fischer Verlag, pp.1-99.

Munro, A.L.S., The pathogenesis of bacterial diseases of fishes. In Microbial diseases
1982   of fish edited, by R.J. Roberts. London, Academic Press, pp.131-49

Peters, G., Pathological effects of stress and environmental conditions on cultured fish.
1984   Bull.Europ.Assoc. Fish Pathol., (4):74-5

_____, Gill structure and blood electrolyte levels of European eels under stress.
1985   In Fish and shellfish pathology, edited by A.E. Ellis. London, Academic Press, pp.109-15

Reichenbach-Klinke, H.H., Krankheiten und Schädigungen der Fische. Stuttgart, Gustav
1980   Fischer Verlag, 205 p.

Selye, H., Stress. München, Piper Verlag .
1974

Sindermann, C.J., Principal diseases of marine fish and shellfish. London, Academic
1970   Press, pp.1-20

Snieszko, S.F., The effect of environmental stress on outbreaks and infectious diseases
1973   of fishes. J.Fish Biol., 6:197-208

Wedemeyer, G.A., The role of stress in the disease resistance of fishes. In A Symposium
1970   on diseases of fishes and shellfishes, edited by S.F. Snieszko. Spec.Publ.Am.Fish Soc., (5)/30-5

Wedemeyer, G.A., and J.W. Wood, Stress as a predisposing factor in fish diseases.
1974   Fish.Leaft.U.S.Fish Wildl.Ser., (38):1-7

Wedemeyer, G.A., F.P. Meyer and I. Smith, Environmental stress and fish diseases.
1976   Neptune, New Jersey, T.F.H. Publications, 180 p.

ENVIRONMENTAL CONDITIONS IN ABU-KIR BAY, EAST OF ALEXANDRIA, DOWNSTREAM FROM "EL-TABIA" EFFLUENT. IMPACT ON THE FISH ASSOCIATIONS IN THE COASTAL ZONE

by

Y. HALIM*, H.H. SALEH** and A. SALIM***

* University of Alexandria Research Centre(UNARC), Alexandria, Egypt.
** Alexandria Institute of Oceanography and Fisheries, Egypt
*** Faculty of Engineering, University of Alexandria, Alexandria, Egypt.

1. INTRODUCTION

Abu-Kir Bay, a shallow semi-circular bay east of Alexandria, extends between Rosetta headland, north east, and Abu-Kir headland, south west. The Bay contributes from 6 to 10% to the total fish and shrimp catch of the Egyptian Mediterranean waters. It is also one of the main spawning and nursery grounds for migratory Mugil spp. and Sparus sp., their fry subsequently restocking the brackish coastal lagoons. The north east Bay receives some fresh water from the Rosetta Nile outlet in winter and the south Bay agricultural runoff through the Maadeya outlet. The south west Bay is impacted by a major effluent, "El-Tabia" pump station. More than 2 million m^3 day^{-1} of industrial waste water from two paper mills, a food-canning plant and of the textile compound at Kafr El-Dawar, mixed with domestic waste water and agricultural drainage are disposed of through this pump station. The mixed waste water with its heavy load of cellulose and lignin particles smothers all life in a vast area around the outfall. Its effect also extends to the coastal zone east and west from the outfall, and appears to disrupt the balance between Mullets and Tilapia fish in the adjacent Edku lake. The aim of the present survey is to bring to light the dramatic effect of waste water disposal on the fish associations of both Abu-Kir Bay and Edku lake.

2. MATERIAL AND METHODS

Fishing was carried out in the coastal strip using a purse-seine net at four localities, from "El-Tabia" outfall to Rosetta headland. The purse-seine net used is about 700 m in length and the area swept, about 3800 m^2 with an average depth of 2-3 m. Locality L1 is immediately downstream from the outfall, locality L2 5 km east, locality L3 about 15 km east next to Maadeya outlet and locality L4, 30 km east, south of Rosetta headland (Fig. 1). The fish catch was analysed for the catch per effort, the species composition, the ratio of demersal to pelagic fish, the length-weight and the age composition. Survival and mortality rates were computed from the abundance of successive age groups by plotting the natural log of the age groups against the corresponding ages (catch curve) as applied by Ricker (1975). The instantaneous total mortality (z) was estimated from the slope of the descending limb of the catch curve. The survival rate (S) was computed from the relation $S = e^{-z}$. The catch per effort is given in kg man^{-1} $hour^{-1}$. The bottom algae, crustacea and bivalves swept by the net were also identified but only the commercial forms are included in the ratio of pelagic to demersal biomass (Table I). Water samples were collected from each site for water quality analysis.

3. RESULTS AND DISCUSSION

A distinct gradation is observed from the outfall eastward. The water quality characteristics along the coast downstream from the effluent, are given in Table II and shown in Fig. 2.

The species composition of the purse-seine catch at the corresponding four localities investigated and the relative importance of pelagic to demersal commercial species are given in Table I. The catch per effort is shown in Table III. In Tables IV and V, respectively, the age composition and the survival and mortality rates for M. capito are given.

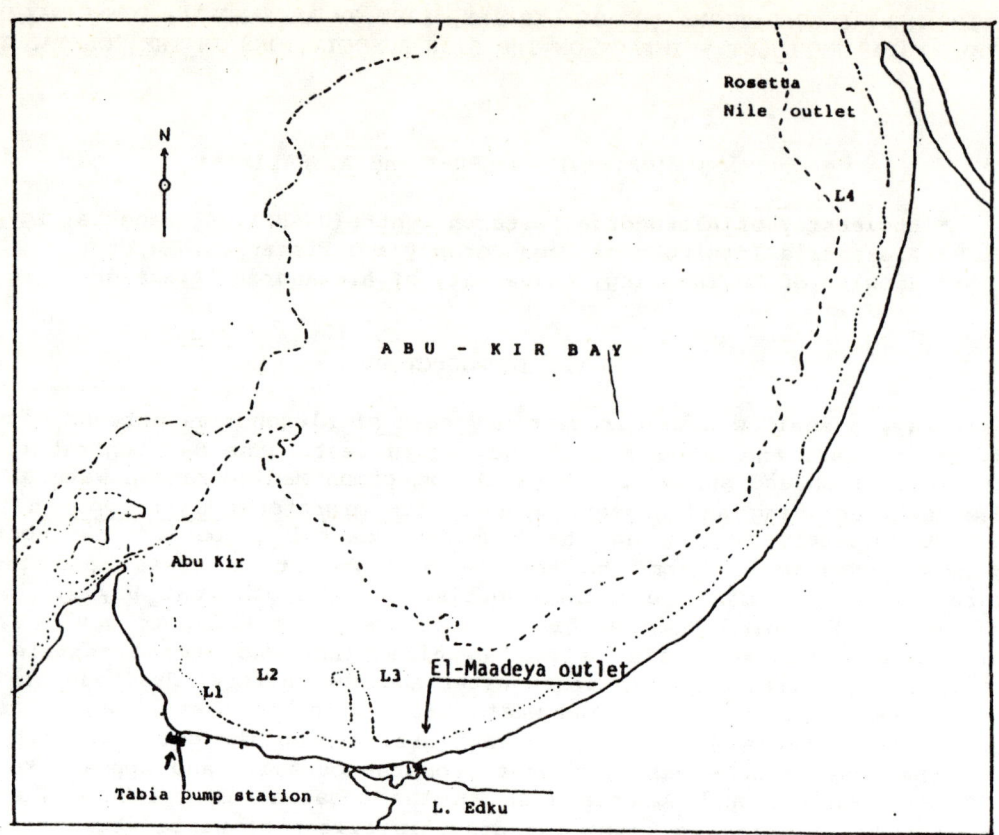

Figure 1. Abu-Kir Bay showing localities.

Table I

The coastal fisheries in Abu-Kir Bay downstream from "El-Tabia" pumping station. Composition and relative importance of pelagic to demersal commercial fish, crustaceans and mussels along a decreasing pollution gradient (% by weight)

Locality 1	
Pelagic : Nil	Demersal : Nil

Locality 2	
Pelagic : 97%	Demersal : 3%
Caranx alexandrinus	Trachypenaeus curvirostris
Julis pavo	Portunus pelagicus (young)
Pagrus mormyrus	
Mugil capito	
Sciaena aquila	
Sardinella sp.	

Locality 3	
Pelagic : 61%	Demersal : 39%
Mugil capito	Trichiurus haumela
Sciaena aquila	Solea solea
Sardinella sp.	Trachypenaeus curvirostris
Julis pavo	Portunus pelagicus
Mugil cephalus	

Table I (Continued)

Locality 4

Pelagic : 46%
Morone labrax
Mugil capito
Sciaena aquila
Sardinella sp.
Pagrus mormyrus
Serranus scriba
Pagrus pagrus

Demersal : 54%
Solea solea
Mullus barbatus
Trichiurus haumela
Trigia hirunda
Metapenaeus monoceros
Metapenaeus stebbingi
Penaeus semiculcatus
Penaeus kerathurus
Trachypenaeus curvirostris
Carcinus mediterraneus
Portunus pelagicus
Donax trunculus

Table II

Results of chemical analysis of effluent and sea water samples.
Effluent: El-Tabia pump station.

Parameter	Effluent	1	2	3	4	Units
Water temp.	25	27	28	28.5	27	°C
Air temp.	29	29	29	29	29	"
pH	7.6	7.55	7.9	7.8	8.2	
Dissolved oxygen	1.1	0.6	6.0	7.0	6.7	mg l^{-1}
Solids, sett.	5.5	3.3	0.8	0.7	0.05	"
Total solids (110°C)	1364	1558	41452	41474	40032	"
Fixed solids (550°C)	1184	1160	35002	34582	33812	"
Volatile solids	180	198	6460	6692	6220	"
Suspended solids	148	236	170	108	9.3	"
Susp.-fixed solids	94	180	140	100	8	"
Volatile Susp. solids	54	56	30	8	1.3	"
COD, conventional	405	558	–	–	–	"
KMnO$_4$	47	45	6.4	5.3	1.6	"
BOD	60	72	10	6	1.6	"
Sulphide	7	9	nil	nil	nil	"
Nitrogen NH$_3$	1.45	1.45	ND	ND	ND	"
NO$_2^-$	nil	nil	0.1	0.1	nil	"
NO$_3^-$	nil	nil	nil	0.88	1.15	"
Phosphorus PO$_4^{3-}$	0.06	0.06	0.01	0.005	0.005	"
Cl$^-$	0.395	0.403	20.6	20.0	20.6	g l^{-1}
SO$_4^{2-}$	161	161	2636	2591	2409	mg l^{-1}
TH (total hardness)	360	356	7000	7000	6700	"
Ca^{++}	64	38	320	360	360	"
Mg^{++}	48	62	1488	1464	1392	"
Alkalinity	256	254	140	128	132	"
Silica (active)	23	18	4.6	9.6	16	"

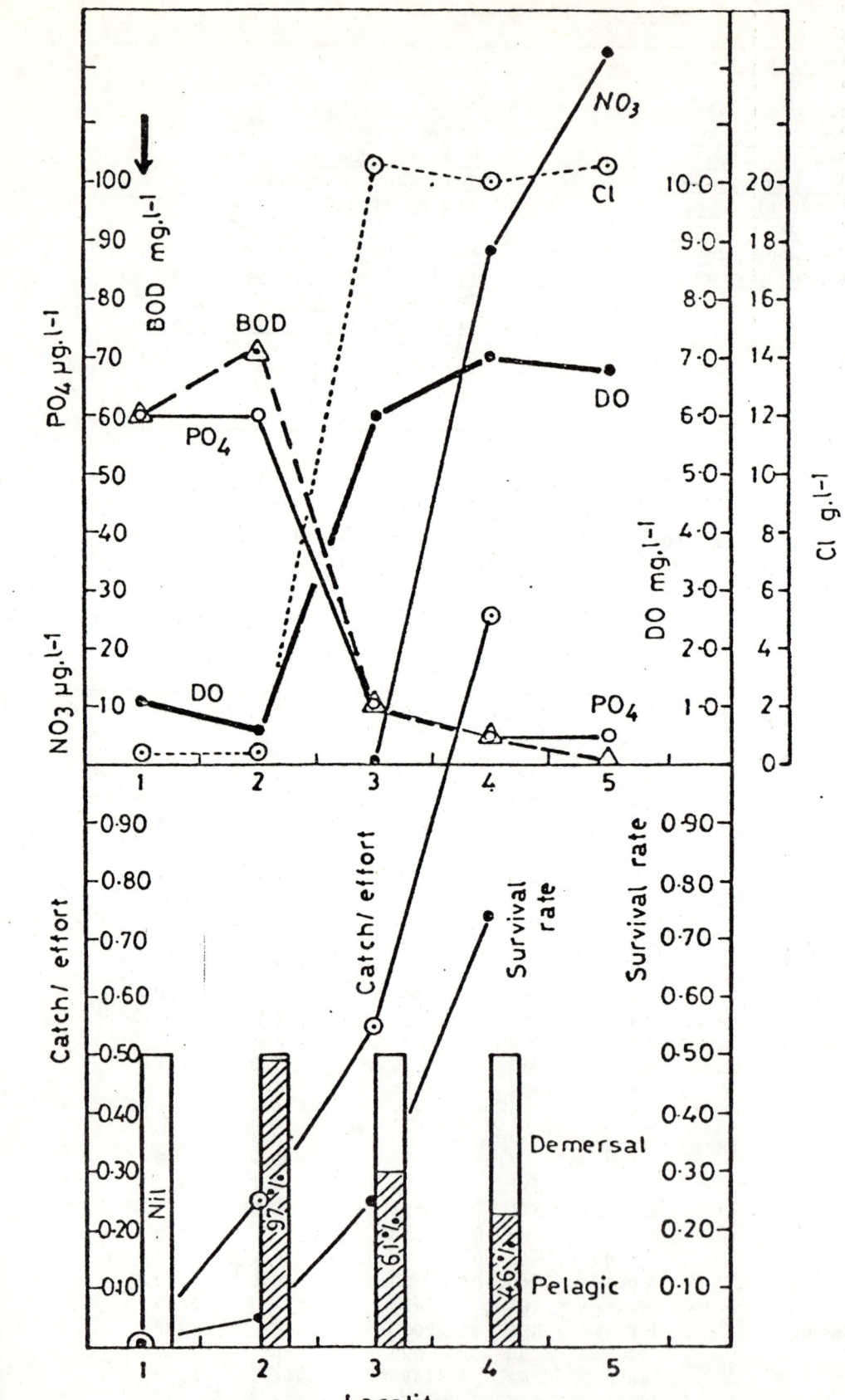

Figure 2. Environmental parameters (upper fig.) and fishery parameters (lower fig.) at five coastal localities in Abu-Kir Bay.

Table III

Catch per effort for the purse-sein net in the four localities investigated (man hour^{-1})

Locality	Catch per effort in Kg
L1	zero
L2	0.25
L3	0.56
L4	1.26

Table IV

Age composition of <u>Mugil Capito</u> from the catch of the purse sein net with decreasing pollution gradient downstream from "El-Tabia" pumping station

Locality	Age of the fish in years				
	1	2	3	4	5
L1	-	-	-	-	-
L2	95%	5%	-	-	-
L3	75%	20%	5%	-	-
L4	30%	30%	20%	15%	5%

Table V

Survival and mortality rates of <u>Mugil capito</u> in the four localities

Locality	Mortality	Survival rate
L1	Complete	-
L2	2.9944	0.0526
L3	1.3541	0.2582
L4	0.2947	0.7448

The outfall area is subjected to multiple pollution. The first locality, directly under the impact of the outfall is azoic with regards to plankton (Dorgham and Osman, pers. comm.) benthic invertebrates and fish. It is characterized by a heavy load of particulate materials, mainly cellulose and lignin particles, imparting a black colour to the water and preventing light penetration, and by a thick layer of sludge on the bottom. Chloride content, nitrate and dissolved oxygen are low. Phosphate, ammonia, COD, BOD, and alkalinity are relatively high. The pattern of circulation has been investigated by Dorgham and Osman (Pers Comm.). The S.W. bay is the site of an anticlock-wise eddy, retarding exchange with the open sea.

Coastal waters in the second locality, 5 km east, are less turbid but comparable as to the other characteristics. Life however, begins to reappear at this transitional zone. The bottom is still muddy and three bottom organisms showed to be extremely abundant: the brown alga Caulerpa prolifera, the brackish amphipod Gammarus aequicauda and a nematode worm. An examination of the gut content of the fish caught from this area showed that they feed abundantly on both the worm and the amphipod. The catch per effort is very low (0.25 kg man^{-1} hour^{-1}). Beside a few specimens of the prawn Trachypenaeus curvirostris, all the commercial catch is made up of pelagic fish. Among the six fish species caught (Table I) Caranx alexandrinus and Julis pavo appear to be characteristic of this locality, the others extend to the coastal strip of the bay up to Rosetta. No molluscs and no fish fry were observed.

The third locality next to the Maadeya lake outlet, approximates the normal conditions of the bay. The chloride content (20.6 g l^{-1}) is much higher. Phosphate, BOD, ammonia, alkalinity, turbidity are much lower (Table II and Fig. 2). The catch per effort is doubled (0.56 Kg man^{-1}hour^{-1}). The demersal fish, completely absent from locality 2, now contribute 39% to the biomass (Table I). A small variety of crabs appears to be attracted to the bottom amphipods and nematods : Macropipus vernalis, Portunus latipes, Portunus(Neptunus) pelagicus, Pachygrapsus marmoratus and Hemigrapsus sp. Caulerpa sp. is less abundant.

Locality 4, just south from Rosetta is entirely free from the effect of the outfall (Table II). The fish diversity is greater and the demersal commercial catch rises to about 54% including 5 prawn species (Table I). The catch per effort is almost three times as much as for locality 3. The locality is characterized by the reappearance of fish fry in large numbers, especially Mugil spp. and Sparus sp. This area also constitutes a rich ground for the edible mussel Donax trunculus, from where it is abundantly fished. Caulerpa sp. is absent. The survival index of Mugil capito was found to be zero directly downstream from the "Tabia" effluent, increasing gradually to a maximum at locality 4. The "instantaneous" fish mortality along the decreasing pollution gradient reflects on the catch per effort, the latter rising from zero at the outfall to a maximum at locality 4. The high mortality index of fish at the outfall, especially demersal forms, is due to the fouling of their gills, reducing their respiration capacity (Saleh, 1983). Recovery of healthy environmental conditions is accompanied by an increased catch/effort, an increased contribution of all year classes to the population, an increased biomass of the demersal fish and invertebrates. Mussels and fish larvae are the most sensitive organisms to pollution, reappearing only where healthy conditions are restored, while Caulerpa sp., Gammarus aequicauda and nematod worms disappear. The continuous disposal of waste water through "El-Tabia" outfall strongly affects the fish stock in Abu-Kir Bay and threatens to destroy its spawning and nursing grounds. The adjacent Edku lake is also affected. The records on Table VI cover the period from 1964 to 1981. They show a drastic drop in the fish catch from this coastal lagoon in the last two decades and the quasi-disappearance of commercial crustaceans (crabs and shrimps) and a steady decrease in the relative abundance of Mugil spp. in favour of the non-migratory Tilapia spp. Both the crustaceans and the mullets spawn in the Bay, their surviving young and fry ascending the lake through Maadeya outlet. The drop in the catch of both mullets and crustaceans from Lake Edku reflects the deterioration of the Bay's environment.

4. CONCLUSIONS

The investigated environment provides a clear-cut cause-effect relationship between a point source of industrial wastewater and the downstream perturbations of the ecosystem. Several indices allow us to identify and quantify this effect and its attenuation along a decreasing pollution gradient:

a. The ratio of pelagic to demersal macrofauna, ranging from 1:0 to about 1:1 along the decreasing pollution gradient.

b. The dominance of certain species, in this case Caulerpa prolifera, Gammarus aequicauda and a nematod worm, gradually disappearing along the decreasing gradient.

Table VI

Annual catch from Lake Edku (in tons), and percentages of the main species in the catch (Data of the Statistical Department, Alexandria Institute of Oceanography and Fisheries). "Tabia" pumping station became functional in 1965.

Year	Total catch in tons	% of Tilapia sp. to total catch	% of Mugil sp. to total catch	% of crabs and shrimps to total catch	% of Anguilla to total catch
1964	4787.5	65.5%	12.2%	5.5%	4.6%
1965	4747.8	66.4%	11.8%	5.5%	3.3%
1966	3400	69.1%	11.8%	4.8%	3.2%
1967	2059	72%	7.9%	3.4%	3.4%
1968	1278.9	70.4%	10.5%	2.8%	5.2%
1971	674.2	84.3%	8.1%	0.1%	1.5%
1975	1059.5	84.1%	5%	-	1.6%
1976	875.5	84%	5.8%	-	1.7%
1977	1279.4	84.7%	4.7%	-	2.2%
1978	632.2	76%	6%	-	4.6%
1979	780.5	76.5%	6.5%	-	4.1%
1980	807.5	76%	5%	-	3.9%
1981	442	83.4%	2%	-	0.9%

c. The extremely low fish yield in the vicinity of the outlet, gradually recovering to normal values.

d. The absence in the vicinity of the outlet of such vulnerable forms as fish-fry and the common bivalve in the Bay, Donax trunculus.

e. The survival index of the common fish, Mugil capito rising from zero to a maximum at the farthest sampling locality. It is only at this last locality that all year-classes contribute to the population of M. capito.

f. The disrupted balance between migratory and non-migratory fish in the adjacent lagoon, "Lake Edku", appears to be a direct consequence of the bay conditions.

5. ACKNOWLEDGEMENTS

The work has been financed by the UNESCO/UNDP Aquatic Environmental Pollution Project, EGY/73/058.

6. REFERENCES

Ricker, W.E., Computation and interpretation of biological statistics of fish
1975   populations. Bull.Fish.Res.Board.Can., (191):382 p.

Saleh, H.H., Mortality of fish exposed to water from the Tabia pumping station, Egypt.
1983   Journ.Etud.Pollut.CIESM, 6(1982):765-70

CHANGES IN THE BENTHIC COMMUNITIES DUE TO VARIOUS POLLUTANTS
IN IZMIR BAY (TURKEY)

by

A. KOCATAS, Z. ERGEN and T. KATAGAN
Ege University, Faculty of Science,
Biology Division, Hydrobiological Department,
Bornova, Izmir, Turkey

1. INTRODUCTION

Izmir Bay and its vicinity is one of the most important regions of Turkey from the point of view of inhabitation, industrialization and marine traffic. During the decade following 1960, the population increment of the region accelerated, industrialization started and in the meantime its marine traffic increased. The domestic and industrial sewage resulting from these activities, even today is discharged directly into the waters of the bay, without being treated in any way. Due to this, quite a few physico-chemical and biological characteristics of the bay waters are changing constantly.

The early studies on the biota of Izmir Bay, which has been under the effects of heavy pollution for nearly 20 years, belong to Forbes(1843) and Colombo(1885). Later, taxonomical and ecological investigations were carried out on algae (Zeybek, 1967; Güner, 1970), molluscs (Geldiay and Uysal, 1971), crustaceans (Geldiay and Kocatas, 1968, 1970, 1972; Kocatas, 1971, 1974, 1975; Geldiay et al., 1971), polychaetes (Geldiay and Ergen, 1972; Ergen, 1976), echinoderms (Unsal, 1973) and sponges (Saritas, 1974). In step with these investigations, in 1970 the first studies on the effects of pollution to the Izmir Bay biota were started by the works of Geldiay and Kocatas (1972a, 1973) on the benthic communities of the Bay. More recently, the following investigations on the pollution-dependent distribution of polychaete species (Ergen, 1979), distribution and seasonal changes of benthic communities (Kocatas, 1978, 1979, 1981; Kocatas et al., 1985), some hydrographical and biological effects of pollution (Kocatas and Geldiay, 1979, 1980) and the pollution-dependent distribution of algae (Güner, 1981) were carried out.

The present study will try to explain the changes observed for a ten-year period (1974-1983) on the benthic communities of the inner bay, which is subject to heavy pollution.

2. DESCRIPTION OF THE STUDY AREA

Izmir Bay can be divided into two sections from the standpoint of its topographical and hydrographical characteristics: an inner bay and an outer bay (Fig. 1).

The inner bay is like a lagoon, which is connected to the outer bay by a narrow channel. The length of its shores is approximately 55 km, its length 13 km, its width 2.5-6.4 km and its area is almost 57 km^2. Shallow shores (0.3-4m) of the northern region extend for quite a large area, whereas the water gets deeper further from the southern shores (11-15m) and its depth reaches 21 metres halfway between these shores.

3. DESCRIPTION OF THE EFFLUENTS

The main polluting sources of Izmir Bay can be given under four headings. There are the domestic sewages, bilge water discharges, industrial effluents and land drainages.

Of these, domestic sewages and industrial effluents are especially important. At present, the whole daily refuse of almost 2 million inhabitants of the city of Izmir is emptied into the bay waters by means of 128 sewers and 10 creeks, without any prior processing or refining. Similarly, daily effluents of approximately 1230 industrial establishments (leathers, food, detergents, textile, alcoholic beverages, chemical plants, etc.) are also discharged directly into the bay waters.

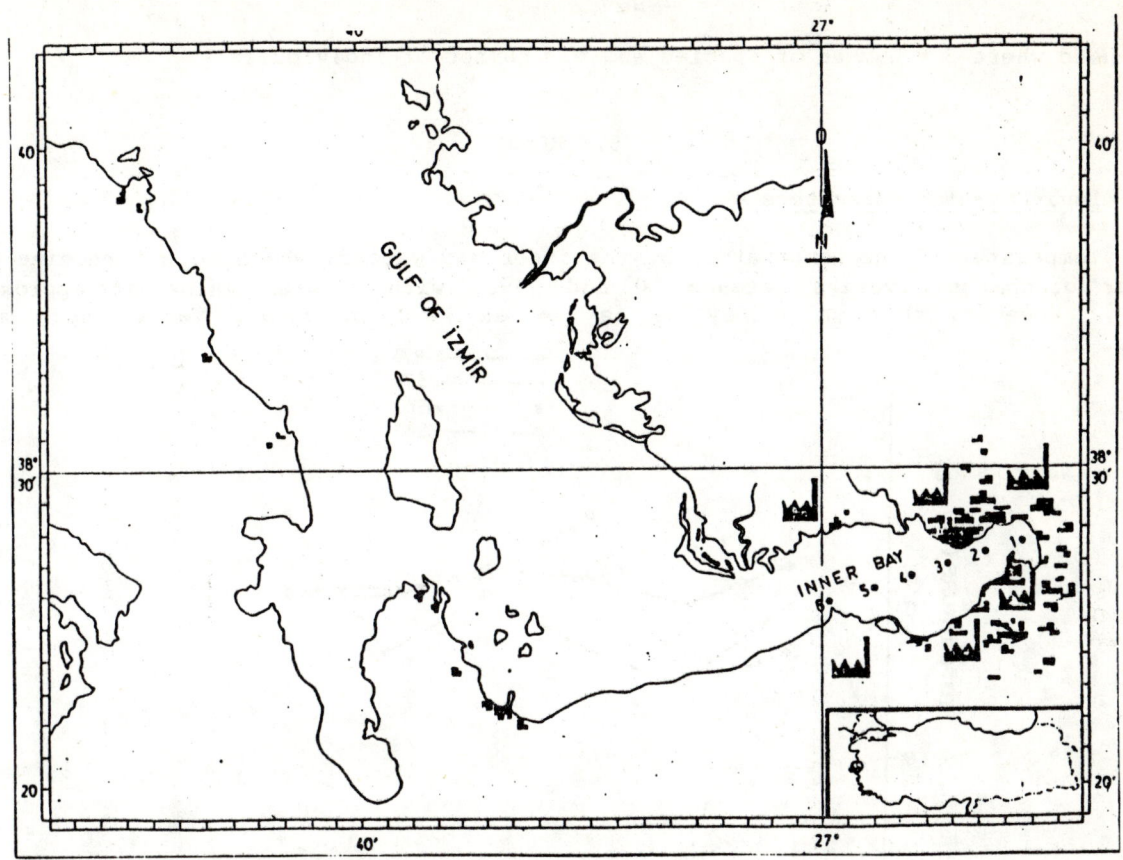

Figure 1. Map of Izmir Bay showing sampling stations in the inner bay.

The relative distribution of the pollutants by source, reaching the inner bay can be summarized as follows:

```
Sewers and industrial effluents ....................... 50%
Rain water drainage from the city ..................... 15%
Artificial fertilizers and pesticides ................. 10%
Rivers ................................................ 10%
Land erosion ..........................................  8%
Bilge water and dock discharges .......................  4%
Untraceable pollution .................................  3%
```

4. MATERIAL AND METHODS

Physico-chemical data were determined monthly and benthic sampling was done simultaneously at the 6 stations chosen in the inner bay.

The following parameters were measured: temperature and salinity (portable temperature-salinity bridge); transparency (Secchi-disc); seston (Millipore HA 0.45-µm filter paper after processing at 60°C in oven); oxygen (Winkler method); pH (pH meter), nitrate and phosphate (portable laboratory kit).

Benthic samples were taken monthly with an orange peel grab (capacity: 4.5 l sediment). Two samples (9 l) of mud were strained through a sieve with a 2mm mesh size. In the laboratory the sample contents were separated into systematic groups. Later, the species of each group were determined.

In order to determine and to compare the diversity of macrofauna, the formula:

$$d = \frac{S - 1}{\log_e N} \quad \text{(Margalef, 1958)}$$

was used where S = number of species and N = number of individuals.

5. RESULTS

5.1 Environmental parameters

Temperature: The temperature of the inner bay waters, which do not receive any hot water discharges, varies between 10 and 27°C, with a mean value of approximately 20°C. However, this value may be as low as 18°C in colder years, such as 1980 (Fig. 2).

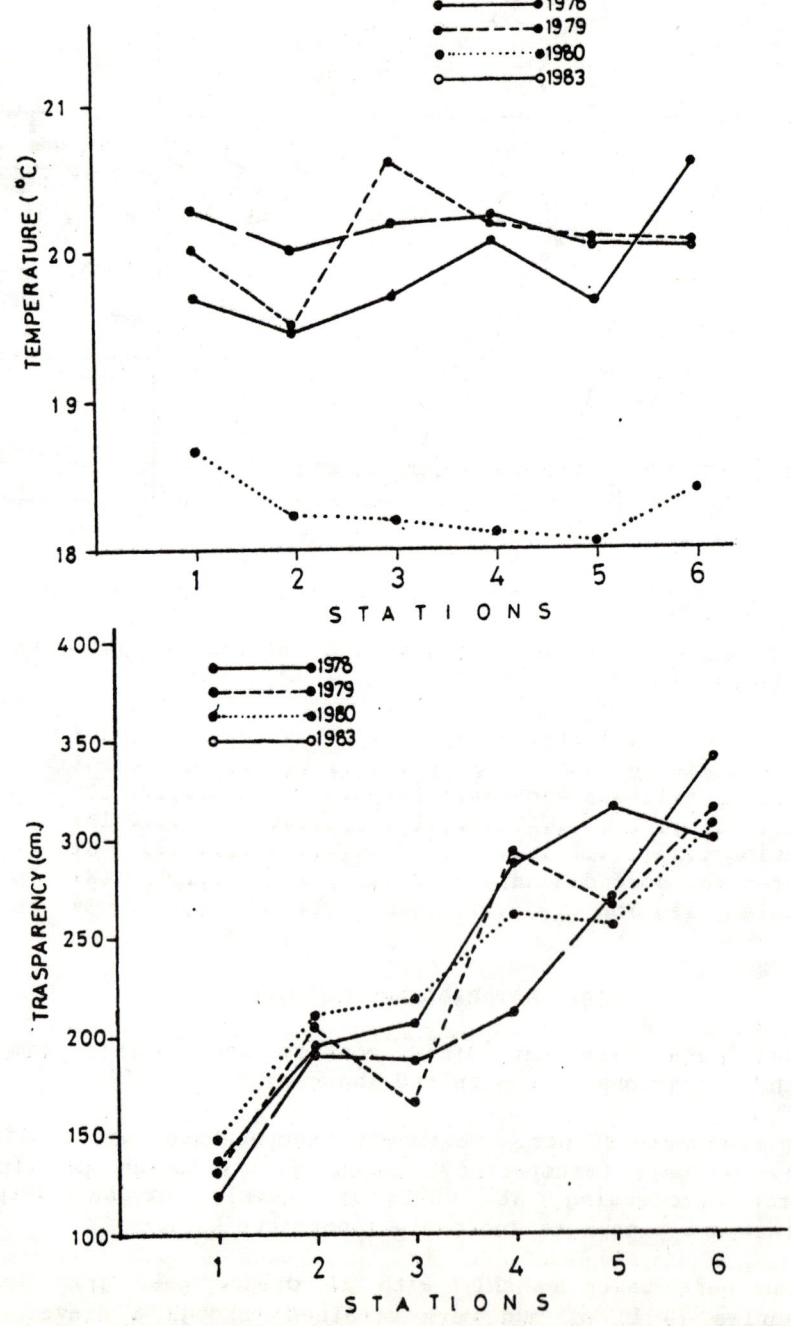

Figure 2. Variation of temperature and transparency yearly means at each station.

Transparency: The transparency of the polluted inner bay waters varies between 70 and 500 cms. No discernible differences have been observed in the yearly means of the stations. However, it can be said that, generally, transparency increases from the inner towards the outer parts of the bay (Fig. 2).

Seston: In the bay waters, this parameter shows a distinct fluctuation from year to year, depending on the station and season of the year. It varies between 4 and 170 mg l^{-1} and its mean value decreases from the inner towards the outer parts of the bay (Fig. 3). The maximal values coincide with the year 1979.

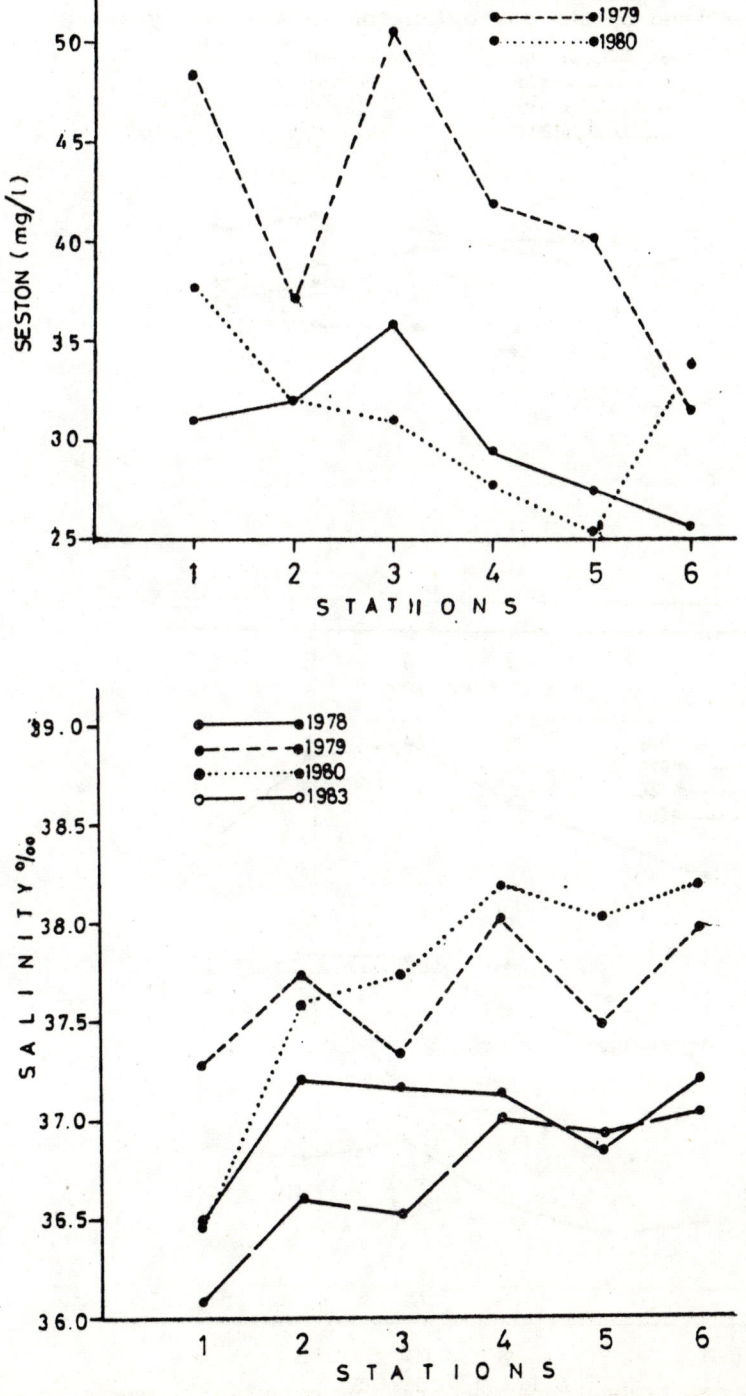

Figure 3. Variation of the seston and salinity yearly means at each station.

Salinity: Salinity of the bay waters varies between 34.4°/oo and 40.5°/oo. However, the mean values are between 36°/oo and 38°/oo (Fig. 3). Yearly salinity variations in the different stations are dependent on temperature and rainfall. During the 10 year survey period, maximum and minimum salinity means were observed in the years 1980 and 1983 respectively.

Dissolved oxygen (DO): The dissolved oxygen content of the surface waters fluctuate significantly from season to season and from year to year in the bay. During the survey period, the values varied between 0.7 mg l^{-1} and 8.0 mg l^{-1}. However, no significant fluctuations are observed in the mean values at different stations in the same year (Fig. 4). Organic matter abundance, temperature and variations in the phytoplankton population play an important role in the seasonal and yearly fluctuation of the DO content. The high DO values observed for the year 1983 are related to the population explosion realised by the phytoplankton in the same year.

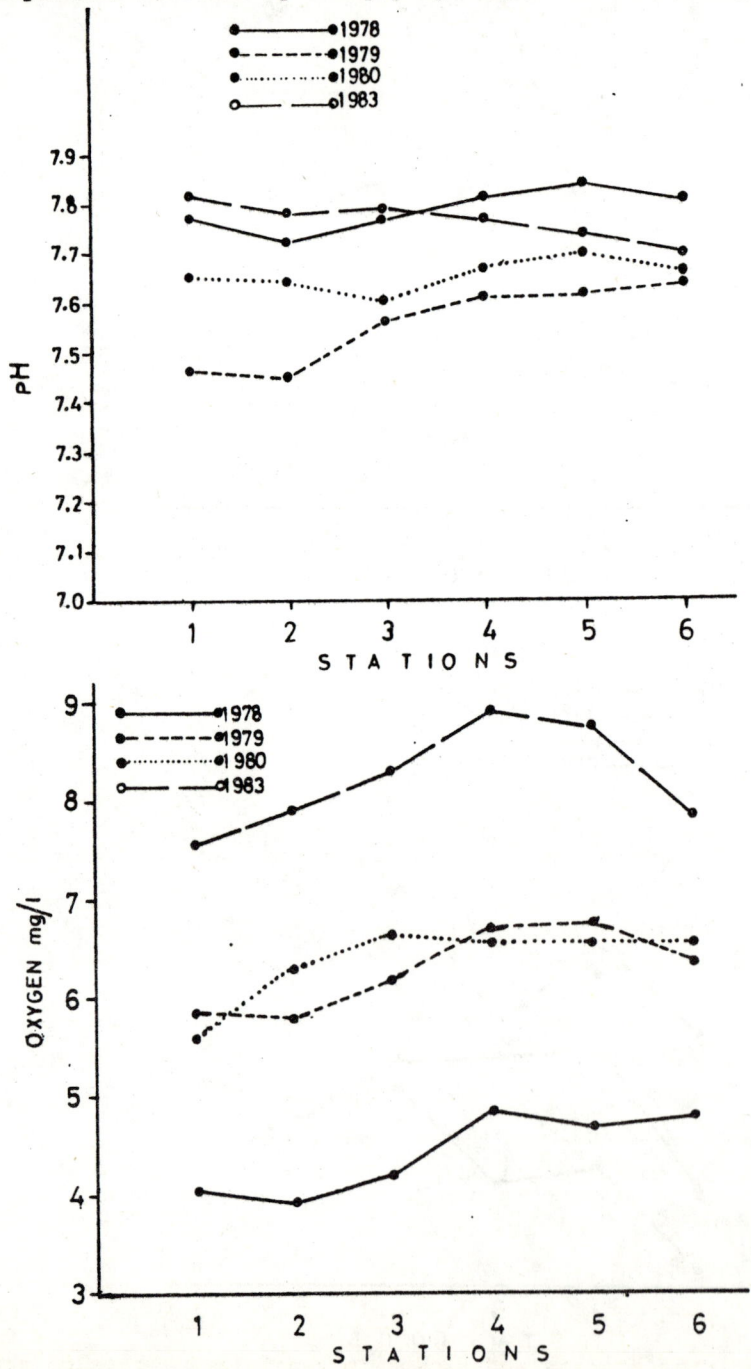

Figure 4. Variation of the dissolved oxygen and pH yearly means at each station.

pH: Bay waters are slightly basic and their pH values vary between 6.5 and 8.6. The yearly means at the stations vary between 7.5 and 7.8 (Fig. 4). The lowest values occurred in 1979.

Nitrate: This nutrient varies seasonally between 3.0 and 166.0 µg l^{-1} in the bay waters. The richest mean values coincide with the year 1983 (Fig. 5).

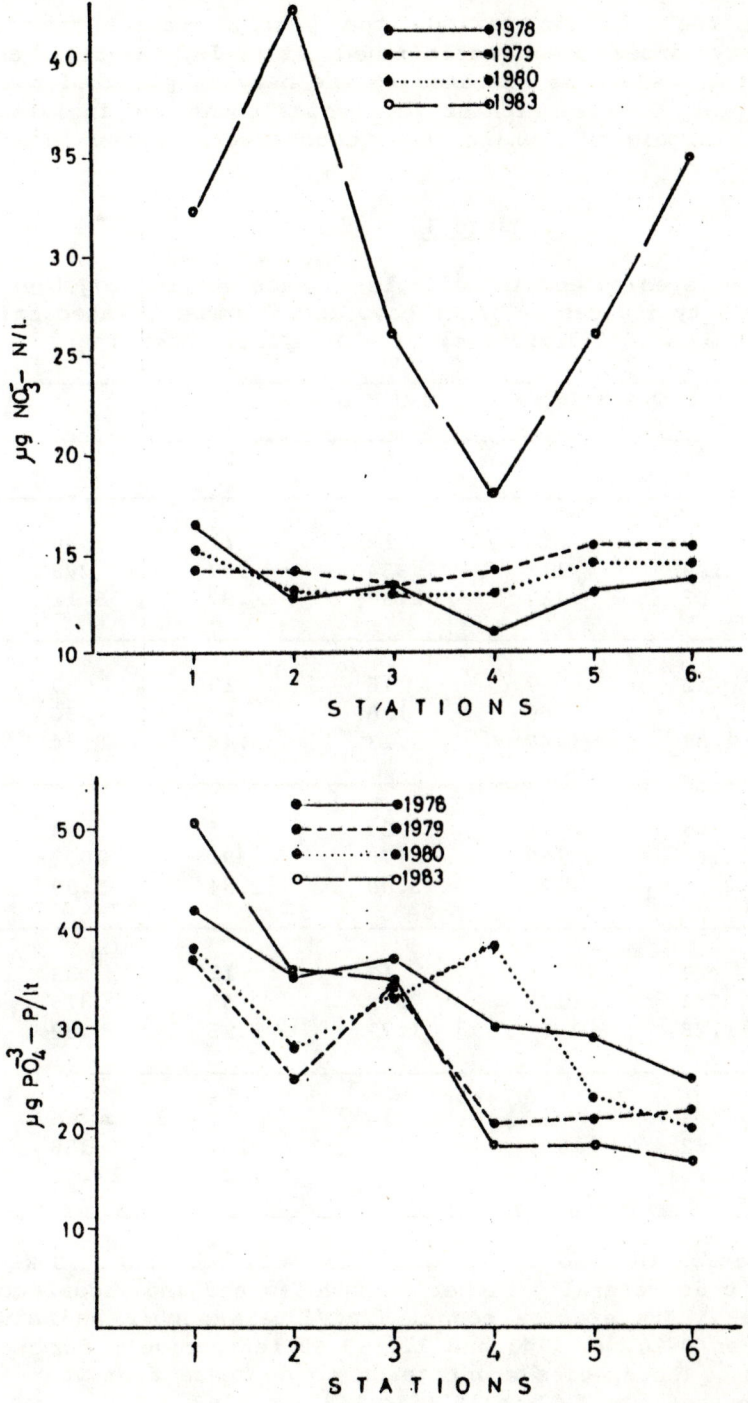

Figure 5. Variation of the nitrate and phosphate yearly means at each station.

5.2 Changes in the benthic communities

Whereas studies on the effects of pollution on the benthic organisms have been started in Izmir Bay during the nineteen seventies, more detailed surveys were initiated in 1974 and are still continuing. According to the data obtained from various investigations carried out in the area during the period 1974-1983, it is apparent that the number of species and individuals as well as the diversity indices change from year to year in the selected stations.

Station 1: During the 10 year period, the yearly species count, number of individuals and diversity index were established at 3-16, 28-1105 and 0.34-2.14 respectively at this station, which is situated in the heavily polluted zone of the bay (Table I, Fig. 6). The year 1974 was richest in species count and individuals' number, the diversity index was also highest, while these counts were lowest in 1980, together with the diversity index.

Table I

Total number of species and individuals at each station together with the diversity indices in inner bay. (SN = Number of species; IN = Number of individuals; DI = Diversity index).

YEAR		STATION NUMBER						
		1	2	3	4	5	6	TOTAL
1974	SN	16	20	12	12	11	30	88
	IN	1105	189	111	354	297	288	3478
	DI	2.14	3.62	1.87	1.87	1.87	5.12	10.66
1978	SN	6	12	7	16	19	50	73
	IN	28	219	59	120	87	356	869
	DI	1.50	2.04	1.47	3.13	3.44	8.34	10.63
1979	SN	12	16	10	21	15	46	59
	IN	761	1149	168	770	391	2072	5311
	DI	1.65	2.12	1.75	3.00	2.34	5.89	6.76
1980	SN	3	9	6	13	11	33	60
	IN	334	553	135	879	165	375	2441
	DI	0.34	1.26	1.01	1.77	1.95	5.39	7.56
1983	SN	13	9	12	10	8	66	75
	IN	784	62	187	173	231	358	1795
	DI	1.80	1.93	2.10	1.74	1.28	11.05	9.87

Station 2: This station is also situated in the polluted zone. However, compared with stations 1 and 3, it is generally richer in species and individual counts and its diversity index is higher. Its species count, individuals number and diversity index were found to vary between 9-20, 62-1143 and 1.26-3.62 respectively during the 10 year period (Table I, Fig. 6). The species count tends to decrease from year to year, i.e. its diversity index increased from 3.62 in 1974 to 1.93 in 1983.

Station 3: This station comprises one of the poorest regions of the inner bay from the point of view of species and individual counts. During the 10 year period, its species count, individual number and diversity index varied between 6-12, 59-287 and 1.01-2.10 respectively (Table I, Fig.6).

Station 4: The species count, individuals number and diversity index varied between 10-21, 173-879 and 1.74-3.13 respectively in this station, which is located in the middle part of the inner bay (Table I, Fig. 6). Except for the years 1978 and 1979, a homogenity is observed at this station from the standpoint of species and individual counts, and the diversity index.

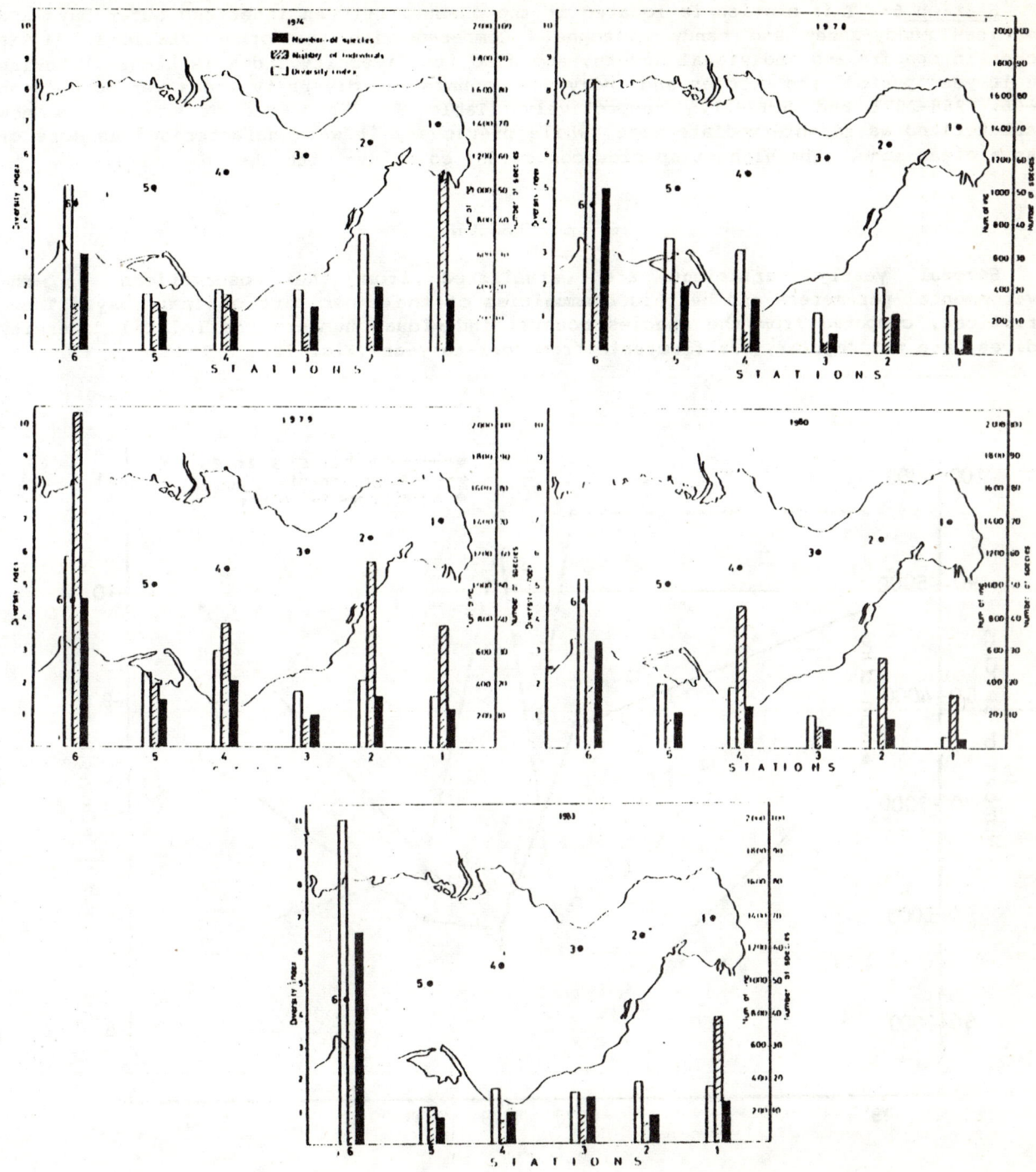

Figure 6. Distribution of the total number of species and individuals at each station and diversity indices during 1974, 1978, 1979, 1980 and 1983.

Station 5: Whereas this station is located near the exit point of the inner bay, still it has a similarity to station 4. During the 10-year period, its species count, individuals number and diversity index varied between 8-19, 87-391 and 1.28-3.44 respectively (Table I, Fig. 6). The observance of only 8 species in 1983 shows that the effects of pollution are gradually increasing at this locality.

Station 6: This station is located at the channel between inner and outer bays and comprises muddy-sandy and sandy biotopes. Compared with the other stations, it is richer in species and individual counts, and also its diversity index is higher. During the 10-year period, its species and individual counts and diversity index varied between 30-66, 288-2072 and 5.12-11.5 respectively (Table I, Fig. 6). This area is now characterized as an intermediate zone, while previously it was characterised as more or less a clear zone. The highest species count (66) coincides with the year 1983.

6. DISCUSSION

Several yearly variations are established from the observations on the environmental parameters and benthic communities of the inner part of Izmir Bay. These variations, computed from the species counts, individual numbers and related diversity indices, are not regular, but fluctuate from year to year (Fig. 7).

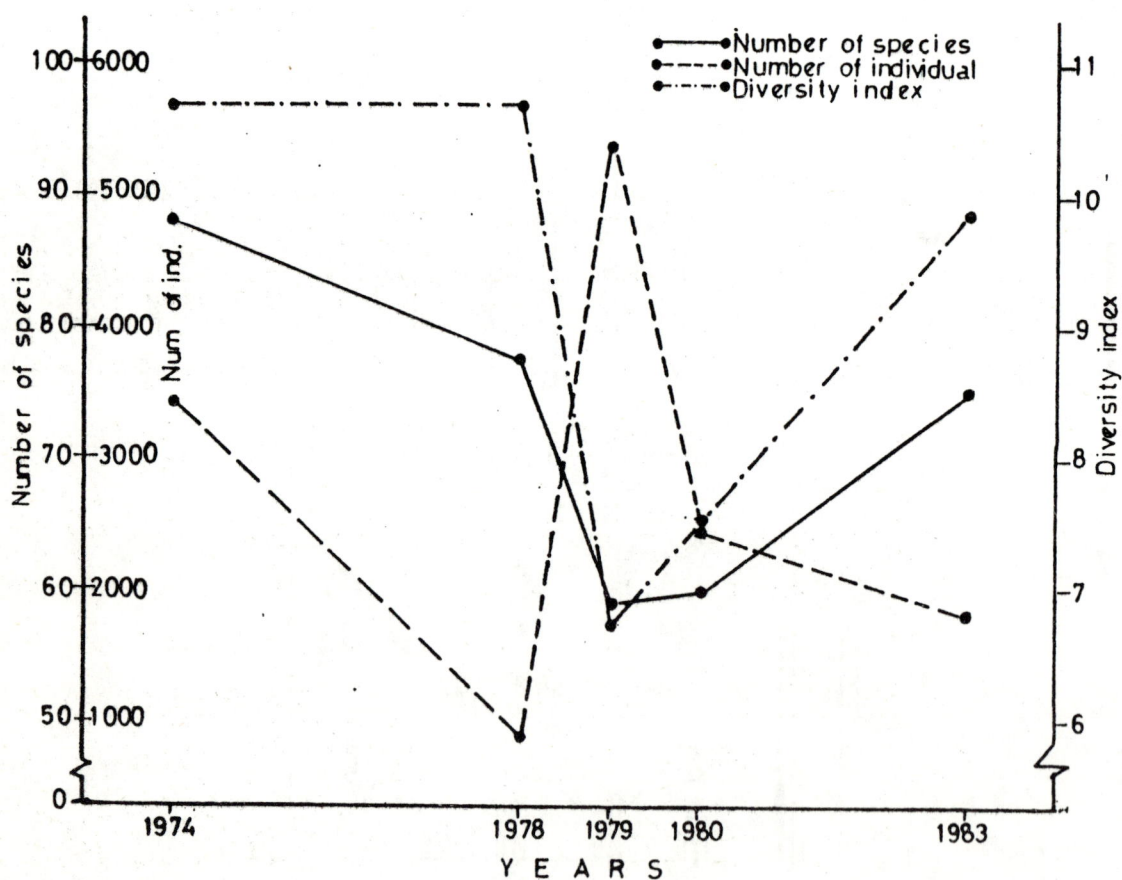

Figure 7. Total number of species and individuals in the inner bay and diversity indices during 1974, 1978, 1979, 1980 and 1983.

The main reasons for these fluctuations are:

(a) Domestic sewage and industrial effluents are discharged into the bay via numerous sewers and creeks in such a way that their local effects vary from season to season and from year to year.

(b) The bottom of the bay, especially at stations 1 and 2, change due to periodic dredging. So, the bottom-dwelling organisms also exhibit periodic changes.

(c) In the last few years, the filling material, used in the construction of the shore highway circling the inner bay, has been causing changes in the bottom structure at numerous localities and this condition affects the bottom-dwellers negatively.

(d) Building of barriers in front of Melez and Halkapinar creeks, which are important pollution sources for the bay, causes less change in the bottom especially at station 1 and, accordingly, a positive effect is observed for the bottom-dwellers.

Depending on these changes observed at the bottom of the bay, a general decrease in species counts is observed at stations 1, 2, 3, 4 and 5 which are more heavily affected by pollution, whereas an increase is evident at station 6 (30 species in 1974 and 66 species in 1983). The reason for this increase is: at the beginning of our survey station 6 was characterised as a clearer zone and included mainly clear zone species but later, as a result of increasing pollution in the inner parts of the bay and the eventual outward spread of this pollution, this locality became characterised as an intermediate zone between clear and polluted zones. Now this locality includes species typical of partly polluted waters, together with clear water species.

7. REFERENCES

Colombo, A., Raccolte zoologiche eseguite dal R. Piroscafo Washington nella campagna
1885   abissale talasso dell'anno 1885. Riv.Marit., (1885): 22-53

Ergen, Z., Investigations on the taxonomy and ecology of polychaeta from Izmir Bay and
1976   its adjacent areas. Sci.Rep.Fac.Sci.Ege Univ., 209(135):1-73 (in Turkish)

_____, The effects of pollution on the distribution of the polychaeta in the Bay
1979   of Izmir. Türk.Ulusal Jeol.Jeod.Bül.Ankara, (11):77-82 (in Turkish)

Forbes, E., Report on the Aegean invertebrates. London, British Association for the
1843   Advancement of Science, pp.130-93

Geldiay, R. and Z. Ergen, Preliminary study on the benthic polychaetes in front of the
1972   Marine Biological Research laboratory. Sci.Rep.Fac.Sci.Ege Univ., 49:1-15
    (in Turkish)

Geldiay, R. and A. Kocatas, Report on a collection of Natantia (Crustacea, Decapoda)
1968   from the Bay of Izmir and its neighbourhood. Sci.Rep.Fac.Sci.Ege Univ.,
    (51):1-46 (in Turkish)

_____, A report on the Anomura collected from the Aegean coast of Turkey
1970   (Crustacea, Decapoda). Sci.Rep.Fac.Sci.Ege Univ., (98):1-35 (in Turkish)

_____, Isopods collected in Izmir Bay, Aegean Sea. Crustaceana, Suppl. 3:19-30
1972

_____, Note préliminaire sur les peuplements benthiques du golfe d'Izmir.
1972a   Sci.Monogr.Rep.Fac.Sci.Ege Univ., 12:1-34 (in Turkish)

_____, Some biological and hydrographical effects of pollution in the Bay of
1973   Izmir. In Proceedings of the Fourth Scientific Congress of TUBITAK, Ankara,
    pp.1-8 (in Turkish)

Geldiay, R. and H. Uysal, A study on the species of Mytilidae living in the Gulf of
1971   Izmir and vicinity. Sci.Rep.Fac.Sci.Ege Univ., 113(72):1-13 (in Turkish)

Geldiay, R., et al., Some littoral Amphipods from the Gulf of Izmir (Aegean Sea,
1971   Turkey). Mem.Mus.Storia Nat.,Verona, 18:369-87

Güner, H., Taxonomische und ökologische Untersuchung über die Strandalgen der Ägäischen
1970   Meeresküste. Sci.Rep.Fac.Sci.Ege Univ., 76(51):1-77 (in Turkish)

_____, Les espèces d'algues recueillies dans les zones polluées de la baie d'Izmir
1981   Rapp.P.-V.Réun.CIESM., 27(2):153-54

Kocatas, A., Investigations on the taxonomy and ecology of crabs "Brachyura" from Izmir
1971   Bay and its adjacent areas. Sci.Rep.Fac.Sci.Ege Univ., 121(76):1-77 (in
         Turkish)

_____, Note préliminaire sur les Amphipodes recueillis dans les horizons
1975   supérieurs de l'étage infralittoral rocheux Golfe d'Izmir. Téthys,
         7(2-3):235-40

_____, Tanaidaces et Isopodes recueillis dans le Golfe d'Izmir. Crustaceana,
1976   30(1):68-72

_____, Contribution à l'étude des peuplements des horizons supérieurs de substrat
1978   rocheux du Golfe d'Izmir (Turquie). Sci.Rep.Fac.Sci.Ege Univ., (12):1-93 (in
         Turkish)

_____, Distribution et évolution des peuplements benthiques du Golfe d'Izmir
1979   (partie intérieure) soumis à des multiples pollutions. Journ.Etud.Pollut.
         CIESM., 4(1978):417-21

_____, Evolution cyclique du benthos dans les zones de pollution du Golfe d'Izmir
1981   (Turquie). Journ.Etud.Pollut.CIESM, 5(1980):643-48

Kocatas, A. and R. Geldiay, Preliminary evaluations on the hyrographical and biological
1979   effects of pollution in the Izmir Bay. Türk.Ulusal Jed.Jeod.Bül.Ankara,
         (11):89-97 (in Turkish)

_____, Effects of domestic pollution in Izmir Bay (Turkey). Helgol.Meeresunters.,
1980   33:393-400

Kocatas, A., et al., Les effets de la pollution sur les écosystèmes benthique et
1985   pelagique dans le golfe d'Izmir (Turquie). Journ.Etud.Pollut.CIESM,
         7(1984):689-98

Margalef, R., Temporal succession and spacial heterogeneity in phytoplankton.
1958   In Perspectives in marine biology, edited by A.A. Buzzati-Traverso. Berkley,
         California, University of California Press, pp.323-49

Saritas, M.U., Taxonomical studies on the siliceous sponges living in the Bay of Izmir.
1974   Ph.D.Thesis. Sci.Rep.Fac.Sci.Ege Univ., (168):54 p. (in Turkish)

Unsal, S., Bio-ecological studies on the Echinoderms living on the Turkish coast of the
1973   Aegean Sea. Ph.D.Thesis, (in Turkish)

Zeybek, N., Some algae collected from the Aegean coast. Sci.Rep.Fac.Sci.Ege Univ.,
1967   (27):1-29 (in Turkish)

MODIFICATIONS DE L'ECOSYSTEME PLANCTONIQUE PAR LA POLLUTION DES
EAUX COTIERES LIBANAISES

par

S. LAKKIS* et R. ZEIDANE**

*Faculté des Sciences, Université Libanaise, Hadeth, Liban
**CNRS/Centre de Recherches Marines, B.P.123, Jounieh, Liban

1. INTRODUCTION

Au cours des dernières années, les écosystèmes marins et plus particulièrement pélagiques, des zones eutrophes, portuaires et polluées, ont fait l'objet de nombreuses études écologiques et hydrobiologiques. En Méditerranée occidentale, plusieurs chercheurs ont travaillé sur l'influence de la pollution sur les systèmes planctoniques (Yamazi, 1964; Groupe EPOPEM, 1979; Patriti et al., 1979; Arfi et al., 1982). Par contre, la Méditerranée orientale, notamment la côte est du Bassin Levantin, est peu étudié de ce point de vue.

Dans le cadre des projets de recherches menés par le PNUE, en collaboration avec la FAO, concernant le programme MED POL Phase II dans tous les pays riverains de la Méditerranée, nous avons eu l'occasion d'éxécuter une étude sur l'écosystème planctonique de quelques secteurs les plus touchés par la pollution marine tels que celui du nord de Beyrouth.

Les travaux relatifs à l'étude de la pollution du milieu marin au Liban sont très maigres. Quelques observations ont été effectuées récemment concernant les zones portuaires et polluées (Kouyoumjian, 1972; Lakkis, 1973; Lakkis et Kouyoumjian, 1975; Basson et al., 1976; Lakkis et Abboud, 1977). Des recherches concernant l'écosystème planctonique côtier ont été menées depuis 1970 (Lakkis, 1971; 1973). Par ailleurs, des observations sur la structure et l'abondance du plancton des eaux portuaires et du large ont fait l'objet de quelques publications (Lakkis et Novel-Lakkis, 1981; 1985; Lakkis et Zeidane, 1983).

Dans cet article, nous allons présenter les résultats de nos observations effectuées entre juillet 1984 et avril 1985 sur l'écosystème planctonique de quelques zones polluées situées au nord de Beyrouth.

2. METHODES, MATERIEL ET ZONE PROSPECTEE

Entre juillet 1984 et avril 1985, des prélèvements réguliers, hydrologiques et planctoniques effectués presque tous les mois dans les 10 stations fixées, ont fourni des données sur les paramètres, température, Salinité, turbidité, oxygène dissous, sels nutritifs et 140 échantillons planctoniques (dont 70 pour le phyto et les protistes). Parmi ces stations, 4 sont situées dans des secteurs très pollués: Z1, G1, DS, DV et les 6 autres sont éloignées des zones d'épandage: Z2S, Z2V, G2S, G2V, G3S, G3V (Fig. 1).

Méthodes de pêche et d'analyse

Les prélèvements planctoniques sont effectués aux filets de type WP2 (UNESCO, 1968), l'un de 200 microns et l'autre de 50 microns de vide de maille, de 130 cm de long et 40 cm de diamètre d'ouverture. Les deux filets sont couplés sur un système rigide de type Bongo et muni chacun d'un volucompteur de type Hydro-Bios calibré pour la mesure de l'eau filtrée à travers les filets. Des traits horizontaux de surface durant 2 minutes sont effectués à partir du bateau de recherche SETA III allant à faible vitesse (1 noeud/h.); des pêches verticales sont aussi effectuées aux mêmes stations. Les échantillons sont fixés à bord au formol neutralisé à 4% pour être analysés.

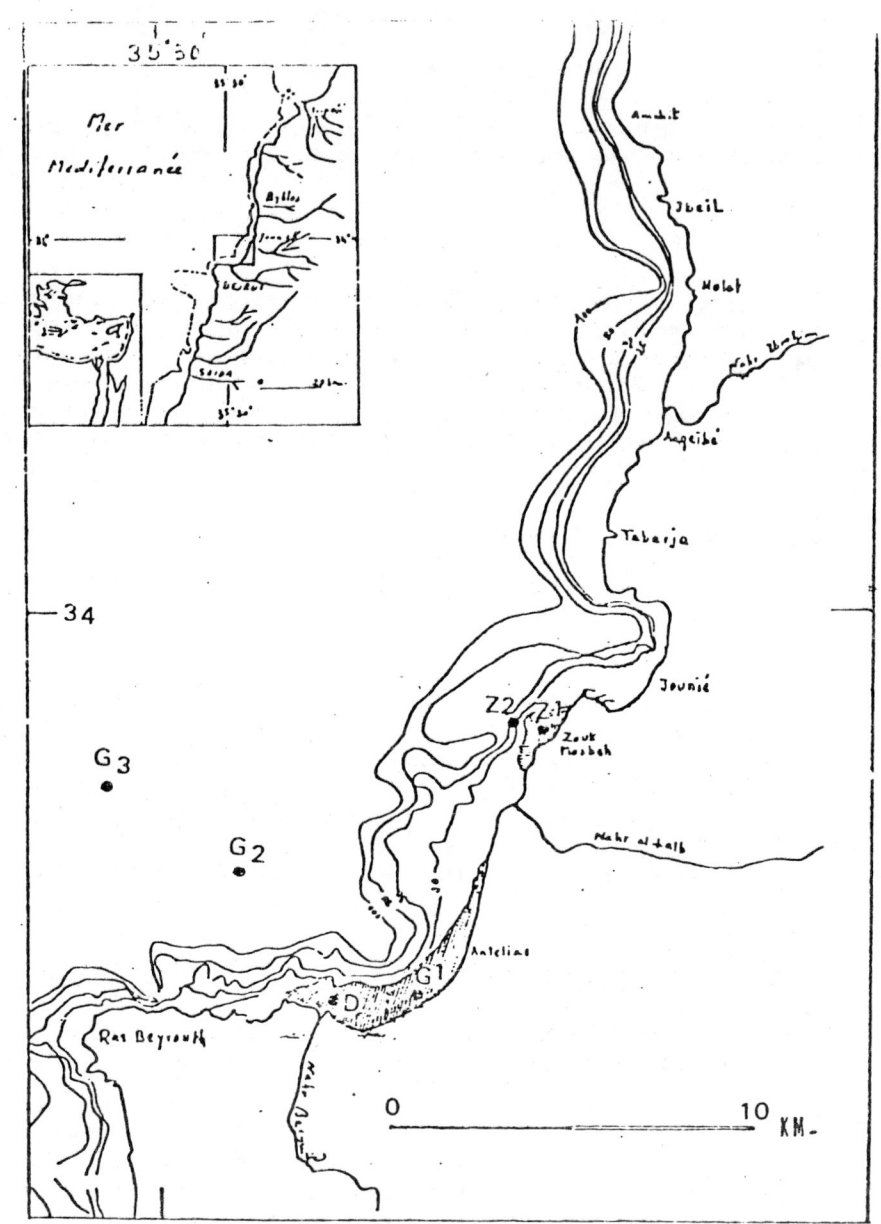

Figure 1. Location des stations sur la côte libanaise visitées entre juillet 1984 et avril 1985. La zone étudiée est située entre Beyrouth et Jounieh; les secteurs hachurés sont fortement pollués. L'emplacement de la zone étudiée est visible sur la carte de la côte du Liban avec une carte de la Méditerranée orientale.

Plusieurs paramètres hydrologiques et de pollution ont été pris en considération: température, salinité, transparence de l'eau, oxygène dissous, nitrites, nitrates et phosphates. Leurs méthodes de mesures classiques sont celles qui sont utilisées dans la plupart des laboratoires et que nous avions utilisées auparavant (Lakkis, 1983). Tous ces paramètres ont été relevés simultanément avec les pêches planctoniques.

Température: thermomètre à renversement calibré Negretti & Zambra, London.

Salinité: l'eau prise aux bouteilles Nansen est mesurée au laboratoire au moyen d'un salinomètre à conductivité de type Beckman calibré à l'eau normale de Copenhague.

Eléments nutritifs: nitrates, nitrates et phosphates sont mesurés au laboratoire selon la méthode colorimétrique préconisée par Strickland et Parsons (1968).

La transparence de l'eau est mesurée in situ au moyen du disque de Secchi.

L'oxygène dissous a été analysé à partir d'échantillons d'eau pris au moyen des bouteilles Nansen et fixés à bord et analysés au laboratoire selon la méthode de Winkler.

Les analyses planctoniques au laboratoire sont effectuées de la façon suivante:

(a) Le Zooplancton: détermination jusqu'à l'espèce et comptage des organismes dans des sous-échantillons fractionnés à la boîte de Motoda (1959) et comptés sous stéréoscope dans une cuvette de Dolfuss en verre rectangulaire de 5 cm sur 10 dont le fond est partagé en 200 carrés de 5 mm de coté par un quadrillage en relief (fournisseur: Leune, Paris). La fraction de récolte examinée varie selon la densité du zooplancton entre 2 et 4. En supposant que la distribution est normale (Frontier, 1973), les résultats sont extrapolés et rapportés en nombre d'organismes par m^3 d'eau de mer.

(b) Le Phytoplancton: Une aliquot de 0,5 cm^3 de l'échantillon étalée dans une cellule graduée et quadrillée placée sous un microscope. La détermination des espèces les plus communes est effectuée ainsi que le comptage des cellules rapporté au nombre de cellules par litre.

Stations fixées et sources de pollution

La région étudiée est située au nord de la ville de Beyrouth, entre celle ci et Jounieh, de latitudes 33°54' et 33°59'N et 35°29' et 35°36'E. Cette zone est fortement polluée à cause des agents polluants déversés en mer dans les rejets urbains et industriels. Les sources de pollution varient selon les stations de la façon suivante:

Z1 : Station côtière située immédiatement au voisinage du rejet de l'eau de refroidissement prise directement en mer par l'usine thermo-électrique de Zouk. L'eau de cette station est donc sous l'influence d'une pollution thermique. Par ailleurs, des dérivés d'hydrocarbure provenant des moteurs et des bateaux en stationnement au bout du quai polluent encore la région. Fond sableux de 3-4 m de profondeur (voir carte Fig. 1). Huit prélèvements.

Z2S: Située à 1 km à l'ouest de Z1, cette station se trouve par 20 m de fond dans une zone éloignée des rejets polluants à eau relativement propre. Station témoin. Huit prélèvements.

Z2V: Même point que Z2S, les prélèvements hydrologiques et planctoniques verticaux sont effectués entre 10 et 0 m. Huit prélèvements.

G1 : Station côtière située par 3-4 m de fond sablo-vasseux dans une zone fortement polluée. Dans ce secteur la densité d'habitation est très grande et plusieurs usines notamment de savon, sucre, d'huile, de tanneries, de batteries, levure de bière, de peinture, de marbre etc. rejettent leurs déchets directement en mer sans aucun traitement. Par ailleurs, des dépôts de carburants et de gaz butane sont aménagés sur la côte avec un long quai de mouillage pour les bateaux. Station visitée 8 fois.

Ds : Station côtière située par 10 m de fond vasseux dans une zone fortement polluée. La source de pollution provient des dépôts d'ordure accumulés sur la côte et rejetés par les boueurs d'une partie de la ville de Beyrouth. Non loin de cette station se jette un cours d'eau (sec en été) qui est le fleuve de

Beyrouth amenant des déchets et des rejets de quelques usines et égouts au cours de route. Les abattoirs de la ville ne sont pas très loin de ce point. Par ailleurs, une pollution par les hydrocarbures affecte ce secteur, provenant du port de Beyrouth et des bateaux en mouillage dans la zone portuaire. Station visitée huit fois entre juillet 1984 et avril 1985.

DV : Station située au même point côtier que DS avec des prélèvements hydrologiques et planctoniques verticaux 5-0 m; visitée 8 fois.

G2S: Station située à 3 milles au large de la baie de St Georges dans le prolongement de la station Gl avec G3. Ce point se trouve par 150 m de fond sablo-vaseux et éloignée des zones polluées; 6 visites.

G2V: Station verticale (50-0 m) au même point que G2S; 6 prélèvements.

G3S: Station au large située à 6 milles de la station Gl et par 300m de profondeur; fond vaseux. Visitée 5 fois, 5 prélèvements.

G3V: Station à prélèvements hydrologiques et planctoniques verticaux (50-0 m) et située dans une zone propre éloignée de la côte au même point que G3S. Station visitée également 5 fois.

Les données hydrologiques et biologiques sont tabulées et traitées de façon à voir les informations nécessaires pour comparer entre les différentes stations. Les moyennes ainsi que les déviations standards et le coefficient de variations ont été calculés suivant les méthodes statistiques classiques. La date et le nombre de prélèvements sont donnés au Tableau I.

Tableau I

Date et nombre de prélèvements planctoniques et hydrologiques effectués aux 10 stations entre juillet 1984 et avril 1985

Date Stat.	Z1	Z2S	Z2V	G1	DS	DV	G2S	G2V	G3S	G3V
19/ 7/84	-	-	-	x	x	x	x	x	x	x
25/ 8/84	x	x	x	x	x	x	x	x	-	-
19/ 9/84	x	x	x	x	x	x	-	-	-	-
9/11/84	x	x	x	x	x	x	-	x	x	x
11/12/84	x	x	x	x	x	x	x	x	x	x
9/ 1/85	x	x	x	x	x	x	x	x	x	x
13/ 2/85	x	x	x	-	-	-	-	-	-	-
27/ 3/85	x	x	x	x	x	x	x	x	x	x
24/ 4/85	x	x	x	x	x	x	x	x	x	x
Nbre Sorties	8	8	8	8	8	8	6	6	5	5
Nbre Ech.Zoo	8	8	8	8	8	8	6	6	5	5
N.Ec.Phyto	8	8	8	8	8	8	6	6	5	5
Tot.Ech.	16	16	16	16	16	16	12	12	10	10
Grand Total	....70 + 70 = 140									

Du point de vue circulation générale, le courant de surface parcourt la côte libanaise parallèlement à celle-ci et portant presque toujours vers le nord; gardant ainsi le même aspect que la circulation générale régnant sur la côte est de la Méditerranée. Ce courant pourrait être modifié localement suivant la configuration de la côte et la topographie du plateau continental, très étroit dans ce secteur. Le mouvement des masses d'eau le long de la côte est associé au système des courants de surface et des facteurs météorologiques. En hiver (décembre-mars), une isothermie verticale due à un mélange des masses d'eau sous-jacentes et une température homogène de minimum 17° gagne toute la colonne d'eau 150-0 m. Par contre, entre avril et novembre, une phase thermique chaude s'installe avec établissement de la thermocline entre 35 et 75 m et une stratification au niveau des couches d'eau superficielles et profondes. La température de surface monte à un maximum de 30°C alors qu'à 300 m elle se stabilise à 15° (Lakkis, 1984).

3. RESULTATS

Paramètres hydrobiologiques

(a) <u>Température</u>: Les valeurs de la température correspondent à la moyenne annuelle, soit autour de 23°C. Les températures à la station Z1 sont affectées directement par l'eau de refroidissement rejetée. Au voisinage immédiat du rejet, la moyenne de la température de l'eau de mer est de 1 à 2° plus que la normale. L'écart entre le minimum (février-mars, 17°) et le maximum (août-sept. 30°) est assez grand en surface; alors qu'à 50 m l'écart est beaucoup plus faible, la thermocline étant très prononcée entre 35 et 75 m. Les variations géographiques locales et saisonnières sont présentées à la Fig. 2A, B, C et Tableau III.

(b) <u>Salinité</u>: Le Bassin levantin en général et les eaux libanaises en particulier sont caractérisés par la salinité la plus élevée en Méditerranée (Lakkis, 1983). Les mesures effectuées ont des valeurs moyennes dans les différentes stations fluctuant légèrement autour de la moyenne générale de 39,10°/oo. A la station Z2S, la salinité de surface est toujours inférieure à celle dans les autres stations à cause de l'influence de l'eau douce en provenance du fleuve de Nahr El Kalb. A la station D, le fleuve de Nahr Beyrouth affecte aussi la salinité du secteur qui descend parfois à 30°/oo (Tableau IV). Les variations géographiques locales de la salinité de surface sont donc faibles malgré les conditions particulières des deux stations Z2S et DS. Ces variations locales sont les plus importantes en juillet (coefficient de variation 9%) et les plus faibles en novembre (0,15%). A la figure 3, nous présentons les variations géographiques locales et saisonnières de la salinité de surface. Le facteur hydroclimatique semble être le plus important qui régit les variations de la salinité. En effet, le minimum saisonnier qui est entre février et mars, correspond à l'apport maximal d'eau de pluie, de la fonte des neiges et des crues des fleuves. Par contre, la période sèche estivale très longue (mai-novembre) induit une salinité maximale en septembre-octobre.

(c) <u>Oxygène dissous</u>: En hiver, le point de saturation en oxygène est enregistré dans les eaux éloignées des zones polluées. Par contre dans les secteurs pollués, la teneur en oxygène de l'eau de mer est sévèrement affectée par les facteurs polluants. En Z1, la moyenne annuelle ne dépasse pas 5,21 ml l^{-1}, l'eau étant toujours chauffée par l'eau de refroidissement de l'usine thermo-électrique. En G1, DS et DV, les valeurs de la teneur en oxygène sont plus basses que la normale à cause de la pollution organique très forte qui exige une consommation en oxygène plus grande. En regardant les résultats présentés aux Tableaux II et V et la courbe d'oxygène de la figure 4, on voit nettement la différence entre les zones polluées (Z1, G1, DS, DV) et les zones non polluées (Z2S, Z2V, G2S, G2V, G3S, G3V).

(d) <u>Transparence</u>: Ce facteur peut être un indicateur de pollution ainsi qu'un indicateur de production primaire. En effet, l'eau très côtière et celle au voisinage des émissaires urbains est toujours plus turbide qu'au large. Les matières organiques en suspension sont beaucoup plus importantes dans les secteurs pollués ce qui réduit beaucoup la transparence de l'eau. Aux Tableaux II et VI, nous présentons les valeurs enregistrées du disque de Secchi. Les stations G1, DS et DV ont toujours les eaux les moins transparentes puisque le disque disparaît à 1 ou 2 m. Par contre l'eau devient plus transparente au fur et à mesure que l'on s'éloigne de la côte, pour atteindre un maximum de transparence à la station G3.

Tableau II

Valeurs des différents paramètres hydrobiologiques aux dix stations étudiées entre juillet 1984 et avril 1985 (moyennes annuelles). Les valeurs de l'indice de diversité sont calculées à partir de la formule de Gleason-Margalef

STATIONS	S°/oo	T°C	O₂ ml/l	Turbid. Secchi	PO₄³⁻ µg/l	NO₃⁻ µg/l	NO₂⁻ µg/l	Zoopl. ind/m³	Ind.Div. bit/ind.	Phytopl. N cel/l	Ind.Div. phyto
Z1	39,00	24,78	5,21	fond	0,10	0,09	0,08	455	5,7	74222	4,9
Z2S	38,58	22,98	6,08	10,4	0,06	2,56	0,07	312	7,2	96889	4,9
Z2V (10 m)	39,15	22,20	5,97	"	0,26	0,00	0,05	1194	5,6	57111	5,6
G1	38,45	23,68	5,28	3,0	1,52	0,80	0,40	401	4,8	62375	3,3
DS	37,52	24,61	4,71	2,3	0,32	0,97	0,11	387	4,4	39375	2,7
DV (5 m)	39,05	23,03	4,55	"	4,76	1,96	0,10	2788	3,6	23500	2,6
G2S	39,01	23,01	6,73	15,6	0,39	0,12	0,03	473	5,7	56000	5,8
G2V (50 m)	39,15	20,93	5,30	"	0,16	0,00	0,03	917	8,5	48333	5,0
G3S	39,05	21,74	6,86	17,5	0,16	0,00	0,03	358	6,9	39667	4,4
G3V (50 m)	39,14	20,66	5,56	"	0,14	0,00	0,005	1165	9,4	36166	4,5

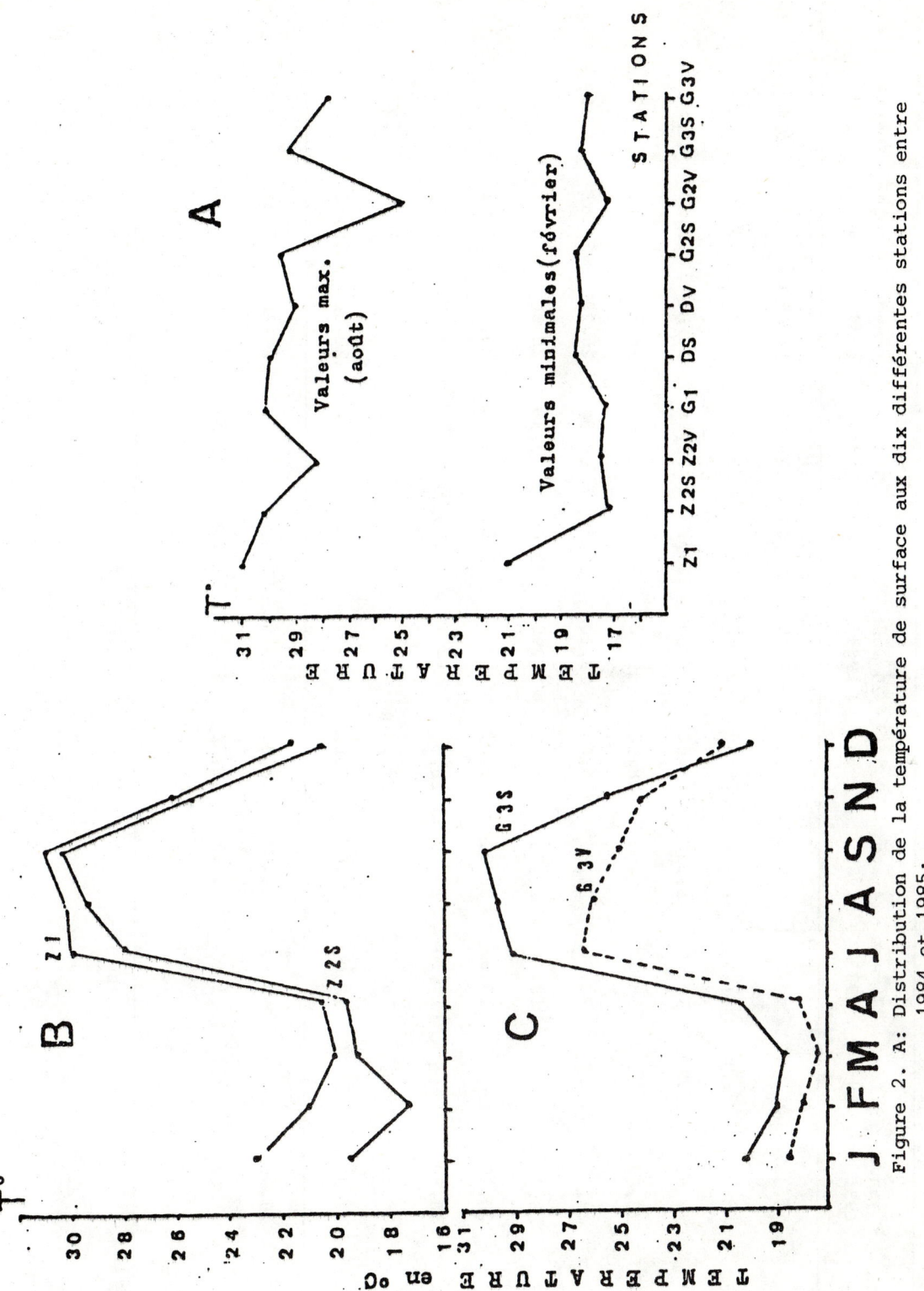

Figure 2. A: Distribution de la température de surface aux dix différentes stations entre 1984 et 1985;

B: Variations mensuelles de la température de surface aux stations Z1 (pollution thermique) et Z2S (côtière non polluée);

C: Evolution mensuelle de la température de surface en G3S (au large) et à 50 m en G3V.

Tableau III

Distributions saisonnières de la température (°C) aux 10 stations visitées entre
19 juillet 1984 et 24 avril 1985

STATIONS	9 janv. 1985	13 fév. 1985	27 mars 1985	24 avr. 1985	19 juil 1984	25 août 1984	19 sep. 1984	9 nov. 1984	11 déc. 1984	Moyenne ann.	Ecart-type	% C.V.
Z1	23,00	21,00	20,05	20,25	30,00	30,00	31,00	26,00	21,80	24,78	4,43	18%
Z2S	19,30	17,20	17,50	19,50	28,00	29,60	30,20	25,20	20,40	22,98	5,25	23%
Z2V (10 m)	19,60	18,20	17,30	19,20	26,00	27,30	28,10	23,90	20,20	22,20	4,15	19%
G1	20,80	-	18,80	19,90	28,80	29,00	30,10	24,00	18,10	23,68	4,97	21%
DS	19,40	-	23,80	20,80	28,60	29,10	30,00	25,00	20,20	24,61	4,26	17%
DV (5 m)	19,00	-	18,20	19,50	27,10	27,90	29,00	23,90	20,00	23,07	4,43	19%
G2S	20,20	-	18,80	20,30	29,00	-	29,50	-	20,30	23,01	4,86	21%
G2V (50 m)	18,80	-	17,10	18,00	26,30	-	25,00	-	20,40	20,93	3,83	18%
G3S	20,60	-	18,50	19,90	29,10	-	-	-	20,60	21,74	4,20	19%
G3V (50 m)	18,90	-	18,00	18,00	27,80	-	-	-	20,60	20,66	4,13	20%
Moyenne gén.	19,96	18,80	18,80	19,53	28,07	28,81	29,10	24,66	20,26	22,76	4,45	19%
Ecart-type	1,27	1,97	1,96	0,93	1,28	1,02	1,86	0,87	0,90	1,40	0,44	1,8%
% Coef.Var.	6%	10%	10%	5%	4%	3%	6%	3%	4%	7%	1%	1%

Tableau IV

Distribution de la salinité dans les stations visitées entre 1984 et 1985

STATIONS	9 janv. 1985	13 fév. 1985	27 mars 1985	24 avr. 1985	19 juil 1984	25 août 1984	19 sep. 1984	9 nov. 1984	11 déc. 1984	Moyenne ann.	Ecart-type
Z1	39,26	37,93	38,76	38,84	-	39,30	39,30	39,42	39,21	39,00	0,49
Z2S	39,24	36,86	36,89	38,65	-	39,24	39,26	39,27	39,28	38,58	1,07
Z2V (10 m)	39,25	38,85	38,99	38,93	-	39,28	39,30	39,28	39,31	39,15	0,19
G1	38,89	-	36,19	37,96	38,94	39,10	39,25	39,24	38,06	38,45	1,04
DS	38,96	-	37,76	38,02	29,45	38,50	39,00	39,26	39,19	37,52	3,30
DV (5 m)	39,29	-	38,54	38,85	39,10	39,10	39,00	39,33	39,12	39,04	0,25
G2S	39,30	-	38,82	38,94	38,53	-	39,25	-	39,24	39,01	0,30
G2V (50 m)	39,29	-	39,04	38,91	39,20	-	39,28	-	39,18	39,15	0,15
G3S	39,28	-	39,01	38,96	38,66	-	-	-	39,32	39,05	0,27
G3V (50 m)	39,28	-	39,10	38,91	39,10	-	-	-	39,31	39,14	0,16
Moyenne	39,20	37,88	38,31	38,69	37,56	39,08	39,20	39,30	39,12	38,80	0,64
Ecart-type	0,15	0,99	1,02	0,38	3,59	0,30	0,12	0,06	0,38	0,51	
% Coef.Var.	0,4%	3%	3%	1%	9%	0,7%	0,3%	0,15%	0,9%	1%	

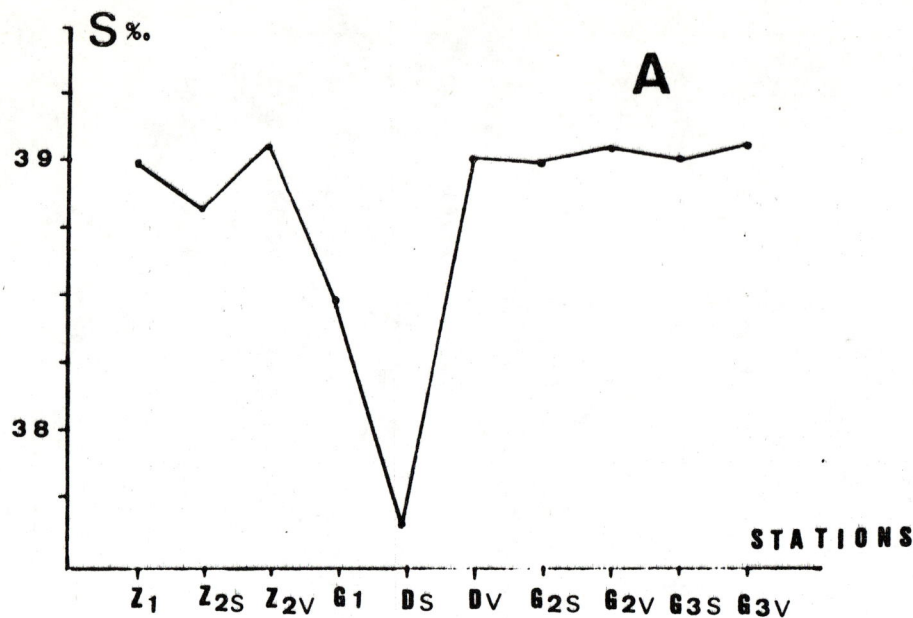

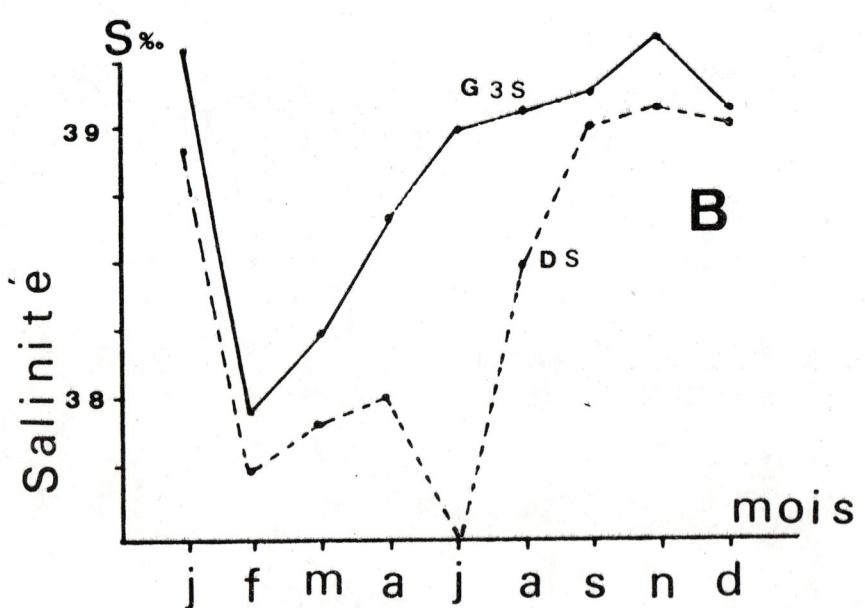

Figure 3. Distributions des valeurs moyennes annuelles de la salinité aux dix stations visitées tous les mois entre juillet 1984 et avril 1985 (courbe A). En B, évolution mensuelle de la salinité à la station côtière DS (zone polluée) et la station au large G3S (zone non polluée), en surface.

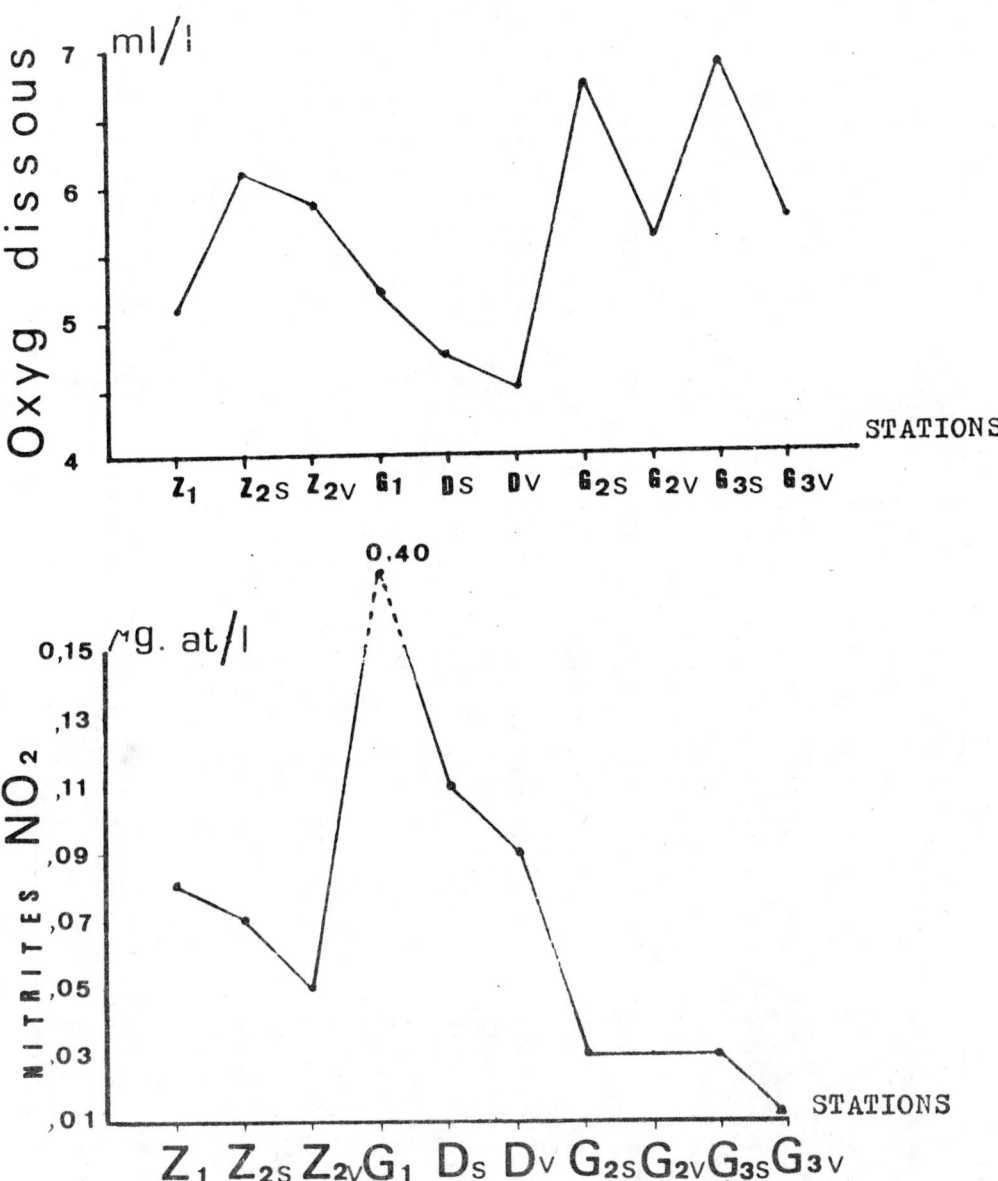

Figure 4. Répartitions des moyennes annuelles de l'oxygène dissous et des nitrites aux dix stations étudiées entre juillet 1984 et avril 1985 (valeurs moyennes)

Tableau V

Valeurs de l'Oxygène dissous (ml/litre) aux différentes stations visitées entre juillet 1984 et avril 1985

STATIONS	9 janv. 1985	13 fév. 1985	27 mars 1985	24 avr. 1985	19 juil 1984	25 août 1984	19 sep. 1984	9 nov. 1984	11 déc. 1984	Moyenne ann.	Ecart-type	% C.V.
Z1	5,5	5,7	5,7	6,1	5,2	4,3	4,1	5,2	5,1	5,21	0,65	12%
Z2S	6,3	6,1	6,4	6,7	6,0	5,6	5,5	5,9	6,3	6,08	0,38	6%
Z2V (10 m)	6,3	6,2	6,1	6,8	5,9	5,3	5,1	6,0	6,1	5,97	0,51	8%
G1	5,1	-	5,7	5,8	4,7	4,8	5,2	5,4	5,7	5,26	0,41	8%
DS	4,0	-	5,2	5,7	4,3	3,9	4,6	5,1	5,0	4,71	0,59	12%
DV (5m)	3,8	-	4,8	5,3	4,2	4,0	4,5	5,2	5,1	4,55	0,56	12%
G2S	6,7	-	7,1	7,2	6,6	-	5,9	-	6,9	6,73	0,43	6%
G2V (50 m)	5,4	-	6,1	5,9	5,0	-	4,7	-	5,1	5,30	0,53	10%
G3S	6,8	-	7,0	7,3	6,8	-	-	-	7,0	6,86	0,33	5%
G3V (50 m)	5,3	-	5,9	5,8	5,3	-	-	-	5,5	5,56	0,27	5%
Moyenne gén.	5,5	6,0	6,0	6,3	5,4	4,6	4,9	5,47	5,78	5,62		
Ecart-type	1,04	0,26	0,70	0,69	0,90	0,70	0,58	0,39	0,76	0,78		
% Coef.Var.	19%	4%	12%	11%	17%	15%	12%	7%	13%	14%		

Tableau VI

Transparence de l'eau (disque de Secchi en m.) aux différentes stations visitées
entre juillet 1984 et avril 1985

STATIONS	Prof.	9 janv. 1985	13 fév. 1985	27 mars 1985	24 avr. 1985	19 juil 1984	25 août 1984	19 sep. 1984	9 nov. 1984	11 déc. 1984	Moyenne ann.	Ecart-type	% C.V.
Z1	4 m	fond	fond	fond	fond	fond	fond	fond	fond	fond	fond		
Z2S	15 m	6	5	9	10	13	fond	fond	12	9	10,4	3,6	34%
Z2V	15 m	=	=	=	=	=	=	=	=	=	=	=	=
G1	4 m	2	-	1	3	fond	-	-	fond	2	3	1,2	40%
DS	7 m	2	-	3	3	3	3	2	2	1	2,3	0,7	32%
DV	7 m	=	=	=	=	=	=	=	=	=	=	=	=
G2S	100 m	9	-	11	14	22	-	23	-	15	15,6	5,7	37%
G2V	100 m	=	-	=	=	=	-	=	-	=	=	=	=
G3S	200 m	10	-	13	15	24	-	25	-	18	17,5	6	34%
G3V	200 m	=	-	=	=	=	-	=	-	=	=	=	=

(e) <u>Sels nutritifs:</u> On observe un net enrichissement en phosphates, nitrates et nitrites dans les stations à forte pollution organique (D et Gl) par rapport aux autres stations des secteurs non pollués (Tableaux VII, VIII, IX et Fig. 5). Les concentrations en sels nutritifs sont très faibles, parfois nulles dans les stations éloignées de la côte. La station Z2 fait exception, car cette station se trouve sur le trajet du courant enrichi par l'eau douce de Nahr El Kalb, plus riche en sels nutritifs. Les concentrations en phosphates varient 0,20 et 0,34 µg-at.l^{-1} en moyenne; à la station DV, la moyenne annuelle étant de 4,76 µg-at.l^{-1} et en septembre, nous avons enregistré la valeur la plus élevée soit 18,71 µg-at.l^{-1}.

Le Phytoplancton et les Protistes

Dans une étude précédente sur le phytoplancton des eaux côtières libanaises (Lakkis et Novel-Lakkis, 1981), nous avons dressé un inventaire de 263 taxa dont 107 espèces et variétés de diatomées et 157 dinoflagellés. Dans cette étude qui ne couvre qu'une région limitée et restreinte de la côte, nous avons identifié 43 espèces de diatomées et 41 dinoflagellés qui sont les plus communes; par ailleurs, 32 espèces de tintinnides sont recensées (Tableau X).

Parmi les diatomées, une quinzaine d'espèces sont plutôt caractéristiques des zones polluées, 31 sont des formes néritiques des eaux non polluées et 11 sont typiquement des eaux de large ou de tendance incertaine. Chez les dinoflagellés, 15 sont rapportées dans les eaux polluées aussi bien que néritiques. Par contre, plusieurs espèces rencontrées dans les eaux néritiques propres uniquement ou océaniques sans être enregistrées dans les eaux polluées. Parmi les espèces indifférentes (trouvées dans les eaux polluées et non polluées avec la même densité) nous citons: <u>Chaetoceros compressus</u>, <u>Hemiaulus hauckii</u>, <u>Leptocylindrus danicus</u>, <u>Licmophora abbreviata</u>, <u>Rhizosolenia alata forma indica</u>, <u>R.hebetata</u>, <u>Skeletonema costatum</u>, <u>Thalassionema nitzschioides</u>, <u>Ceratium arietinum</u>, <u>C.furca</u>, <u>C.macroceros</u>, <u>C.tripos</u>, <u>Peridinium brochi</u>. Parmi les espèces trouvées en abondance dans les eaux polluées, signalons: <u>Chaetoceros decipiens</u>, <u>Ch.teres</u>, <u>Streptotheca thamesis</u>, <u>Striatella unipunctata</u>, <u>Licmophora abbreviata</u>. Parmi les espèces de diatomées communes dans les eaux néritiques et au large et évitant les zones polluées, signalons: <u>Biddulphia elegans</u>, <u>B.vesiculosa</u>, <u>Asterionella janonica</u>, <u>Bacteriastrum sp.</u>, <u>Cerataulina pelagica</u>, <u>Chaetoceros affinis</u>, <u>Ch.brevis</u>, <u>Ch.anastomosans</u>, <u>Hemiaulus sinensis</u>, <u>Melosira juergensii</u>, <u>Thalassiothrix mediterranea</u>. Chez les dinoflagellés nous citons: <u>Ceratium carriense</u>, <u>C.egyptiacum</u>, <u>C.extensum</u>, <u>C.karsteni</u>, <u>C.pulchellum</u>, <u>Peridinium oceanicum</u>.

En ce qui concerne les tintinnides, parmi les 32 espèces rencontrées dans toute la région étudiée, quelques unes sont abondantes dans les eaux polluées telles que: <u>Codonella aspera</u>, <u>Coxliella sp.</u>, <u>Epiplocylis blanda</u>, <u>Eutintinnus fraknoi</u>, <u>E.lusus-undae</u>, <u>Favella sp.</u>, <u>Tintinnopsis campanula</u>, <u>Tintinnus inquilinus</u>, etc. (voir Tableau X).

Les effectifs des diatomées sont souvent plus élevés que ceux des dinoflagellés et sont maximaux dans les eaux côtières et polluées. Si l'on regarde les résultats présentés aux Tableaux II et XI, ainsi que dans les courbes Fig. 6 et 7, on voit que la densité maximale moyenne du phytoplancton est enregistrée en Z2S, suivie de Z1 et de Gl; alors qu'en DV et G3V (au large, traits verticaux) la densité était la plus faible. Les variations saisonnières d'abondance, montrent que la maximum de la production phytoplanctonique se situe en avril (300,000 cel./l à la station Z2S) pour toutes les stations; en août la densité devient très faible et tombe à une moyenne de 10,000 cel./l. La diversité spécifique calculée au moyen de la formule de Gleason-Margalef

$$I = \frac{S-1}{\log N}$$

permet de connaître l'évolution spatio-temporelle des populations phytoplanctoniques. En général, la diversité est faible dans les zones polluées et côtières et devient plus élevée au large et dans les prélèvements verticaux. Ainsi, l'indice le plus faible correspond avec la densité la plus forte (en avril). La diversité la plus forte est enregistrée en février lorsque le phytoplancton est encore faible. Du point de vue locale, l'indice de diversité générale le plus fort est enregistré en Z2V et G2S et le plus faible en DV, DS, Gl (zones polluées).

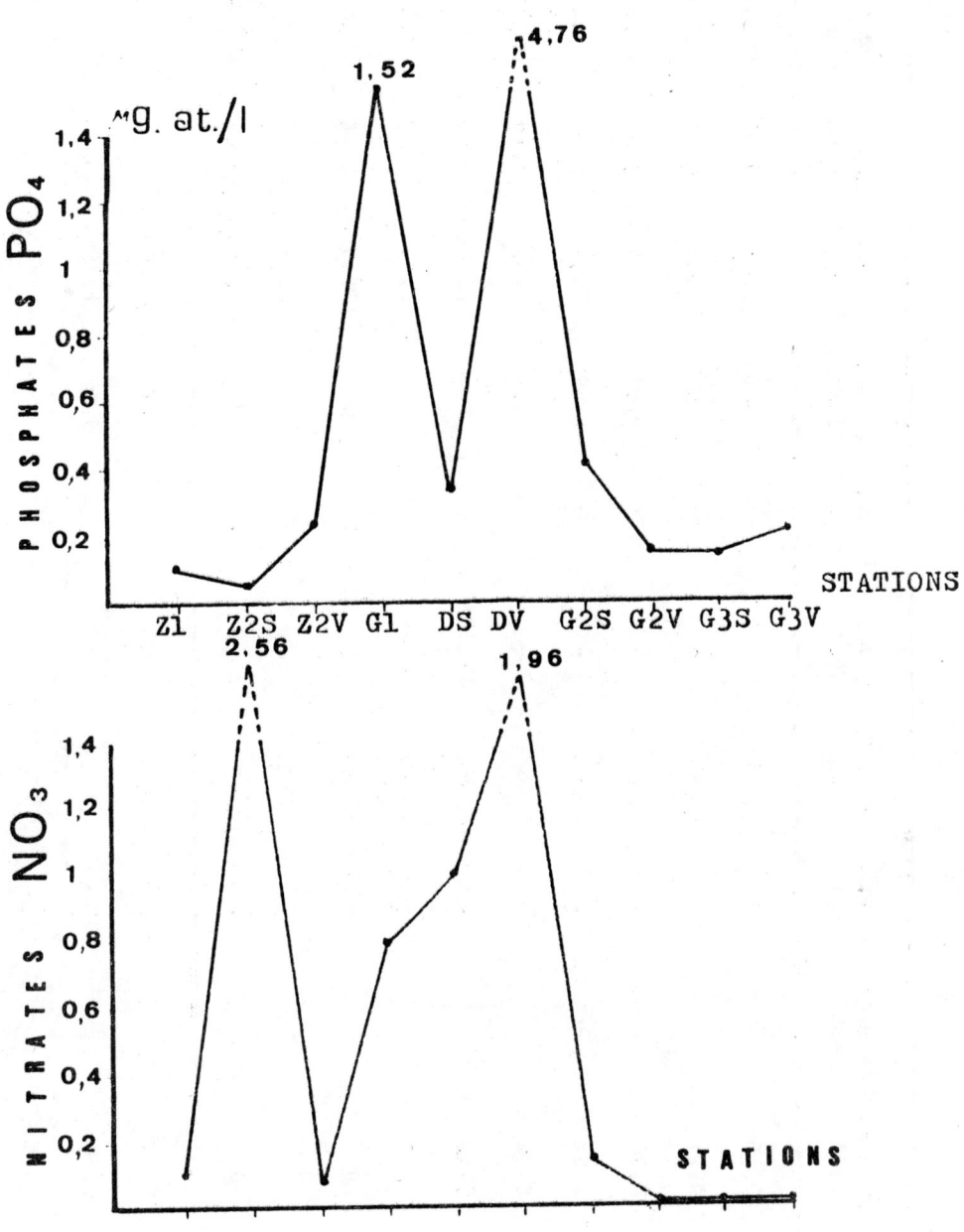

Figure 5. Distributions des sels nutritifs (Phosphates et Nitrates) aux dix stations prospectées entre juillet 1984 et avril 1985

Tableau VII

Valeurs des phosphates (PO_4 µg.at./l) effectuées aux différentes stations

STATIONS	9 janv. 1985	13 fév. 1985	27 mars 1985	24 avr. 1985	19 juil 1984	25 août 1984	19 sep. 1984	9 nov. 1984	11 déc. 1984	Moyenne ann.	Ecart-type	% C.V.
Z1	0,19	-	0,15	-	-	-	-	0,01	0,05	0,01	0,08	84%
Z2S	0,09	-	0,21	-	-	-	0	0	0	0,06	0,09	154%
Z2V (10 m)	0,14	-	0,07	-	-	-	-	0,04	0,01	0,26	0,05	21%
G1	0,45	-	1,04	-	-	-	-	-	3,09	1,52	1,38	91%
DS	0,42	-	0,32	-	-	-	-	-	0,21	0,32	0,10	33%
DV (5m)	0,23	-	0,12	-	-	-	18,71	-	0	4,76	9,29	195%
G2S	0,18	-	0,07	-	-	-	0,14	-	0	0,39	0,07	20%
G2V (50 m)	0,16	-	0	-	-	-	0	-	0	0,16	0,08	50%
G3S	0,14	-	0	-	-	-	-	-	0,02	0,16	0,07	47%
G3V (50 m)	0,14	-	0	-	-	-	-	-	0	0,14	0,08	58%
Moyenne gén.	0,21	-	0,20					-	0,34	0,78		
Ecart-type	0,12	-	0,30						0,96	1,45		
% Coef.Var.	58%	-	151%						285%	1,87		

Tableau VIII

Valeurs des nitrates (NO_3 µg-at./l) effectuées aux différentes stations

STATIONS	9 janv. 1985	13 fév. 1985	27 mars 1985	24 avr. 1985	19 juil 1984	25 août 1984	19 sep. 1984	9 nov. 1984	11 déc. 1984	Moyenne ann.	Ecart-type	% C.V.
Z1	0,04	-	0,13	-	-	-	-	0	0	0,09	0,06	68%
Z2S	0	-	0	-	-	-	12,83	0	0	2,56	5,74	224%
Z2V (10 m)	0	-	0	-	-	-	-	0	0	0		
G1	0,91	-	0	-	-	-	-	-	1,50	0,80	0,75	94%
DS	2,91	-	0	-	-	-	-	-	0	0,97	1,68	173%
DV (5 m)	0,14	-	0	-	-	-	7,70	-	0	1,96	3,82	195%
G2S	0	-	0	-	-	-	0,47	-	0	0,12	0,23	196%
G2V (50 m)	0	-	0	-	-	-	0	-	0	0		
G3S	0	-	0	-	-	-	-	-	0	0		
G3V (50 m)	0	-	0	-	-	-	-	-	0	0		

Tableau IX

Valeurs des nitrites (NO_2^- µg.at./l) effectuées aux différentes stations

STATIONS	9 janv. 1985	13 fév. 1985	27 mars 1985	24 avr. 1985	19 juil 1984	25 août 1984	19 sep. 1984	9 nov. 1984	11 déc. 1984	Moyenne ann.	Ecart-type	% C.V.
Z1	0,02	-	0,12	-	-	-	-	0,12	0,06	0,08	0,05	61%
Z2S	0,06	-	0,03	-	-	-	0,07	0,10	0,12	0,07	0,03	50%
Z2V (10 m)	0,04	-	0	-	-	-	-	0,06	0,11	0,05	0,04	91%
G1	0,19	-	0,27	-	-	-	-	-	0,75	0,40	0,30	75%
DS	0,15	-	0,04	-	-	-	-	-	0,14	0,11	0,06	55%
DV (5 m)	0,06	-	0,02	-	-	-	0,15	-	0,17	0,10	0,07	71%
G2S	0,05	-	0	-	-	-	0,04	-	0,04	0,03	0,02	73%
G2V (50 m)	0,01	-	0	-	-	-	0,07	-	0,04	0,03	0,03	100%
G3S	0,02	-	0	-	-	-	-	-	0,06	0,03	0,03	100%
G3V (50 m)	0,01	-	0	-	-	-	-	-	0,01	0,006	0,005	96%
Moyenne gén.	0,06	-	0,05	-	-	-	0,08	0,09	0,15	0,09		
Ecart-type	0,06	-	0,08	-	-	-	0,05	0,03	0,22	0,11		
% Coef.Var.	100%	-	172%	-	-	-	59%	34%	144%	126%		

Tableau X

Inventaire des espèces phytoplanctoniques et des tintinnides:
P(zones polluées), N(néritique), O(océanique), I(indéterminée),
A(abondante), C(commune), R(rare)

ESPECES	P	N	O	I
DIATOMEES				
Achnanthes longipes	R	C	-	-
Actinocyclus octonarius	R	R	-	-
Asterionella japonica	-	A	-	-
Asterolampra marylandica	-	R	-	-
Bacteriastrum delicatulum	-	-	C	-
B.elegans	-	-	R	-
Biddulphia vesiculosa	-	-	R	-
Ceratauline pelagica	-	-	A	-
Chaetoceros affinis	-	A	-	-
Ch.anastomosans	-	C	-	-
Ch.brevis	-	A	-	-
Ch.compressus	A	A	-	-
Ch.curvisetus	X	A	-	-
Ch.decipiens	A	X	-	-
Ch.didymus	X	A	-	-
Ch.peruvianus	C	C	-	-
Ch.pseudocurvisetus	X	A	-	-
Ch.teres	C	X	-	-
Chaetoceros sp.	X	A	C	-
Coscinodiscus sp.	X	C	-	-
Dactyliosolen mediterraneus	-	X	C	-
Girosigma balticum	-	R	-	-
Hemiaulus hauckii	R	C	C	-
H.sinensis	-	X	C	-
Leptocylindrus danicus	C	A	R	X
L.minimus	X	A	X	X
Licmophora abbreviata	R	R	-	-

suite 2

ESPECES PHYTOPLANCTONIQUES	P	N	O	I
Melosira juergensii	-	R	-	-
Nitzchia seriata	R	C	A	X
Rhabdonema adriaticum	-	R	-	-
Rhizosolenia alata forma alata	-	-	C	-
R.alata forma indica	R	R	C	-
R.calcar-avis	-	C	C	-
R.castracanei	-	R	-	-
R.hebetata var.semispina	R	R	R	-
R.stolterfothii	X	R	R	-
R.styliformis	X	R	C	-
Skeletonema costatum	C	A	R	-
Streptotheca thamesis	C	R	X	-
Striatella unipunctata	C	X	-	-
Thalassionema nitzschioides	A	A	R	-
Thalassiothrix frauenfeldii	X	X	R	-
Th.mediterranea	-	-	C	-
DINOFLAGELLES				
Amphisolenia clavipes	-	-	X	-
Ceratium arietinum	C	C	X	-
C.candelabrum	-	X	R	-
C.carriense forma volans	-	X	C	-
C.declinatum forma declinatum	-	X	R	-
C.egyptiacum	-	R	R	-
C.extensum forma extensum	-	X	R	-
C.furca	C	C	A	-
C.fusus var.seta	X	R	C	-
C.karsteni	-	R	R	-
C.macroceros var.gallicum	C	C	C	-
C.massiliense var.massiliense	X	C	R	-

suite 3

ESPECES PHYTOPLANCTONIQUES	P	N	O	I
Ceratium pentagonum forma tenerum	R	R	C	-
C.pulchellum forma semipulchellum	-	C	X	-
C.teres	-	X	R	-
C.tripos forma ponticum	C	C	C	-
Ceratocorys armata	X	R	R	-
C.gourreti	X	R	R	-
Dinophysis caudata	R	A	R	-
D.tripos	R	C	X	-
Diplosalis sp.	X	C	X	-
Gonyaulax sp.	R	C	R	-
Ornithocercus sp.	X	R	R	-
Peridinium brochi	C	C	R	-
P.claudicans	R	C	C	-
P.depressum	R	C	A	-
P.diabolus	C	A	C	-
P.divergens	R	R	C	-
P.globulus var.quarnerense	X	C	R	-
P.grani	X	R	C	-
P.mediterraneum	X	C	C	-
P.oceanicum	-	R	A	-
P.piriforme	R	C	C	-
Phalacroma sp.	X	R	X	-
Podolampas bipes	X	R	X	-
Prorocentrum micans	R	C	R	-
HAPTOPHYCEAE, CRYSOPHYCEAE, EBRIIDEAE				
Calcisolenia	-	X	C	-
Dictyocha fibula	X	C	C	-
Ebria tripartita	-	R	R	-
Hermesinum adriaticum	-	X	X	-

suite 4

TINTINNIDES	P	N	O	I
Amphorella quadrilineata	R	C	C	-
Codonella aspera	R	R	X	-
Codonellopsis morchella	X	C	C	-
Coxliella sp.	C	C	R	-
Cyttarocylis sp.	R	R	C	-
Dadayiella ganymedes	X	X	R	-
Dictyocysta elegans	-	X	R	-
Dictyocysta sp.	-	R	X	-
Epiplocylis undella	X	C	R	-
E.blanda	R	R	X	-
Eutintinnus fraknoi	C	A	R	-
E.lusus undae	R	R	R	-
Favella ehrenbergi	C	C	R	-
F.markuzowskyi	A	A	R	X
F.serrata	A	A	R	X
Helicostomella subulata	X	R	R	-
Petalotricha ampulla	R	C	R	-
P.major	R	A	C	-
Proplectella claparedei	X	R	C	-
Rhabdonella conica	R	C	C	-
R.spiralis	R	C	R	-
Salpingella attenuata	X	R	R	-
Tintinnidium neopolitanum	X	R	X	-
Tintinnopsis beroidea	R	C	X	-
T.campanula	A	C	R	-
T.cylindrica	C	C	R	-
Tintinnus inquilinus	R	R	X	-
Undellopsis subangulata	R	C	X	-
Xystonella longicauda	X	R	X	-
X.treforti	R	R	-	-
Xystonellopsis cymatica	X	R	X	-
X.paradoxa	X	X	X	-

Tableau XI

Distribution numérique des protistes planctoniques (phytoplancton, protozoaires) rapportée au nombre de cellules/litre aux dix stations visitées entre 1984 et 85. Le nombre d'espèces (N.sp.) par échantillon est donné (prélèvements au filet 50 microns)

STATIONS	N/l=cellu- les/litre N.sp.=esp.	9 janv. 1985	13 fév. 1985	27 mars 1985	24 avr. 1985	19 juil 1984	25 août 1984	19 sep. 1984	9 nov. 1984	11 déc. 1984	Moyenne ann.	Ecart- type	% C.V.
Z1	N/l.	30000	58000	110000	200000	70000	15000	20000	120000	45000	74222	59930	80%
	N sp.	68	74	62	51	33	36	51	62	65	56	14	25%
Z2S	N/l.	45000	52000	130000	300000	80000	30000	27000	140000	68000	96889	86165	89%
	N sp.	71	75	65	50	40	41	48	66	67	58	13	23%
Z2V	N/l.	20000	25000	75000	200000	45000	12000	17000	90000	30000	57111	59951	105%
	N sp.	75	72	58	52	58	50	55	68	70	62	9	15%
G1	N/l.	100000	—	55000	250000	65000	15000	12000	67000	25000	62375	79469	127%
	N sp.	34	—	24	32	38	42	38	40	46	37	7	18%
DS	N/l.	20000	—	40000	130000	35000	5000	10000	50000	25000	39375	39590	100%
	N sp.	28	—	25	19	32	29	31	37	40	30	7	22%
DV(5-0)	N/l.	10000	—	24000	75000	18000	3000	8000	35000	15000	23500	23083	100%
	N sp.	25	—	31	24	27	18	26	32	33	27	5	18%
G2S	N/l.	23000	—	55000	150000	50000	—	28000	—	30000	56000	47795	85%
	N sp.	75	—	70	55	50	—	66	—	70	64	10	15%
G2V	N/l.	20000	—	50000	100000	45000	—	40000	—	35000	48333	27325	57%
	N sp.	65	—	56	43	54	—	53	—	67	56	9	16%
G3S	N/l.	15000	—	45000	95000	35000	—	28000	—	20000	39667	29132	73%
	N sp.	60	—	48	40	45	—	38	—	56	48	9	18%
G3V	N/l.	10000	—	40000	75000	38000	—	31000	—	23000	36166	21958	61%
	N sp.	63	—	45	38	48	—	42	—	55	48	9	18%
Moy.gén.	N/l.	20300	45000	62400	157500	48100	13333	22100	83666	31600	53364		
	N sp.	56	74	48	40	42	36	45	51	57	50		
Ec.type	N/l.	10843	17578	33400	77540	18729	9605	10386	40775	15247	21155		
	N sp.	20	2	17	12	10	11	12	16	13	13		
% C.V.	N/l.	53%	39%	53%	49%	39%	72%	47%	49%	48%	40%		
	N sp.	35%	2%	35%	31%	24%	31%	27%	32%	23%	26%		

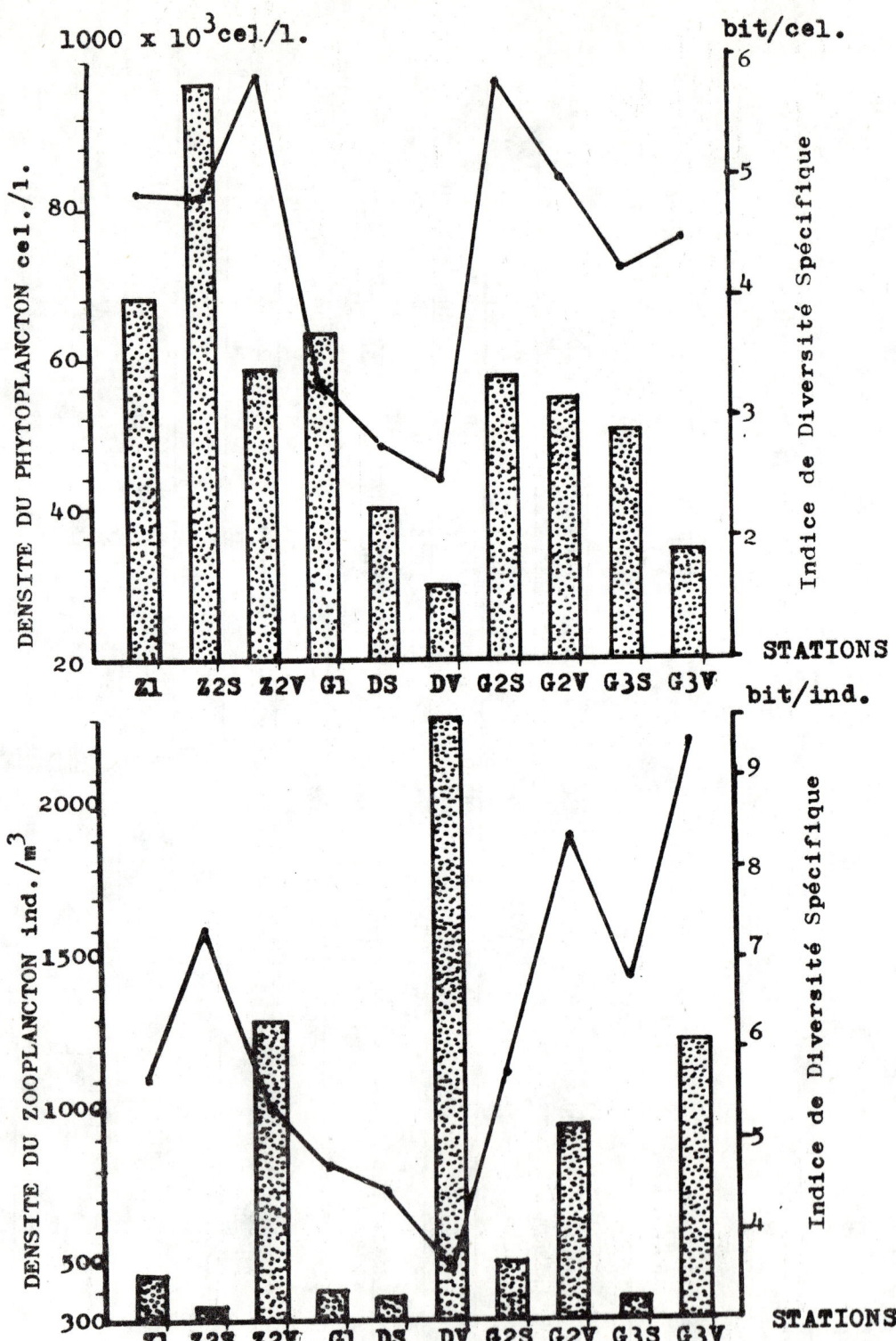

Figure 6. Variations de la densité (colonnes pointillées) et de l'indice de diversité chez le phytoplancton et le zooplancton dans les dix stations visitées entre juillet 1984 et avril 1985

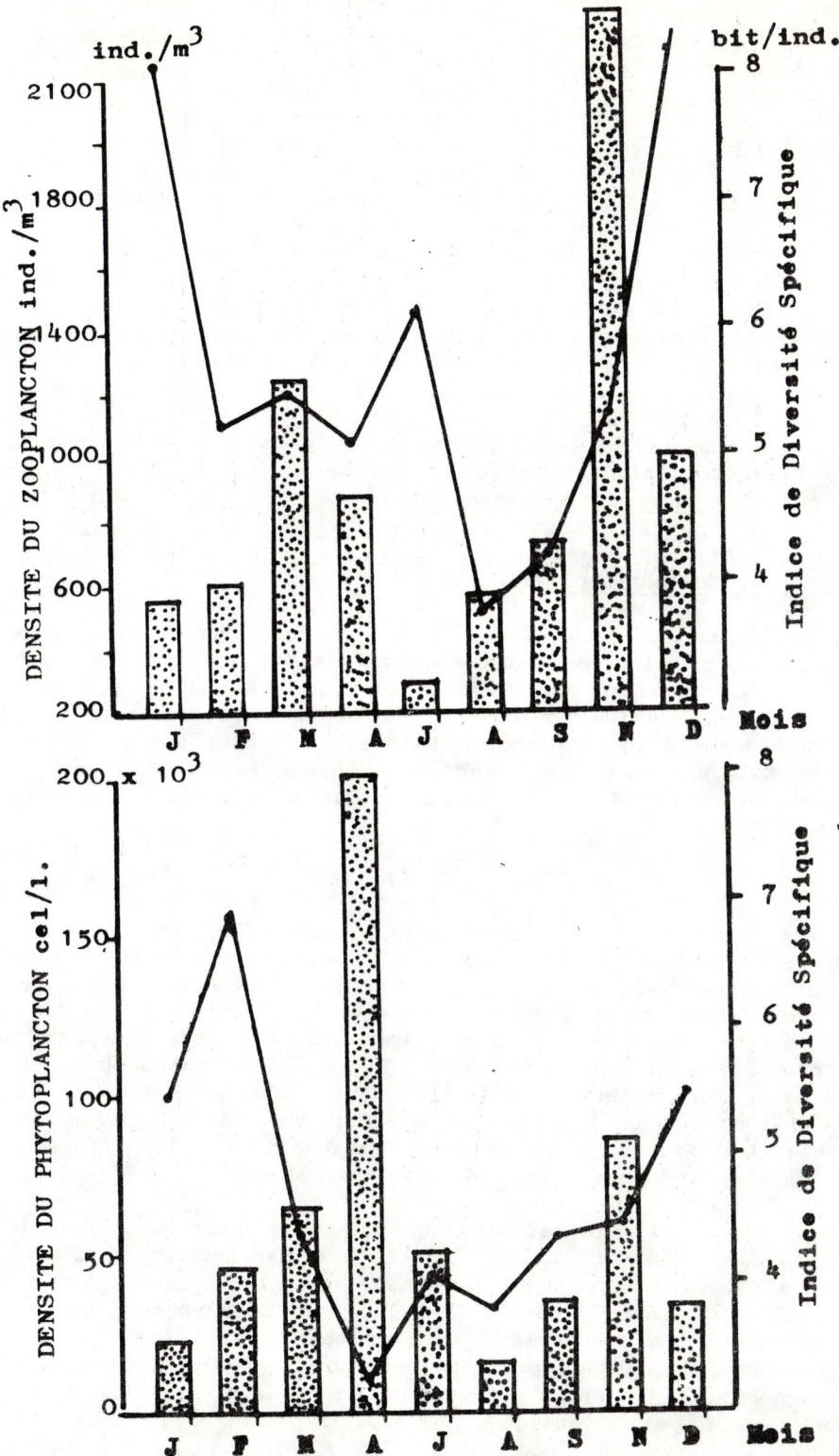

Figure 7. Evolution mensuelle de la densité du phytoplancton et du zooplancton (collones pointillées) ainsi que l'indice de diversité spécifique dans la zone étudiée (valeurs combinées pour les 10 stations).

Le Zooplancton

D'une façon générale, le zooplancton atteint son maximum d'abondance au printemps, quelques semaines après la poussée du phytoplancton (Lakkis, 1983). Mais dans le cas de nos observations présentes, le maximum est enregistré entre novembre et décembre à la station polluée DV (Tableau XII and XIII). Cette poussée automnale est due principalement au développement exceptionnel de Paracalanus parvus, P.crassus et Labidocera pavo; ce dernier semble être un bon indicateur de pollution. Dans ces zones polluées, la densité chez le phytoplancton et celle du zooplancton ne sont pas synchrones, comme c'est le cas dans d'autres stations comme Z2V, G2V, G3V. Du point de vue distribution géographique, la station DV est la plus riche en effectif, suivie de Z2V, G3V et G2V. En DS (même point que DV), ainsi qu'en Z2S (propre), la densité du zooplancton était la plus faible. Parmi les groupes qui abondent dans les eaux polluées, à part les copépodes, signalons les larves de polychètes, les appendiculaires, les larves cypris. Par contre, plusieurs groupes zooplanctoniques sont, sinon absents, du moins très rares dans les zones polluées (Tableau XIV). Parmi ceux-ci, signalons: les hydroméduses, les siphonophores, les ostracodes, les chaetognathes, les hétéropodes, les ptéropodes, les ostracodes, les cladoderes, les amphipodes et les appendiculaires. Parmi les espèces caractéristiques des eaux polluées signalons: Acartia clausi, A.discaudata, Var.mediterranea, A.josephinae, A.negligens, Calocalanus pavo, C.styliremis, Labidocera pavo, L.orsinii, Paracalanus crassus, Corycaeus flaccus, Euterpina acutifrons, Oithona nana, O.plumifera, larves de polychètes (Polydora ciliata), Leucifer typus, Echinoplutei (Fig. 8).

L'étude de la diversité spécifique, a été suivie pour chaque prélèvement afin de caractériser l'évolution et les successions écologiques des populations. Les résultats sont présentés à la Fig. 7 et aux Tableaux II et XII. En hiver (décembre-janvier), durant la période thermique froide et d'homothermie, la diversité est la plus forte, correspondant avec une densité de zooplancton faible. Du point de vue géographique, la station G2S, suivie Z2V, montrent la diversité moyenne la plus élevée (5,8 et 5,6 bit/ind.); par contre en DV et DS (les stations les plus polluées), la diversité était la plus faible (2,6 et 2,7).

4. DISCUSSION ET CONCLUSION

L'aire prospectée comprend donc 3 secteurs pollués (Z1, G1, D) et 3 secteurs éloignés des zones polluées dont un côtier (Z2S, Z2V) et les deux autres au large (G2 et G3). Les caractéristiques physico-chimiques et hydrologiques sont différentes entre ces secteurs. A la station Z1, mise à part une légère augmentation (1 ou 2°C) de la température de l'eau de mer au voisinage immédiat du rejet de refroidissement ($20 m^3/sec$), les caractéristiques hydrologiques (S°/oo, sels nutritifs, turbidité) ne sont pas exceptionnels. Par ailleurs, l'anomalie dans cette augmentation de température disparaît à quelques 100 m de l'émissaire du rejet d'eau. L'écosystème ne semble pas être modifié ou perturbé, surtout que les mouvements d'eau et la houle dispersent ce réchauffement local faible et renouvellent l'eau en permanence.

Dans ce secteur, les variations saisonnières de l'abondance du phytoplancton et du zooplancton sont presque normales, malgré une légère diminution en concentration de phosphates et nitrates (0,10 et 0,09 µg-at.l^{-1}, moyenne annuelle). La poussée phytoplanctonique en avril avec 200,000 cel/l., ne correspond pas avec le pic zooplanctonique saisonier qui coïncide en novembre (2100 ind./m^3), malgré un développement assez important du zooplancton en avril (500 ind./m^3). Ce phénomène n'est pas toutefois normal par rapport à ce que l'on observe dans des secteurs côtiers normaux. En effet, en Z2 et à quelques 1000 m de Z1, le maximum saisonnier de l'abondance du phytoplancton et du zooplancton est concordant (voir Tableau XII). Ce décalage dans le timing de la poussée phyto-zooplancton est dû probablement à l'influence de cette anomalité thermique présente en Z1. En août, lorsque la température au voisinage de l'émissaire monte à 31°, le phytoplancton est effecté par cette hausse anormale de la température et voit son développement chuter au minimum (effectifs 10,000 cel./l). Il en est de même pour la densité du zooplancton qui tombe à 20-30 ind./m^3 début septembre. Du point de vue structure des populations, on n'observe vraiment pas un changement significatif par rapport à la station Z2S, si ce n'est l'action hydrodynamique et les mouvements des masses d'eau qui affectent en permanence le déplacement des essaims planctoniques. Au niveau du phytoplancton,

- 149 -

Tableau XII

Répartition du nombre d'organismes zooplanctoniques (N/m3), protozoaires non compris, ainsi que du nombre d'espèces par prélèvement aux différentes stations étudiées

STATIONS		9 janv. 1985	13 fév. 1985	27 mars 1985	24 avr. 1985	19 juil 1984	25 août 1984	19 sep. 1984	9 nov. 1984	11 déc. 1984	Moyenne ann.	Ecart-type	% C.V.
Z1	N/m³	197	537	226	437	110	150	16	2093	335	455	635	140%
	N sp.	31	31	49	35	27	24	14	51	59	36	14	40%
Z2S	N/m³	234	19	605	480	66	653	109	352	296	312	230	74%
	N sp.	64	25	56	40	29	31	20	52	60	42	16	39%
Z2V	N/m³	425	363	2065	2897	110	1796	627	1216	1244	1194	922	77%
	N sp.	47	43	39	38	25	29	28	55	62	41	12	30%
(10-0 m)	N/m³	111	-	176	490	48	50	66	1146	1126	401	475	118%
G1	N sp.	45	-	29	27	15	17	18	42	48	30	13	44%
DS	N/m³	502	-	38	59	146	38	1097	997	224	387	435	112%
	N sp.	39	-	15	15	22	25	32	32	40	27	10	36%
DV	N/m³	1806	-	2373	620	150	120	200	12107	4935	2788	4106	147%
	N sp.	38	-	18	17	31	27	30	40	43	30	10	32%
(5-0 m)	N/m³	54	-	588	179	27	-	1393	-	596	473	516	109%
G2S	N sp.	47	-	32	30	19	-	32	-	57	36	13	37%
G2V	N/m³	290	-	2128	548	600	-	1311	-	625	917	683	74%
	N sp.	68	-	60	57	47	-	53	-	64	58	7	13%
(50-0 m)	N/m³	69	-	414	958	221	-	-	-	129	358	360	100%
G3S	N sp.	50	-	37	33	35	-	-	-	54	42	9	23%
G3V	N/m³	669	-	3357	771	710	-	-	-	322	1165	1237	106%
	N sp.	69	-	58	62	66	-	-	-	79	67	8	12%
(50-0 m)													
Moy.gén.	N/m³	436	472	1197	744	226	468	602	2985	983	845		
	N sp.	50	33	39	35	34	25	28	45	57	41		
Ec.-type	N/m³	521	156	1170	799	240	689	587	4503	1438	762		
	N sp.	13	9.2	16.2	15	14	5	12	9	11.2	13		
% C.V.	N/m³	119%	33%	98%	107%	111%	147%	97%	151%	146%	90%		
	N sp.	26%	28%	41%	43%	42%	19%	43%	19%	20%	31%		

Tableau XIII

Répartition de la biomasse du zooplancton (exprimée en cm^3/m^3) de plancton sédimenté, pêché au filet WP2 de 200 microns aux différentes stations sur la côte libanaise entre 1984 et 1985

STATIONS	9 janv. 1985	13 fév. 1985	27 mars 1985	24 avr. 1985	19 juil 1984	25 août 1984	19 sep. 1984	9 nov. 1984	11 déc. 1984	Moyenne ann.	Ecart-type	% C.V.
Z1	0,27	0,63	0,90	0,81	0,10	0,30	0,02	0,90	0,36	0,48	0,34	71%
Z2S	0,54	0,10	0,54	1,81	0,05	0,36	0,50	0,27	17,20	2,37	5,58	235%
Z2V (10-0)	1,66	0,80	2,50	3,00	0,05	0,36	0,83	3,33	4,00	1,84	1,42	77%
G1	0,54	-	1,81	0,54	0,27	0,18	0,01	0,90	1,00	0,66	0,58	87%
DS	2,72	-	déch.	0,10	0,36	0,06	1,09	0,63	0,18	0,73	0,95	130%
DV (5-0)	6,00	-	1,81	3,33	0,80	0,30	0,80	2,33	3,33	2,33	1,87	80%
G2S	0,27	-	0,81	1,10	0,03	-	0,23	-	0,54	0,50	0,40	80%
G2V (50-0)	0,60	-	2,00	3,33	0,61	-	0,33	-	1,33	1,37	1,14	83%
G3S	0,18	-	0,45	0,09	0,10	-	-	-	0,18	0,20	0,15	73%
G3V (50-0)	0,66	-	3,00	3,33	0,36	-	-	-	0,33	1,54	1,50	97%
Moyenne gén.	1,28	0,51	1,53	1,74	0,27	0,26	0,48	1,39	2,90	1,20		
Ecart-type	1,91	0,36	0,90	1,39	0,26	0,12	0,40	1,18	5,19	0,80		
% Coef.var.	149%	71%	59%	79%	97%	45%	85%	85%	179%	66%		

Tableau XIV

Inventaire des groupes et espèces zooplanctoniques: P(zones polluées), N(zones néritiques non polluées), O(zones au large non polluées), I(indéterminées), A(espèce abondante). C(commune), R(rare), X(présente)

ESPECES ZOOPLANCTONIQUES	P	N	O	I
Foraminifères	R	R	C	-
Acanthaires	R	C	C	-
Radiolaires	X	R	C	-
COPEPODES Calanoides Total	C	A	A	-
Acartia clausi	A	A	R	-
A.discaudata var.mediterranea	A	C	R	-
A.grani	R	R	C	-
A.italica	X	R	-	-
A.josephinae	C	X	-	-
A.negligens	C	R	R	-
Calanus minor	R	C	C	-
C.tenuicornis	X	R	R	-
Calanopia elliptica	-	X	R	-
C.media	-	-	X	-
Calocalanus contractus	C	C	R	-
C.pavo	C	R	R	-
C.styliremis	A	C	R	-
C.elegans	R	R	X	-
Candacia bispinosa	R	C	R	-
C.aethiopica	X	R	R	-
Centropages kröyeri	R	C	C	-
C.violaceus	R	C	R	-
Clausocalanus furcatus	A	C	R	-
Cl.lividus	C	C	R	-
Cl.mastigoforus	X	R	C	-
Cl. parapergens	-	X	R	-
Cl.paululus	X	C	A	-
Euaetideus giesbrechti	-	X	R	-
Euchaeta marina	-	-	R	-

suite 2

ESPECES ZOOPLANCTONIQUES	P	N	O	I
Isias clavipes	-	R	C	-
Labidocera pavo	A	C	C	-
L.madurae	C	X	R	-
L.orsinii	C	R	R	-
Lucicutia flavicornis	R	C	R	-
L.ovalis	-	-	R	-
Mecynocera clausi	-	X	R	-
Paracalanus crassus	A	C	R	-
P.nanus	X	R	C	-
P.parvus	A	A	C	-
P.pygmaeus	R	C	C	-
Pleuromamma abdominalis	-	-	R	-
P.gracilis	-	-	R	-
Scolecithricella dentata	-	-	R	-
Temora stylifera	R	A	C	-
Copépdites calanoides	C	A	A	-
COPEPODES Cyclopoides Total	C	A	C	-
Clytemnestra rostrata	X	R	C	-
Copilia mediterranea	X	X	X	-
Corycaeus brehmi	R	R	C	-
C.clausi	X	R	C	-
C.flaccus	C	R	R	-
C.giesbrechti	-	R	C	-
C.limbatus	X	R	R	-
C.ovalis	-	R	C	-
C.typicus	A	C	A	-
Corycella rostrata	R	C	A	-
Euterpina acutifrons	A	C	C	-
Oithona nana	A	C	R	-
O.plumifera	A	C	C	-
Oncaea media	C	C	R	-
O.mediterranea	C	C	A	-
O.minuta	X	X	R	-
O.neobscura	C	C	X	-

suite 3

ESPECES ZOOPLANCTONIQUES	P	N	O	I
Sapphirina gemma	R	X	X	-
S. metallina	X	R	C	-
Vettoria sp.	-	X	X	-
Copepodites cyclopoides	C	A	C	-
HYDROMEDUSES	C	A	A	-
Aglaura hemistoma	-	C	R	-
Liriope tetraphylla	X	C	A	-
Obelia sp.	C	A	R	-
Phialidium hemisphaericum	R	C	X	-
Rhopalonema velatum	R	R	X	-
Méduses jeunes	X	R	C	-
SCYPHOMEDUSES				
Cotylorhiza tuberculata	-	X	R	-
Rhizostoma pulmo	-	C	R	-
SIPHONOPHORES	R	C	C	-
Physophora	-	R	C	-
Bassia bassensis	X	R	C	-
Eudoxoides spiralis	C	C	R	-
Lensia sp.	R	R	C	-
Monophies irregularis	-	X	R	-
Sphaeronectes sp.	-	X	R	-
CTENOPHORES (larves et adultes)	R	C	C	-
POLYCHETES PELAGIQUES (adultes)	-	X	C	-
Larves de Polychètes	A	C	C	-
CLADOCERES				
Evadne spinifera	C	A	C	-
E. tergestina	C	C	R	-
OSTRACODES	X	R	R	-

suite 4

ESPECES ZOOPLANCTONIQUES	P	N	O	I
CIRRIPEDES (nauplii et Cypris)	C	A	C	-
AMPHIPODES	-	R	R	-
EUPHAUSIACES (larves et adultes)	A	A	C	-
DECAPODES (larves)	C	C	R	-
SERGESTIDES (Leucifer typus)	A	C	C	-
MOLLUSQUES (larves)	A	A	C	-
ECHINODERMES (larves)				
Ophiopluteus	-	X	C	-
Echinopluteus	C	C	R	-
PTEROPODES				
Creseis acicula	-	-	R	-
C.virgula	R	C	A	-
Limacina inflata	C	C	R	-
Styliola subula	X	R	C	-
HETEROPODES	-	X	R	-
CHAETOGNATHES				
Sagitta enflata	R	A	R	X
S.friderici	R	A	C	-
APPENDICULAIRES				
Fritilaria borealis et F.pellucida	R	C	C	-
Oikopleura cophocerca	-	R	C	-
O.dioica	X	C	C	-
O.fusiformis	-	X	R	-
O.longicauda	-	C	A	-
Megalocercus abyssorum	-	X	R	-
Stegosoma magnum	-	-	X	-
THALIACES (Dolioles & Salpes)	C	C	R	-
OEUFS & LARVES DE POISSONS	R	C	C	-

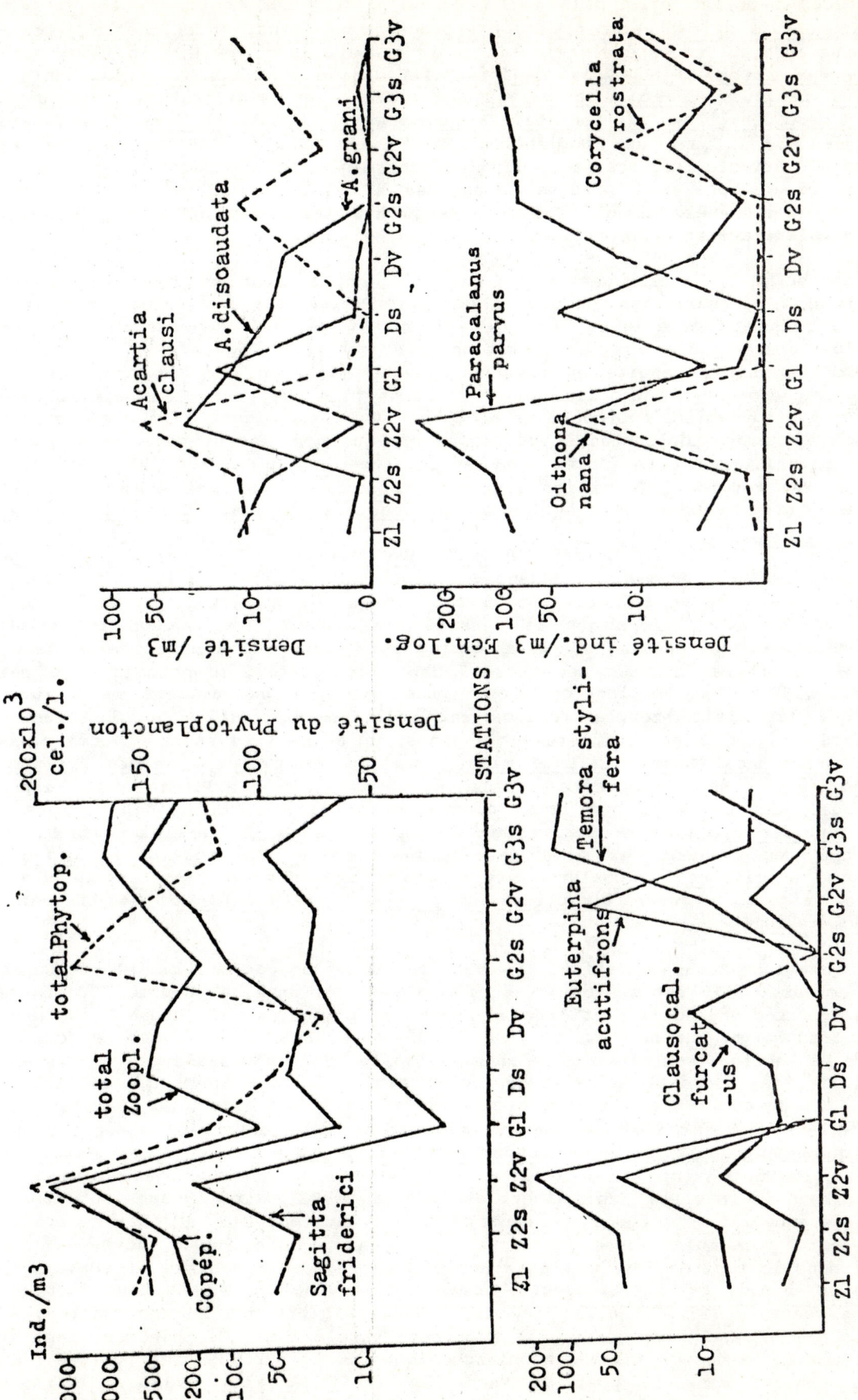

Figure 8. Fluctuations en abondance des populations zooplanctoniques et phytoplanctoniques ainsi que quelques espèces typiques aux dix stations étudiées entre 1984 et 1985.

quelques espèces semblent être plus rares en Z1 qu'en Z2S, telles que: Nitzschia seriata, Rhizosolenia sp., Chaetoceros sp., Prorocentrum micans, Asterionella japonica. Par contre plusieurs autres espèces semblent avoir une abondance normale en Z1, c'est le cas de Ceratium furca, Dinophysis caudata, Peridinium sp., Skeletonema costatum, Leptocylindrus danicus, Chaetoceros curvisetus. Chez les tintinnides, Tintinnopsis campanula, T.beroidea, Favella serrata, F.markuzowskii, Coxliella sp., sont aussi abondants qu'en Z2. Quant au zooplancton du secteur Z1, il ne semble pas être particulièrement affecté par cette pollution thermique; l'abondance ainsi que la structure des populations sont presque identiques à celles de la station Z2. On y observe toutefois un développement plus important chez les copépodites et les nauplii des copépodes calanoides et cylopoides.

Dans le secteur G1, l'action des agents polluants sur le système planctonique semble être plus évidente. Malgré un degré de pollution très élevé à cause des rejets industriels, urbains et organiques, et qui se traduit par une concentration en sels nutritifs plus élevée qu'aux autres stations (PO_4 = 1,52 µg-at.l^{-1}, NO_3 = 0,80 mg-at.l^{-1}) et une teneur réduite en oxygène dissous (5,28 ml/l.), il ne semble pas qu'il existe un phénomène d'eutrophisation. D'autant plus que ce secteur est largement ouvert au fort hydrodynamisme des courants et de la houle qui aident à la dispersion des agents polluants. Le phytoplancton se développe presque normalement dans cette zone et des poussées printanières très fortes sont enregistrées en avril-mai (250,000 cel/l.) rendant l'eau très turbide. Parmi les espèces qui abondent particulièrement en cette saison en G1, nous avons: Leptocylindrus danicus, L.minimus, Nitzschia seriata, Prorocentrum micans, Cocconeis pseudomarginata, Chaetoceros sp., Ceratium sp., Peridinium sp. Chez les tintinnides, Favella markuzowskii, F.serrata, Tintinnopsis campanula sont particulièrement abondants. Dans ce secteur pollué, le maximum saisonnier du zooplancton ne coïncide pas avec celui du phytoplancton. Celui-ci montre son pic maximal en avril alors que la densité maximale du zooplancton tombe entre novembre et décembre. Ce décalage dans le timing saisonnier qui a été observé aussi à la station Z1, n'est pas un phénomène normal dans l'écosystème planctonique de cette région (Lakkis, 1983). Le zooplancton, étant plus sensible aux facteurs de pollution, on observe que la densité zooplanctonique est relativement faible en G1, plusieurs groupes et espèces sont, sinon complètement absents, du moins très rares. Parmi ceux-ci signalons: Paracalanus parvus, Calanus minor, Centropages typicus, Temora stylifera, Clausocalanus sp., des cladocères, siphonophores, etc. Parmi les espèces caractéristiques de ce secteur, nous signalons particulièrement des larves de polychètes, Leucifer typus, des nauplii de copépodes, larves de lamellibranches. Des espèces de copépodes abondent dans ce secteur telles que: Acartia negligens, Labidocera pavo, Euterpina acutifrons, Calocalanus styliremis, Clausocalanus furcatus; à côté de ceux-ci, nous signalons aussi Oikopleura longicauda et O.fusiformis et en moindre quantité Evadne spinifera.

Dans le secteur de Daoura (stations DS, DV), la pollution est la plus élevée de tous les autres secteurs; elle est surtout d'origine organique, urbaine (dépotoir d'ordures, émissaires d'égoûts d'une bonne partie de Beyrouth) et portuaire. Ce degré très fort de pollution se traduit par une très forte odeur nauséabonde (H2S et autre qui se dégage en permanence de ce secteur, une eau peu agitée, à l'abri des forts brassages connus en G1 et une turbidité très grande. Du point de vue chimique, la plus forte concentration en phosphates (4,76 µg-at.l^{-1}) et en nitrates (1,96 µg-at.l^{-1}) et la plus faible teneur en oxygène (4,55 ml/l.) ont été mesurées. L'écosystème marin est très perturbé dans ce secteur et qui se traduit par des fonds azoiques de vase noire. Ce secteur est comparable à celui qui existe dans le port de Marseille et notamment au voisinage de l'émissaire urbain de la ville (Champalbert et Patriti, 1982). Ces auteurs ont constaté qu'au voisinage immédiat de l'égoût, la densité du zooplancton, bien qu'elle ne soit pas très faible, est 2 fois moins forte que la zone la moins polluée. Nous avons d'ailleurs observé ceci dans le port de Beyrouth dans une étude précédente (Lakkis et Kouyoumjian, 1975; Lakkis et Abboud, 1977). La densité moyenne du phytoplancton est plus faible que dans d'autres secteurs (23,000 cel/l. moyenne annuelle) alors que le zooplancton, s'il est très faible en surface, par contre, il est très dense à 5 m. Ce phénomène peut être expliqué par une fuite des organismes zooplanctoniques de la surface très polluée. Ceci a été aussi observé dans le port de Marseille (Arfi et al., 1982). La poussée phytoplanctonique enregistrée en avril est plus faible qu'en G1 et Z1 (Fig. 9). Parmi les espèces caractéristiques signalons: Asterionella japonica, Chaetoceros curvisetus,

Nitzschia seriata, Rhizosolenia stolterfothii ainsi que Favella sp., Tintinnopsis campanula. Le zooplancton est caractérisé par une densité très faible en surface, compensée par une concentration très grande vers les 5 m., mais la diversité spécifique est très faible. Parmi les espèces caractéristiques, nous avons: Acartia grani, Calocalanus styliremis, Euterpina acutifrons, Acartia discaudata, Clausocalanus paululus, Oithona nana, Paracalanus parvus, Labidocera pavo, Evadne spinifera, Obelia sp. Plusieurs larves d'annélides, des appendiculaires et des larves d'échinodermes abondent aussi dans la couche subsuperficielle. On remarque d'une façon générale que les formes herbivores sont les plus dominantes alors que les zooplanctontes carnivores et de grande taille évitent ces eaux polluées.

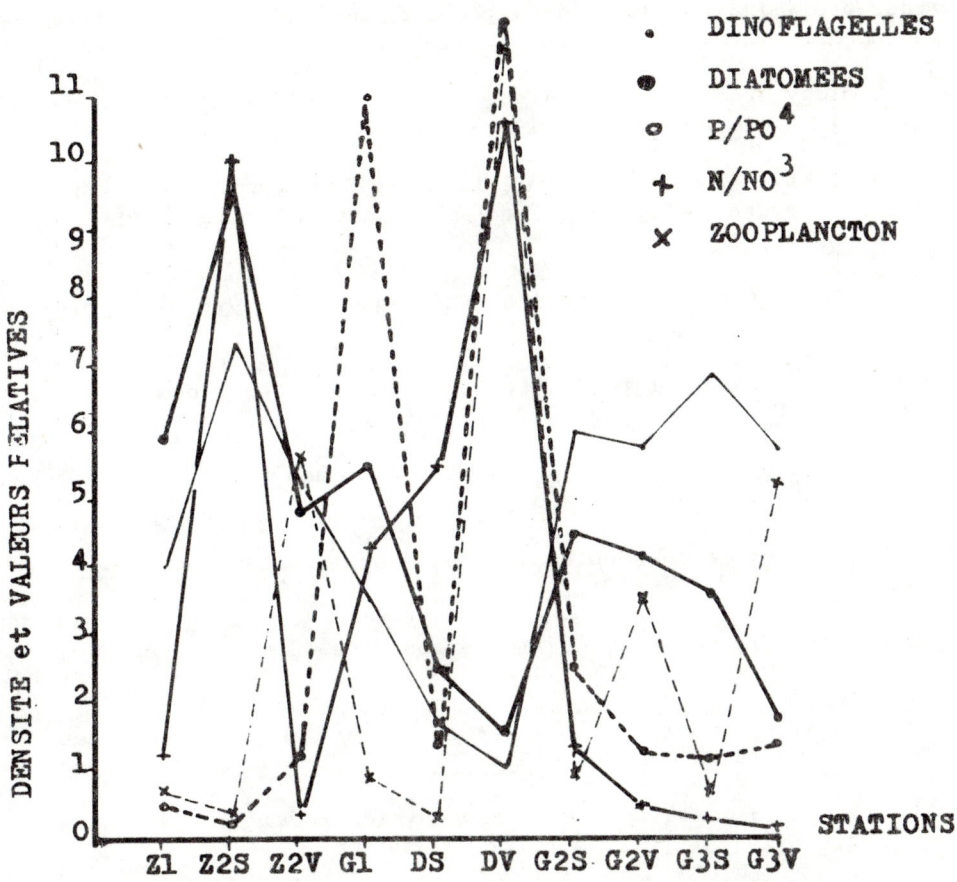

Figure 9. Evolution comparée des moyennes hydrobiologiques dans la zone étudiée (10 stations) entre juillet 1984 et avril 1985.

En conclusion, on peut dire qu'il existe une certaine modification quantitative et qualitative au niveau de l'écosystème planctonique des zones polluées. Les perturbations dans les facteurs physico-chimiques ne peuvent qu'affecter le développement et la distribution des organismes planctoniques. La teneur des produits polluants a tendance à augmenter avec la profondeur et à diminuer avec la diffusion depuis la côte vers le large. Il est certain que les apports de polluants étant irréguliers, les prélèvements ponctuels ne reflètent qu'imparfaitement le degré de pollution global qui, en fait, détermine l'évolution, l'abondance et la structure des communautés. Si la pollution locale contribue à l'appauvrissement des populations planctoniques, lequel aboutit à une diminution de la diversité spécifique, cette situation n'existe que dans les aires fermées, semi-fermées et portuaires. Là dans ces conditions, une forte eutrophisation crée des populations qui s'adaptent à ces milieux par la prolifération d'organismes d'un nombre limité d'espèces. Par ailleurs, la pollution des aires ouvertes aux mouvements permanents des eaux et des courants, ne peut affecter l'écosystème planctonique d'une façon dramatique. Si la pollution affecte beaucoup plus les écosystèmes benthiques, les systèmes pélagiques peuvent s'adapter plus facilement ou même s'y échapper.

5. REMERCIEMENTS

Nous remercions Dr. H. Kouyoumjian dont la collaboration nous a aidé à l'éxécution de ce travail. Nous remercions également Mlle J. Yazbeck pour avoir aimablement effectué l'analyse des sels nutritifs. L'aide de Mrs. M. Khachan et E. Tarek à bord du bateau SETA III est apprécié.

6. REFERENCES

Arfi, R. et al., Etude préliminaire comparée du plancton du vieux port de l'avant-port
1982 et du golfe de Marseille. Téthys, 10(3):211-7

Basson, P.W. et al., Ecology of marine macroalgae in relation to pollution along the
1976 coast of Lebanon. Acta Adriat., 18(19):307-24

Champalbert, G. et G. Patriti, Impact de la pollution sur les communautés zooplancto-
1982 niques dans la zone d'épandage de l'émissaire urbain de Marseille. Hydrobiologia, 89:17-27

Frontier, S., Etude statistique de la dispersion du zooplancton. J.Exp.Mar.Biol.Ecol.,
1973 12(2):229-62

Groupe EPOPEM, Système planctonique et pollution urbaine. Un aspect des populations
1979 zooplanctoniques. Oceanolog.Acta, 2(4):379-88

Kouyoumjian, H., A preliminary study of bacterial pollution and physiography of Beirut
1972 coastal waters. Rev.Int.Océanogr.Méd., 26:5-27

Lakkis, S., Contribution à l'étude du zooplancton des eaux libanaises. Mar.Biol.,
1971 11(2):138-48

_____, Situation actuelle de la pollution le long des côtes libanaises.
1973 Journ.Etud.Pollut.CIESM, 1(1972):53-4

_____, Note préliminaire sur la présence et la répartition des copépodes dans les
1973 les eaux superficielles libanaises. Rapp.P-V.Réun.CIESM, 21:459-64

_____, Contribution à la connaissance de l'écosystème néritique des eaux
1983 Libanaises (Médit.orientale). Thèse Doctorat d'Etat., Université de Paris 6, 505 p.

_____, On the presence of some rare copepods in the Levantine basin. Crustaceana,
1984 Suppl.7:286-304

Lakkis, S. et H. Kouyoumjian, Observations sur la composition et l'abondance du
1975 zooplancton aux embouchures d'effluents urbains des eaux de Beyrouth. Rapp.P-V.Réun.CIESM, 22(9):107-8

Lakkis, S., et M. Abboud, Zooplancton et pollution du secteur libanais en Méditerranée
1977 orientale. Rapp.P-V.Réun.CIESM, 23(9):79-81

Lakkis, S., et V. Novel-Lakkis, Composition, annual cycle and species diversity of the
1981 phytoplankton in Lebanese coastal waters. J. Plankton Res., 3(1):123-36

_____, Les Tintinnides (Tintinnina) des eaux côtières libanaises: composition,
1985 distribution et cycle annuel. Lebanese Sci.Bull., 1(1):43-58

Lakkis, S., et R. Zeidane, Caractéristiques écologiques et dynamiques du zooplancton
1983 des eaux côtières libanaises. Rapp.P-V.Réun.CIESM, 28(9):215-6

Motoda, S., Devices of simple plankton apparatus. Mem.Fac.Fish.Hokkaido Univ.,
1959 (7):73-94

Patriti, G. et al., Système planctonique en milieu portuaire (Port de Marseille).
1979 Structure et fonctionnement - étude chronologique. Téthys, 9(2):137-48

Strickland, J.D.H., and T.R. Parsons, A manual of sea-water analysis.
1968 Bull.Fish.Res.Board Can., (125):311 p.

Unesco, Zooplankton sampling. Unesco Monogr.Oceanogr.Methodol., (2):174 p.
1968

Yamazi, I., Structure of the netted plankton communities in the inner area of the Gulf
1964 of Naples in September 1962. Pubbl.Staz.Zool.Napoli, (34):98-136

SEASONAL FLUCTUATIONS OF A POLYCHAETE COMMUNITY IN A SMALL BRACKISH
ENVIRONMENT (MICROHABITAT)

by

C. LARDICCI* and A. CASTELLI**

*Istituto di Biologia Marina
Università degli Studi di Pisa
Pisa, Italy

**Istituto di Zoologia
Università degli Studi di Modena
Modena, Italy

1. INTRODUCTION

The analysis of a benthic community structure is a good method in the study of environmental modifications caused both by natural and anthropogenic perturbations (Crema and Bonvicini Pagliai, 1980). Moreover this kind of analysis can help also to find out the effects of certain factors which have acted in the past. By contrast this is not possible by chemico-physical analysis, which gives us only temporary information. The present paper is a preliminary study of a benthic community in a brackish "microhabitat" which are typical unpredictable environments (Curini-Galletti et al., 1985).

The study area, located in the Bay of Portoferraio (Elba Island), is characterized by a great variability of chemico-physical parameters such as: O_2, T°, pH and salinity, due to tidal stress and fresh water inputs.

The "microhabitat", on account of their peculiar characteristics (small dimensions, low number of species, facility of monitoring) are particularly interesting in this kind of study, especially concerning the community changes related to environmental modifications.

Among the benthic organisms the Polychaetes are generally the most representative group, therefore the analysis of the distribution of only this taxa gives good information about environmental characteristics.

2. MATERIAL AND METHODS

The study area is a small drain channel 100m long and 5m wide, which opens into a small confined area with a blocking toward the open sea. The mouth of the channel is about 100m from the sea. The samples were collected in April 1983, October 1983, October 1984, and April 1985 with a small grab in two sites, within and outside the channel (Fig. 1).

For each sample 10^{-2} m^3 of sediment were taken and sieved through a 0.5 mm mesh. The salinity was measured by a Beckman salinometer. The structural analysis was performed using multivariate ordering techniques (Fresi and Gambi, 1982; Zurlini, 1983). Data were elaborated by factorial analysis of correspondence (Benzecrì et al., 1973) on a set of species ordinated in three kinds of matrices:

(a) "matrix des états", describing the seasonal conditions of the samples on the basis of simultaneous variation of the species;

(b) "matrix des histoires", describing the "history" of each species in the two samples;

(c) "matrix des états (II)", in which the vector of the species change in abundance is split into several vectors, to emphasize the relationship between the abundance changes of such species.

Significance of the axes were tested by the method of Lebart.

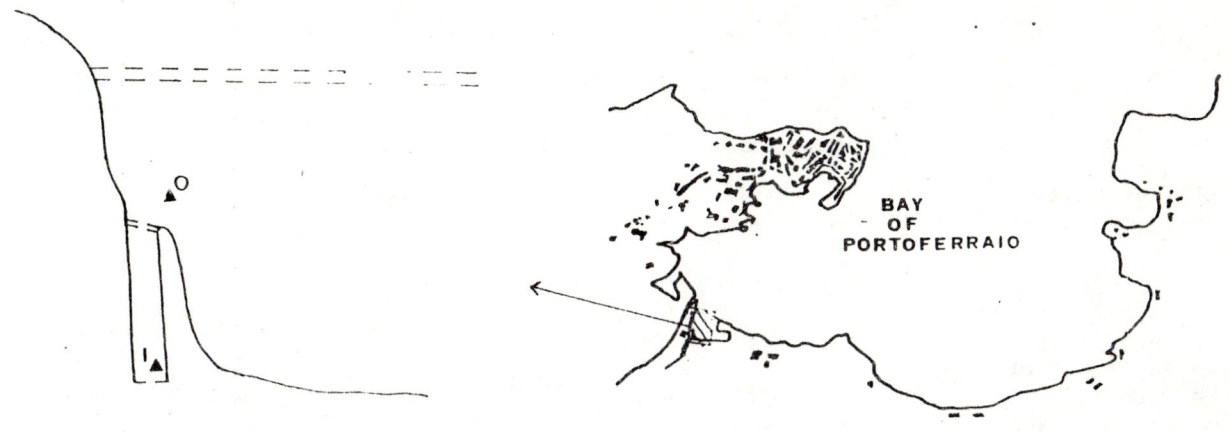

Figure 1. Study area

3. RESULTS

The trend of salinity (Table I) shows low values in the spring months, especially inside the channel because of the big freshwater inputs (0.5°/oo) (in April 1985 the salinity values were very low outside the channel also) whereas in the autumnal months the values of salinity were very similar to those of the open sea (37°/oo).

Table I

Salinity values (°/oo) inside (I) and outisde (O) the channel

	APR. 1982		OCT. 1983		OCT. 1984		APR. 1985	
	I	O	I	O	I	O	I	O
	0.5	33.0	37.8	38.0	38.0	37.9	0.6	14.6

Altogether 20 species belonging to 8 families were collected (Table II). Some species such as N.diversicolor and S.shrubsolii were very abundant in the inner zone while, in the outer station, species such as E.assimilis, B.oculata, P.cultrifera, C.tentaculata and S.shrubsolii were well represented. Species richness inside the channel was always lower than outside during the whole period of study.

In the analysis performed on the "matrix des états", only the first two axes were significant and explain 81.14% of the total variance. The ordination model obtained on the plane of the two factors is shown in Fig. 2. The first axis sharply separates the inner samples from the outer ones over the whole period of study. On the second axis the seasonal fluctuations are developed particularly as far as the inner samples are concerned.

Table II

Abundance of species inside (I) and outside (O) the channel

	APR. 1983		OCT. 1983		OCT. 1984		APR. 1985	
	I	O	I	O	I	O	I	O
Eusyllis assimilis	-	-	-	-	-	+++	-	-
Sphaerosyllis hystrix	-	+	-	+	-	+	-	-
Brania pusilla	-	-	-	-	-	+	-	-
Brania oculata	-	-	-	-	-	+++	-	-
Parapionosyllis elegans	-	+	-	+	-	-	-	-
Exogone gemmifera	-	+	-	+	-	+	-	-
Nereis diversicolor	++	-	++	-	+	-	++	++
Nereis caudata	-	+	-	-	-	-	-	+
Perinereis cultrifera	-	++	-	++	-	++	-	+++
Perinereis oliveirae	-	-	-	-	-	+	-	+
Marphysa sanguinea	-	-	-	-	-	+	-	-
Cirrophorus furcatus	-	-	-	-	-	++	-	-
Malacoceros sp.	-	++	-	-	-	++	-	++
Spiophanes bombyx	-	-	-	-	-	+	-	-
Microspio meczinikowianus	-	+	-	+	-	-	-	+
Streblospio shrubsolii	+++	+++	++	++	++	+++	++	+++
Cirriformia tentaculata	-	+++	+	+++	+	+++	-	+
Capitella capitata	-	++	-	++	+	++	-	++
Heteromastus filiformis	-	+	-	+	-	-	-	-
Fabricia sabella	-	+	-	+	-	-	-	-

In the analysis performed on the "matrix des états (II)", also only the first two axes are significant and explain 56.26% of the total variance. As far as the station points are concerned, the ordination model obtained (Fig. 3) is very similar to that previously described. The trend of abundance vectors of the species points is well related to the station ones. N.diversicolor, linked to the inner stations, shows a variable trend along the second axis. Such trend is linked to time variability. In fact this species has an increase in abundance when the salinity is lower (spring samples). S.shrubsolii, C.tentaculata and C.capitata show some fluctuations in abundances along the first axis. This one is therefore interpreted, in both the models, as a gradient reflecting the changes of abiotic parameters from outside to inside the channel.

In the analysis performed on the "matrix des histoires" only one factor was taken out and all the variance of the system is linked to this axis. Therefore, an axis perpendicular to this including the sampling periods, was placed to better represent the data (Fig. 4). Four groups of species points were obtained:

(a) The first, consisting of S.shrubsolii, the only species always located in the intermediate zone between the inner and outer stations;

(b) The second, consisting of N.diversicolor, is linked to the inner zone, except for April 1985, in which it lies in the intermediate zone;

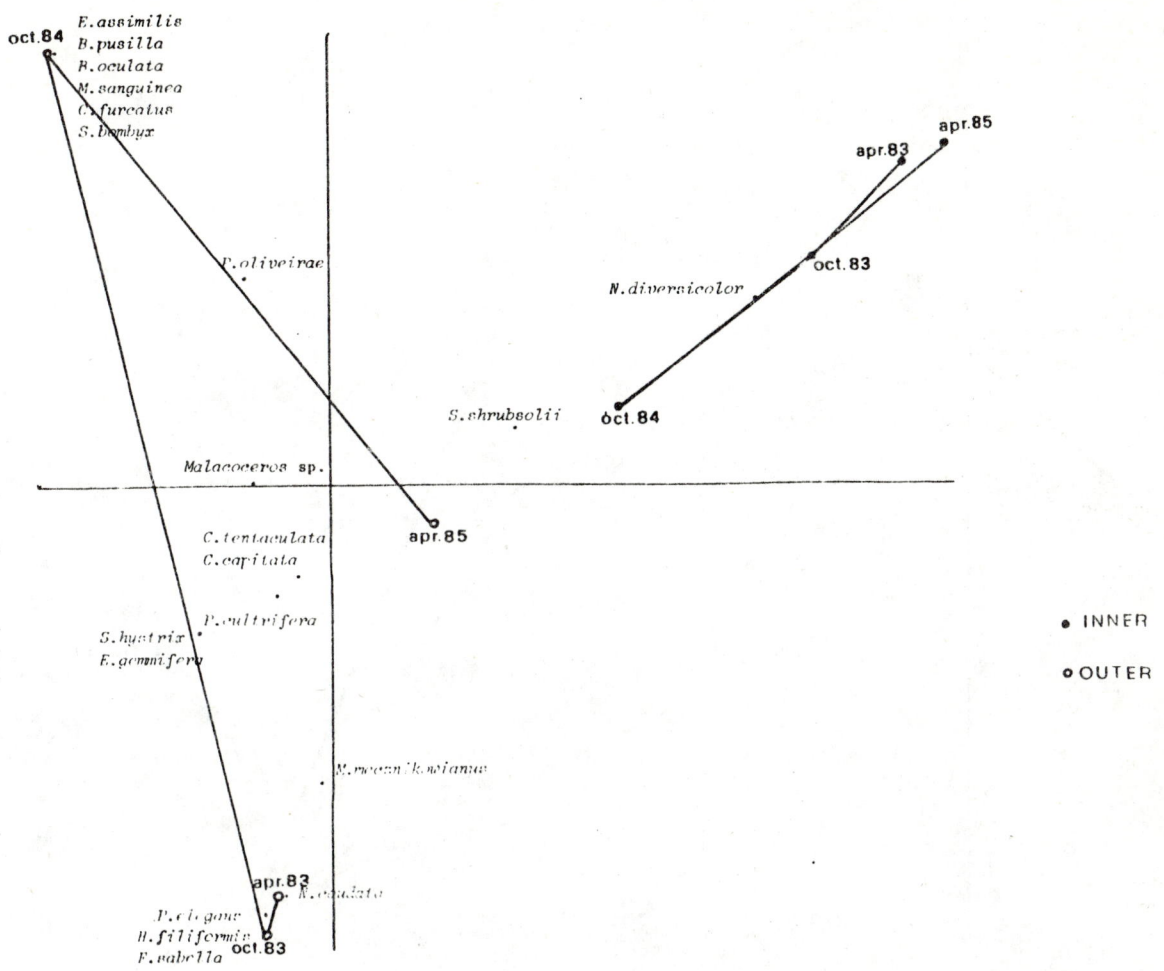

Figure 2. Analysis performed on the "Matrix des états"

(c) The third, consisting of C.capitata and C.tentaculata, is fluctuating from the inner to the outer zones in autumn, whereas it is linked to the outer zone in spring periods;

(d) Finally, the fourth, consisting of the remaining species, is completely linked to the outer zone.

4. DISCUSSION AND CONCLUSIONS

All the above considerations can be summarized as follows:

(a) A sharp discontinuity from inside to outside samples exist during the whole period of study;

(b) Species richness is always higher in the outer zone;

(c) The species moving between the inner and the outer zone are mainly three:

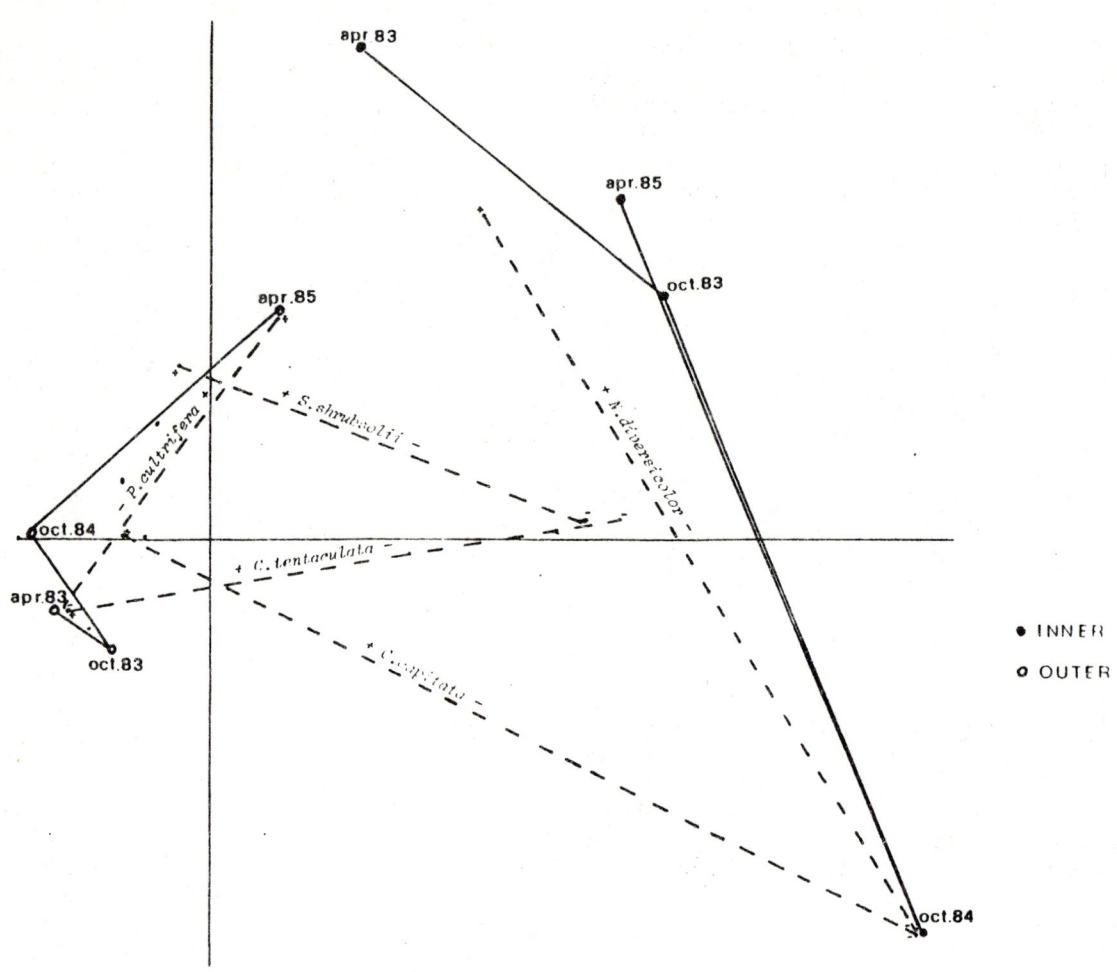

Figure 3. Analysis performed on the "Matrix des états II"

N.diversicolor: this is a typical euryhaline species tolerating low oxygen levels, high concentrations of heavy metals, big temperature variations (Muus, 1967; Bryan and Hummerstone, 1971), but which does not tolerate a high degree of competition (this species was present in the outer zone only in April 1985 when environmental conditions were very similar to those of the inner zone). Therefore we can assume that N.diversicolor colonized this kind of environment by a high "tolerance" of abiotic factors.

C.capitata: this species is a typical r-selected species. It can reproduce both by plactonic larvae and by benthic larvae, has a short life-cycle, and reaches maturity from the egg in about three weeks. It can therefore continuously repopulate sediments subjected to pollution from organic matter. This species does not use "tolerance" as an adaptive strategy, but adapts to continuous disturbance by continuous reproduction (Gray, 1981). C.capitata does not tolerate low salinity values and therefore can colonize the inner zone only when a uniformity of this parameter exists along the whole study area.

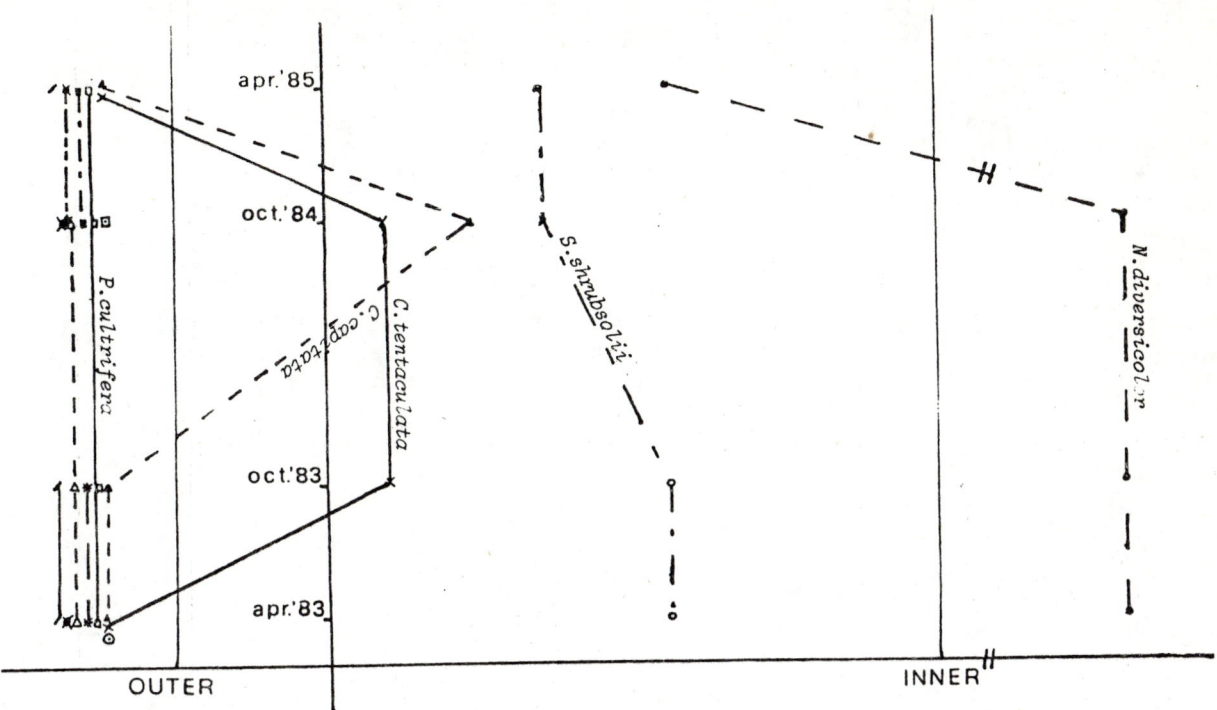

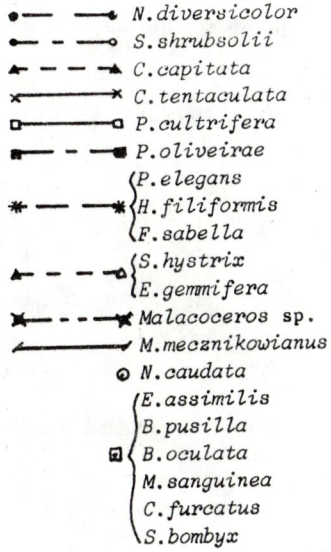

Figure 4. Analysis performed on the "Matrix des histoires"

<u>C.tentaculata</u>: this is a saprobic species, typical of environments with high organic content, such as harbours or polluted zones (Bellan, 1968; Cognetti and Taliercio, 1969). Its life history and strategy are similar to that of <u>C.capitata</u>.

Examining the trend of each species in the study period, three patterns are recognizable as a response to the environmental changes:

(a) Species characterizing the seasonal fluctuation of the whole community by their spatial movement (N.diversicolor, C.capitata, C.tentaculata);

(b) Species which are present both in the inner and outer zones without changes in abundance (S.shrubsolii); this species appears to tolerate both competition and environmental variability;

(c) Stenohaline species linked to the outer zone (P.cultrifera, Malacoceros sp.). To this group also belong some sea species which appeared only after the summer period (E.assimilis, B.oculata, B.pusilla, C.furcatus).

The community studies therefore show a trend related to environmental variability which results in the fluctuation of the species between the inner and the outer zone.

The study of colonization and of community changes in environments subject to changes over time is very interesting in order to identify the effects on the entire community, as well as the adaptive strategies of the single species.

5. ACKNOWLEDGEMENTS

We thank Dr. C.N. Bianchi for his invaluable contribution to the statistical analysis of data.

6. REFERENCES

Bellan, G., Contribution à la connaissance des peuplements de substrat meuble établis
1968   dans les zones polluées de la région de Marseille. Rapp.P-V.Réun.CIESM, 5:19-21

Benzecrì, J.P. et al., L'analyse des données. Vol 2. L'analyse des correspondances.
1973   Paris, Dunod, 619 p.

Bryan, G.W. and L.G. Hummerstone, Adaptation of the Polychaete N.diversicolor to
1971   estuarine sediments containing high concentrations of heavy metals. 1. General observations and adaptations to copper. J.Mar.Biol.Assoc.U.K., 40:845-63

Cognetti, G., and P. Taliercio, Policheti indicatori dell'inquinamento delle acque.
1969   Pubbl.Staz.Zool.Napoli, (37): 149-54

Crema, R., and A.M. Bonvicini Pagliai, The structure of benthic communities in an area
1980   of thermal discharge from a coastal power station. Mar.Pollut.Bull., 2:221-7

Curini-Galletti, M., A. Castelli and C. Lardicci, La fauna dei microhabitat salmastri
1985   delle Isole d'Elba e di Capraia. Oebalia, 11(3):803-5

Fresi, E., and M.C. Gambi, Alcuni aspetti importanti dell'analisi matematica di
1982   ecosistemi marini. Nat.Sicil., 6:449-65

Gray, J.S., The ecology of marine sediments. Cambridge, U.K.,Cambridge University
1981   Press, Cambridge studies in modern biology, 177 p.

Muus, B.J., The fauna of Danish estuaries and lagoons. Distribution and ecology of
1967   dominating species in the shallow reaches of the mesohaline zone. Medd.Dan.Fisk.Havundersog., 5:1-316

Zurlini, G., Applicazioni di metodi multivariati all'analisi ecologica. In Atti
1983   Convegno ENEA "Un esempio di analisi ecologica del sistema marino-costiero da Capo Circeo all'Isola d'Ischia", June 1983, edited by G. Zurlini and V. Damiani, ENEA, Serie Simposi pp.13-34

ECOLOGICAL OBSERVATIONS OF A LOCALLY LIMITED SUMMER BLOOM

by

I. MARASOVIC and T. PUCHER-PETKOVIC
Institute of Oceanography and Fisheries
Split, Yugoslavia

1. INTRODUCTION

The first data on the phytoplankton of the Kastela Bay (vicinity of Split, middle Adriatic) can be traced back to some fifty years ago. Systematic studies of phytoplankton, together with some other biotic and abiotic parameters, have been carried out ever since 1956. The last 15 years have shown some changes such as a gradual increase in primary production, density, change of seasonal cycle as well as changes in the taxonomic structure of phytoplankton (Pucher-Petkovic and Marasovic, 1980; Marasovic and Pucher-Petkovic, 1983, 1985). For the last five years these changes have become rather serious since minor or more severe red tides occurred on several occasions (Marasovic and Vukadin, 1982), accompanied by the mortality of marine organisms from time to time. Summer is the most critical season since the sea water temperature is high, both the vertical and horizontal circulations very poor and the levels of nutrients high (waste water discharges). This causes extra summer phytoplankton blooms. The paper describes two non-toxic red tide blooms which occurred simultaneously in two different parts of the Kastela Bay in 1984.

2. MATERIAL AND METHODS

Phytoplankton samplings were carried out by Nansen bottle from six permanent stations of which five are in the Kastela Bay and one out of the bay in the vicinity of the town port (Fig. 1). During the bloom, a series of additional stations were sampled from time to time.

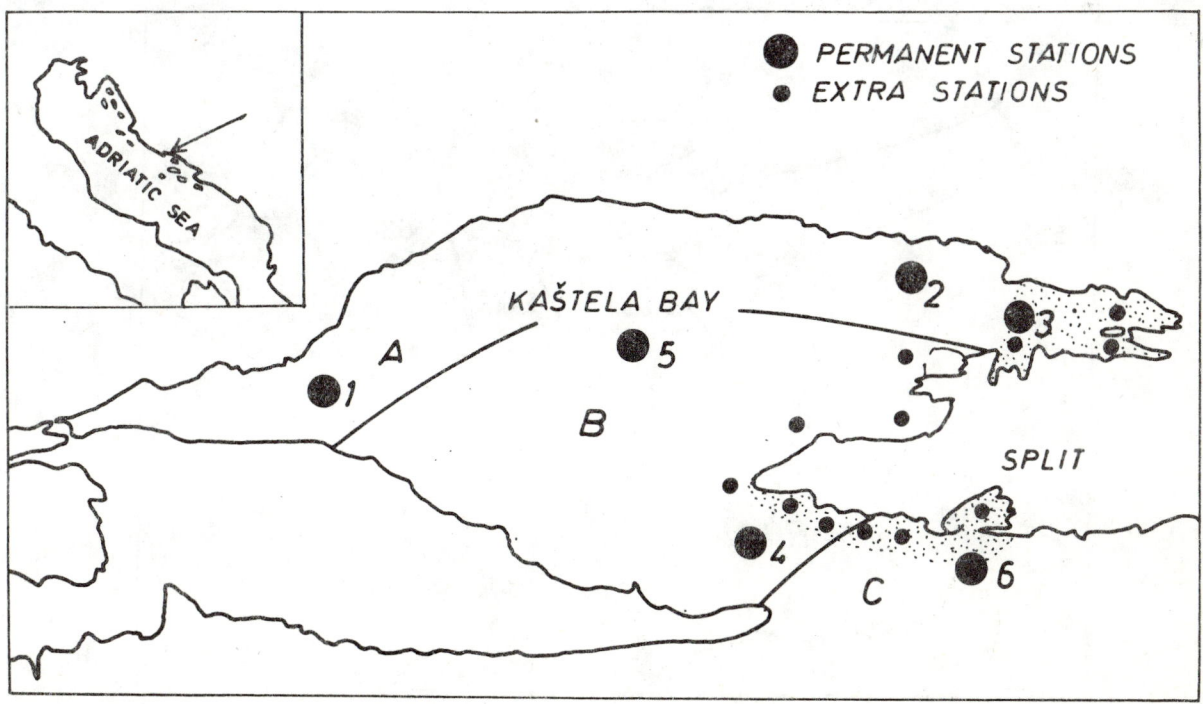

Figure 1. Study area (dotted fields are areas under red tide)

The phytoplankton biomass was determined by the method of sedimentation and counting using a Utermöhl microscope as well as by chlorophyll a fluorescence using a Turner 111 fluorimeter. Chlorophyll a quantities are expressed in fluorescence units.

Nutrient levels were determined by standard methods using an Autoanalyzer, temperature and salinity by CSTD probe, and oxygen by Winkler's method. Climatic data were obtained from the Meteorological Institute of the Republic of Croatia and some climatic measurements were performed at the Institute of Oceanography and Fisheries, Split.

3. RESULTS AND DISCUSSIONS

Earlier investigations have already shown that the taxonomic structure of phytoplankton in the eastern part of the Kastela Bay differs considerably from that of the other parts of the bay and particularly outside the bay (Pucher-Petkovic and Marasovic, 1979, 1980; Marasovic and Pucher-Petkovic, 1983, 1985; Marasovic and Vukadin, 1982). This is particularly evident during summer when the differences in circulation are best marked. They cause considerable differences between the physical and chemical environmental conditions of the inner part of the bay and those of the other parts of the bay (Figs. 2, 3, 4, 5, 6). These situations are also well illustrated by the comparison of the phytoplankton composition from different parts of the bay (A and B respectively) and from the area outside the bay (C) (Fig. 1, Table I) during August 1984. It is evident from Table I that dinoflagellates are predominant in the area A during summer, making up more than 90% of the phytoplankton populations. Diatom species are, however, predominant in the A and C areas with percentages ranging from 65 to 85. Index of similarity between diatoms and dinoflagellatestes was calculated on the basis of these data (SORENSEN, 1948). The similarity index for the A/B areas was 38.9 and even lower, not exceeding 35.3, for the A/C areas. Similarity index for B/C areas is much higher (69.6%).

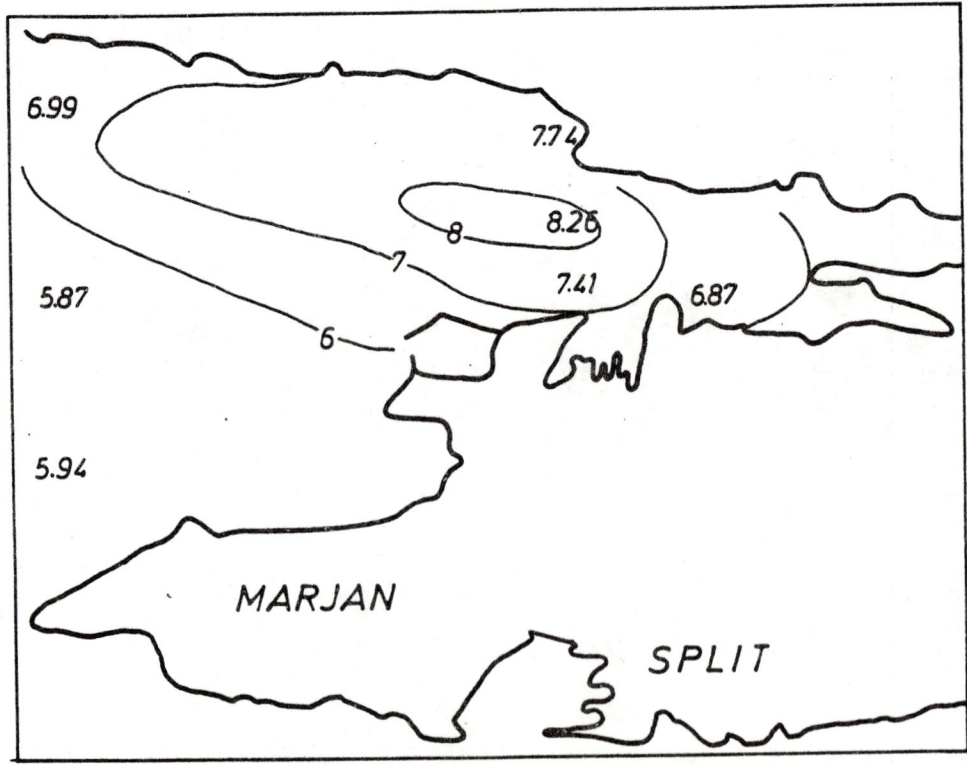

Figure 2. Oxygen quantities (ml O_2 l^{-1}) in the eastern part of the bay during summer (September) (Institut za Oceanografiju i Ribarstvo, 1976)

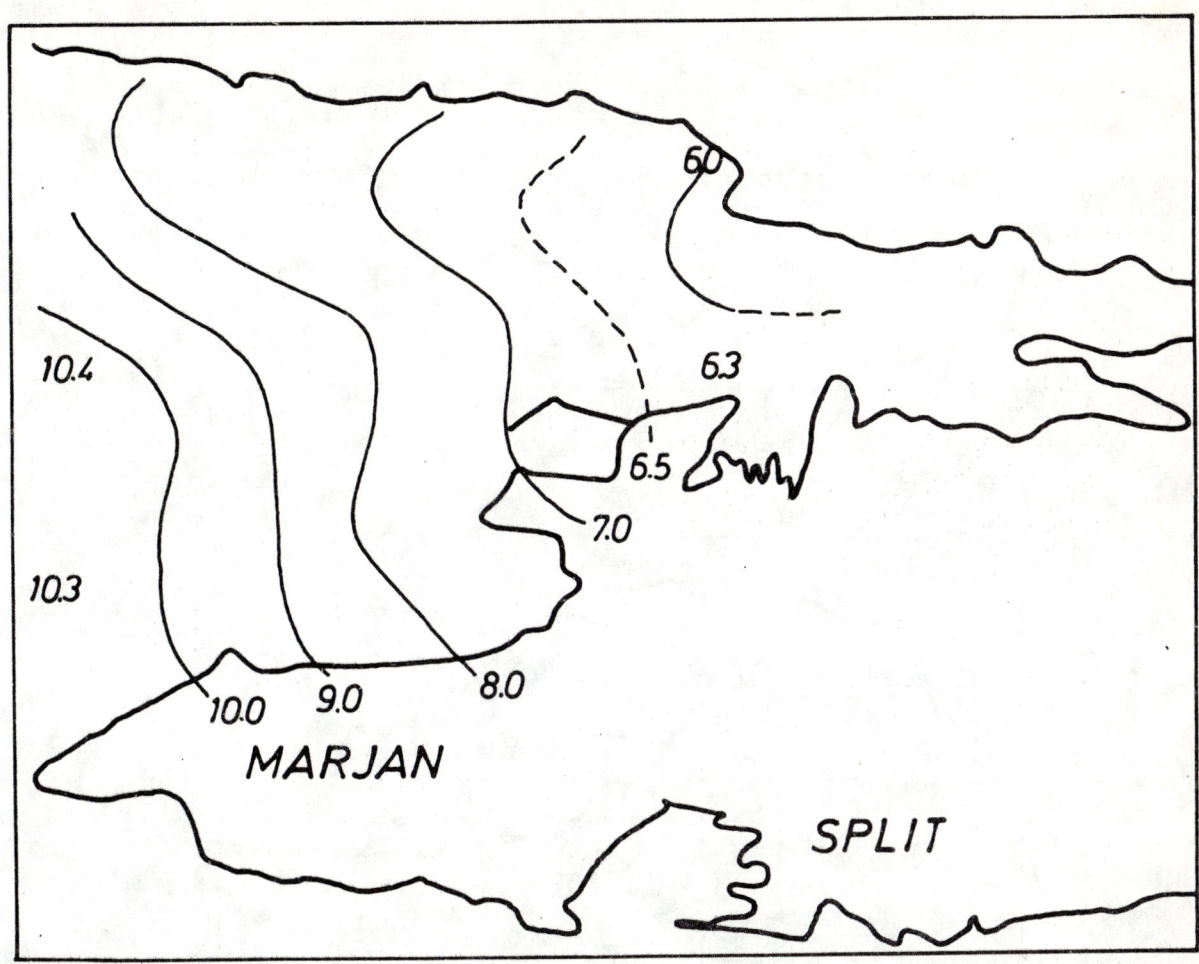

Figure 3. Transparency (m) in the eastern part of the bay (annual mean) (Institut za Oceanografiju i Ribarstvo, 1976)

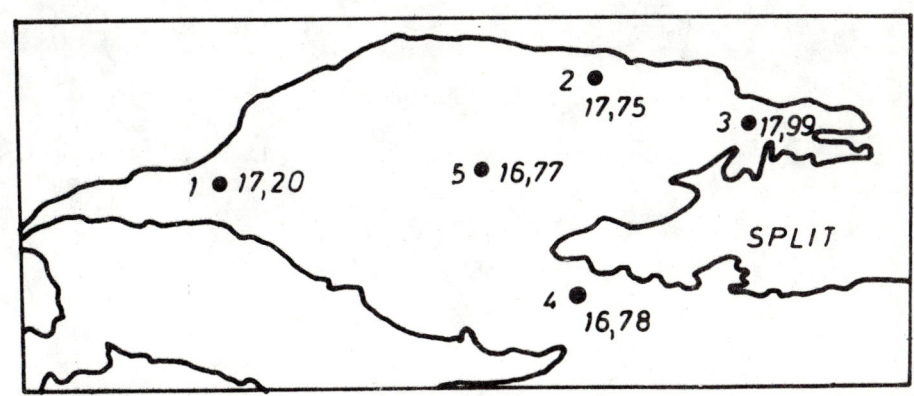

Figure 4. Temperature (°C) at studied stations of the Kastela Bay (long-term means, 1982, 1983 and 1984)

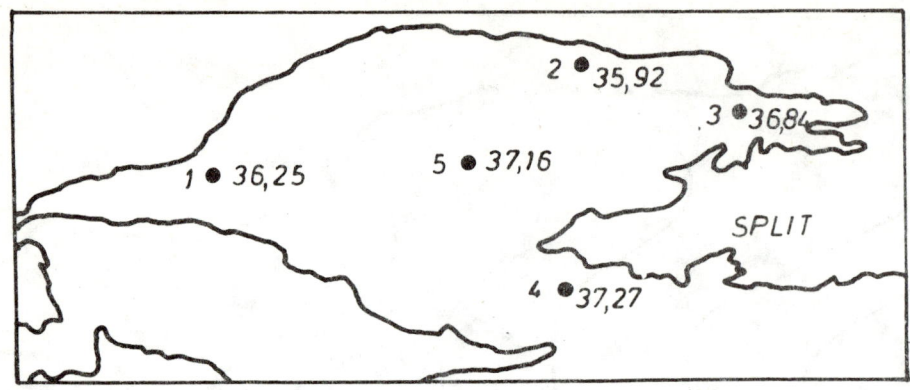

Figure 5. Salinity (S x 10^{-3}) at studied stations of the Kastela Bay (long-term means, 1982, 1983 and 1984)

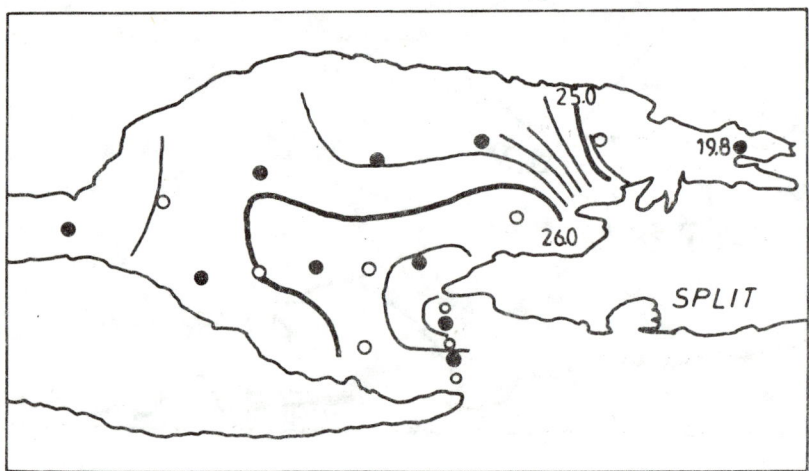

Figure 6. Sea water density (σ_t) in the Kastela Bay (annual mean) (Gacic, M. and T. Smircic, 1971)

An intensive diatom bloom with the predominance of three species: <u>Leptocylindrus danicus</u>, <u>L.adriaticus</u> and <u>Nitzschia seriata</u>, occurred in the beginning of June 1984. This was an ordinary spring bloom which occurred with a slight delay after an unusually rainy first half of the year. Namely, precipitation quantities during the first six months of 1984 considerably exceeded those during the same period of the preceding years:

January-June	Precipitations $1 \, m^{-2}$
1951-84	397.6
1983	301.7
1984	442.6

At the end of June a sudden, intensive monospecific bloom of the diatom <u>Skeletonema costatum</u> occurred. Owing to the sudden increase of the sea water temperature (from 18.5 to 24.4°C in less than a fortnight) this bloom interfered with the ordinary dinoflagellate bloom. The sea became covered by rag-like patches floating on the surface looking like dirt. These patches were composed of a large number of <u>Skeletonema costatum</u> cells within which different dinoflagellates were retained since this dense mass filtered the sea water like a net. The bloom proceeded and by the middle of July about 50 different diatom and dinoflagellate species were included in this bloom. At the end of July when the sea surface temperature reached a value as high as 25°C and the meteorological situation was very stable, a red tide caused by <u>Gonyaulax polyedra</u>

(1×10^7 cells l^{-1} on July 23 at the surface) occurred in the inner part of the bay. Codominant species were <u>Eutreptiella pascheri</u> (1.5×10^5 cells l^{-1}) and <u>Calciosolenia murrayi</u> (2.2×10^5 cells l^{-1}). The same organisms caused red tide at the end of summer 1980 when a severe red tide was accompanied with mass mortality of marine organisms in the sea. That mass mortality was, to a considerable extent, due to an anoxic situation which occurred as a consequence of an intensive decomposition of a huge quantity of organisms (1.8×10^7 cells l^{-1}). However, it was assumed to be also caused by pure mechanical effects of phytoplankton on gills of marine organisms.

Table I

List of phytoplankton species recorded on August 30, 1984 and relative abundance (%) of phytoplankton groups

SPECIES	RELATIVE ABUNDANCE
AREA A	
<u>Rhizosolenia alata f.gracillima</u> <u>Thalassionema nitzschioides</u>	Diatoms 1%
<u>Prorocentrum micans</u> <u>P.minimum</u> <u>P.triestinum</u> <u>Dinophysis sacculus</u> <u>Gyrodinium sp.</u> <u>Glenodinium sp.</u> <u>Scrippsiella trochoidea</u> <u>Gonyaulax polyedra</u> <u>Ceratium karstenii</u> <u>Oxytoxum longiceps</u>	Dinoflagellates 94%
<u>Calciosolenia murrayi</u> <u>Rhabdosphaera longistylis</u> <u>Dictyocha speculum</u> <u>Eutreptiella pascheri</u>	Coccolithophorids + "Microflagellates" 5%
AREA B	
<u>Skeletonema costatum</u> <u>Leptocylindrus adriaticus</u> <u>Rhizosolenia alata f.gracillima</u> <u>Rh.fragilissima</u> <u>Rh.setigera</u> <u>Rh.stolterfothii</u> <u>Chaetoceros curvisetus</u> <u>Ch.danicus</u> <u>Ch.pseudocurvisetus</u> <u>Eucampia cornuta</u> <u>Thalassionema nitzschioides</u> <u>Navicula sp.</u> <u>Nitzschia delicatissima</u> <u>N.longissima</u> <u>N.seriata</u>	Diatoms 65%

Table I (Continued)

SPECIES	RELATIVE ABUNDANCE
AREA B	
Prorocentrum micans	
P.triestinum	
Gyrodinium pingue	
Glenodinium sp.	
Heterocapsa triquetra	
Scrippsiella trochoidea	
Peridinium diabolus	
P.pyriforme	
Gonyaulax polyedra	Dinoflagellates 6%
Syracosphaera pulchra	
Syracosphaera sp.	
Calciosolenia murrayi	
Eutreptiella pascheri	Coccolithophorids +
Carteria sp.	"Microflagellates" 29%
AREA C	
Skeletonema costatum	
Leptocylindrus adriaticus	
Rhizosolenia stolterfothii	
Rhizosolenia styliformis var.longispina	
Chaetoceros curvesetus	
Ch.danicus	
Ch.pseudocurvisetus	
Ch.sp.	
Thalassionema nitzschioides	
Amphora sp.	
Nitzschia delicatissima	
N.longissima	
N.seriata	Diatoms 81%
Prorocentrum micans	
P.minimum	
P.triestinum	
Gymnodinium sp.	
Glenodinium sp.	
Heterocapsa triquetra	
Scrippsiella trochoidea	
Gonyaulax polyedra	
Oxytoxum viride	Dinoflagellates 13%
Syracosphaera sp.	
Rhabdosphaera longistylis	
Calciosolenia murrayi	Coccolithophorids +
Eutreptiella pascheri	"Microflagellates" 6%

Long-term analyses of samples have shown that G.polyedra is, to a larger or smaller extent, always present in the phytoplankton crop of this area throughout the period April-October. A stable meteorological situation (warm and sunny weather) and temperature of the sea water above 22°C may, besides other convenient factors, cause the monospecific bloom of this organism, the duration and intensity of which is dependent on the duration of these weather conditions.

Owing to the fact that red tide occurs repeatedly in the area, it may be explained by the theory of resting cysts acting as seed populations for the development of red tides (Steidinger and Haddad, 1981). According to this theory the red tide is caused by sudden excystment of organisms from resting cysts which survived in sediment of the area in question. Analysing several samples of surface sediment we did not succeed in establishing the presence of G.polyedra cysts. We plan to analyse sediment samples all the year round to provide evidence for this assumption.

In addition, simultaneously with G.polyedra bloom in the inner part of the Kastela Bay, another monospecific bloom caused by Olisthodiscus luteus, an organism not recorded from the Adriatic so far, was recorded from the coastal waters out of the bay and at the entrance to the bay. The sea became an intransparent brownish-red colour all along the coast of the town up to 50-100m offshore. Margins of this patch were clearly distinguished from the adjacent sea which at the same time showed a very high transparency. The analysis of samples showed that a monotype bloom of the flagellate Olisthodiscus luteus ($3-5 \times 10^7$ cells l^{-1}) had developed in the surface layer. Accompanying species were Eutreptiella pascheri, E.lanowii and Carteria sp.

Intensive bloom of O.luteus lasted for ten days. It stopped owing to the change of meteorological conditions (south wind). Analysing these red tide blooms we tried to establish the conditions favourable for the development of characteristic red tide species so as to provide the basis for a forecast of the bloom.

The first requirement to be fulfilled is a sufficiently high level of nutrients (phosphates and nitrates). Even though they do not provoke the bloom, they are its limiting factor. Calm and stable weather is one of these conditions too, since it causes water stratification and concentration of organisms on the surface.

However, the temperature seems to be the most important factor which probably initiates the occurrence of red tide. Many authors believe that temperature significantly affects the emergence and disappearance of red tide.

The majority of authors suggest that the mass blooms of Skeletonema costatum and Olisthodiscus luteus were mutually dependent blooms. Namely, mass bloom of O.luteus always develops after an intensive bloom of S.costatum. Tomas (1980) held that O.luteus has no particular competitive advantage over S.costatum due to their considerable physiological similarity and requirements for similar conditions. However, whereas S.costatum is the food preferred by copepods (Acartia clausi), it was found that O.luteus was only slightly or not at all taken as food. Therefore, the author believes that grazing indirectly has an important role in these successions, since, during the decrease of S.costatum population and increase of O.luteus, the population of Acartia clausi also shows an increase. This, to a certain extent, may account for relatively low quantities of phytoplankton out of the area under red tide.

In our case, we assume as well that the initial impulse is given by temperature. The limiting factor, besides nutrients and light, is the bloom of Skeletonema population. Therefore, only after the reduction of this diatom by grazing, Olisthodiscus may freely develop the bloom.

4. CONCLUSION

From the long-term data it may be stated that the amplitudes of mass blooms are in positive correlation with the process of eutrophication. The increase in the level of nutrient salts results in an increase of primary production and phytoplankton density. The increase becomes apparent through the bloom of so called opportune species and under conditions of further eutrophication causes monospecific red tides. Due to the actual conditions in the Kastela Bay (high levels of nutrients, remarkable quantities of trace metals and organic wastes) during summer, each degree of temperature may act as a "trigger" which will initiate an uncontrolled reproduction (or perhaps excystment) which will, in an unexpectedly short time, result in the emergence of red tide.

5. REFERENCES

Gacic, M. and T. Smircic, Statisticka analiza dinamike povrsinskog sloja Kastelanskog
1971   zaljeva. Hidrogr.God., 1971: 89-102

Institut za Oceanografiju i Ribarstvo, Studije i elaborati Instituta za oceanografiju i ribarstvo. Split, Institut za Oceanografiju i Ribarstvo, 24 p.
1976

Marasovic, I. and T. Pucher-Petkovic, Irregular phytoplankton blooms in the Kastela Bay
1983   (Central Adriatic). Rapp.P.-V.Réun CIESM, 28(9):97-8

_____, Effects of eutrophication on the structure of coastal phytoplankton
1985   community. Rapp.P.-V.Réun CIESM, 29(9):137-9

Marasovic, I. and I. Vukadin, "Red tide" in the Vranjic basin (Kastela Bay). Bilj.
1982   -Notes Inst.Oceanogr.Ribar., Split, (48):7 p.

Pucher-Petkovic, T. and I. Marasovic, Projekti zastite mora u uzoj regiji Splita.
1979   Utjecaj eutrofikacije na fitoplankton Kastelanskog zaljeva. II Konferencija o zastiti Jadrana. Zb.Ref., 2:263-9.

_____, Dévéloppement des populations phytoplanctoniques
1980   caractéristiques pour un milieu eutrophisé (Baie de Kastela, Adriatique Centrale). Acta Adriat., 21(2):79-93

Sørensen, T., A method of establishing groups of equal amplitude in plant sociology
1948   based on similarity of species content and its application to analyses of the vegetation on Danish commons. Biol.Skr., 5(4):1-34

Steidinger, K.A. and K. Haddad, Biologic and hydrographic aspects of red tides.
1981   BioScience, 31(11):814-7

Tomas, C.R., Olisthodiscus luteus (Chrysophyceae). 5. Its occurrence, abundance and
1980   dynamics in Narragansett Bay, Rhode Island. J.Phycol., 16:157-66

LEVELS OF GENETIC DIVERSITY AND RESISTANCE TO POLLUTION IN
MARINE ORGANISMS

by

E. NEVO, R. NOY, B. LAVIE and S. MUCHTAR

Institute of Evolution, University of Haifa,
Haifa 31999, Israel

1. INTRODUCTION

Is the level of genetic diversity related to the resistance against pollution? We know from diverse studies on genetic diversity (reviewed in Nevo, 1983 and Nevo et al., 1984) that the levels of genetic diversity in both plants and animals are positively correlated with ecological heterogeneity. Is the general pattern also true for the effects of pollution? We tested three pairs of species of marine gastropods belonging to three different genera, ranging along the Mediterranean Coast of Israel, in an attempt to answer this question. Each pair consisted of one member with high, the other species with lower genetic diversity, associated with the niche breadth of each. Here we present evidence that in all three pairs, the species characterized by the high levels of genetic diversity was more resistant to all pollutants than the one with low levels of genetic diversity.

2. MATERIALS AND METHODS

The six species of marine gastropods Monodonta turbinata, M. turbiformis, Littorina punctanta, L. neritoides, Cerithium scabridum and C. rupestre are widespread and abundant species of the marine littoral zone along the Mediterranean Sea coast. These species were chosen for the present study because of their abundance and small size, which permitted the testing of relatively large samples.

The pollutants

We employed in the present study three types of pollutants as follows:

Heavy metals: $CuCl_2$, $ZnCl_2$, $PbCl_2$, $HgCl_2$, $CdCl_2$, $HgCl_2$ together with $CdCl_2$,

Salinity: 8.5% NaCl,

Organic: detergent and crude oil.

The nonionic detergent used was the commercially available Marlophen 89 (Hülls, West Germany). This product is a nonylphenol ethoxylate having an approximate molecular weight of 643, the molar ratio between the ethylene oxide and the nonylphenol being 9.55:1 respectively. This nonionic detergent is of the "hard" type, that is, resistant to biological degradation. The crude oil used ("Sonol 2", obtained from the Israeli Oil Refineries Ltd. in Haifa) had the following specifications:

Density at 15°C: 0.9152
Distillation Range (%wt): up to 150, 8.3%; 150-250, 17.7%; 250-400, 21.2%;
Res. > 400, 51.9%
Kinematic Viscosity: at 122F, 15.88cP.
Total Sulfur (%wt): 1.4
Salt PTB: 11
Water (%V): 0.05
Carbon Residue (%wt): 5.3; Asphaltenes (%wt): 0.8.

Laboratory experiments

Samples of hundreds of individuals from each species were collected from the rocky shores in Haifa region and introduced in batches of 50-150 individuals into partly filled aquaria, 80 litres each (70 x 30 x 40 cm) at the Institute of Evolution, University of Haifa. The two Monodonta species were collected near Shikmona, south of the Haifa Bay, the two Cerithium species near Akko, north of Haifa Bay and the two Littorina species at both Shikmona and at Akko. The two Littorina species tend to emerge above water level. In order to avoid this, they were placed into quandrangular cages (25 x 25 x 5 cm) made of perspex and subdivided into small interconnected cells (5 x 5 x 5 cm) by a plastic net, so that water currents could pass freely through all cells. Each cell held two animals. Fresh sea water for the experiments was pumped from 30m depth at the Shikmona National Institute for Oceanography. Conditions in all aquaria were identical (22 degrees Celsius; pH = 8.1; constant aeration and no food was provided). The duration of the test varied depending upon the concentrations of pollutants (Tables III, IV and V). All tests conducted simultaneously were matched with one control. Survival in the controls was almost always 100% otherwise all experiments were discarded. The experimental organisms were observed daily; dead animals were removed and deep frozen (-80 degrees Celsius) as were all survivors at the termination of the tests.

The two Monodonta species were observed for all 5 heavy metals mentioned and organic pollutants. All 6 species were tested in mercury and cadmium pollution (the pollutants that affected the Monodonta species in the smallest concentration). The Cerithium and Littorina species, apart from being tested for cadmium and mercury separately, were also tested for the interaction of cadmium and mercury, and the Littorina species for salinity too. Table I shows the pollutants to which each species was subjected.

Table I

The pollutants to which each species was subjected

POLLUTANTS	SPECIES					
	Monodonta		Littorina		Cerithium	
	turbinata	turbiformis	punctata	neritoides	scabridum	rupestre
$CuCl_2$	+	+				
$ZnCl_2$	+	+				
$PbCl_2$	+	+				
$HgCl_2$	+	+	+	+	+	+
$CdCl_2$	+	+	+	+	+	+
$HgCl_2 + CdCl_2$			+	+	+	+
detergent	+	+			+	+
detergent + oil	+	+			+	+
salinity 8.5‰			+	+		

3. RESULTS

Table II presents the results obtained in previous studies, for genetic variability of the species tested. The species M. turbiformis, L. neritoides and, C. scabridum are the species showing a higher genetic variability than their cogeneric partners distinguished by three genetic criteria, namely, mean number of alleles per locus (A), proportion of polymorphic loci per population (P) and mean heterozygozity per locus per individual (H).

Table II

Genetic variability of the species observed

Parameter/ species	Number of loci tested	Genetic indices A	P	H	Niche breadth	Reference
M. turbiformis	26	2.025	0.571	0.019	Broad	Lavie and Nevo, 1981
M. turbinata	26	1.965	0.417	0.078	Narrow	Lavie and Nevo, 1981
L. neritoides	17	2.681	1.000	0.142	Broad	Noy et al., (in press)
L. punctata	17	1.745	0.529	0.028	Narrow	Noy et al., (in press)
C. scalridum	25	2.680	0.760	0.190	Broad	Lavie and Nevo, 1986
C. rupestre	25	1.480	0.200	0.030	Narrow	Lavie and Nevo, 1986

Table III presents the survivorship results for the two Monodonta species. All 20 comparisons show significantly higher survivorship for the more polymorphic species M. turbiformis. The fact that the number of concentrations varied from one pollutant to another reflects the fact that for most pollutants the concentration ranges were not the same for both species (Nevo et al., 1983) and only concentrations shared by both species are presented in Table III. On all occasions that the concentration ranges were not identical for the two species, M. turbimala had the lower concentration range, because for M. turbinata even a low concentration rendered LD50 in less than a week. Survival for both species is calculated on the day that one species passed LD50. For each pollutant the most impressive difference in survivorship was presented diagramatically in Fig.1. Similarly, L. neritoides proved significantly more resistant than L. punctata and C. scabridum proved significantly more resistant than C. rupestre.

Table III

Survivorship results for the two Monodonta species.

Pollutant	Concentration (mg/l)	Days until LD50	Survival M. turbiformis	M. turbinata
Hg	2	3	0.7282	0.3434*
	3	4	0.2667	0.2369
Cu	12	3	0.7755	0.4050
	16	4	0.5980	0.1948
	20	2	0.8750	0.3750*
Zn	600	2	0.6733	0.1578*
Pb	600	2	0.6122	0.3594*
	1000	4	0.4375	0.2738
Cd	25	7	0.1818	0.0244
	26	6	0.5439	0.3030
	28	5	0.7961	0.4444
	30	5	0.7792	0.3917
	32	5	0.8810	0.4298
	34	5	0.5918	0.3526
	35	3	1.0000	0.3793*
Nonionic detergent, Marlophen 89	4	2	0.6567	0.2167*
	7	2	0.4369	0.3056
Crude oil and 7mg/l detergent	7	3	0.9417	0.3750
	10	2	0.8837	0.1268*
	14	2	0.7652	0.4493

* Data presented in Fig.1

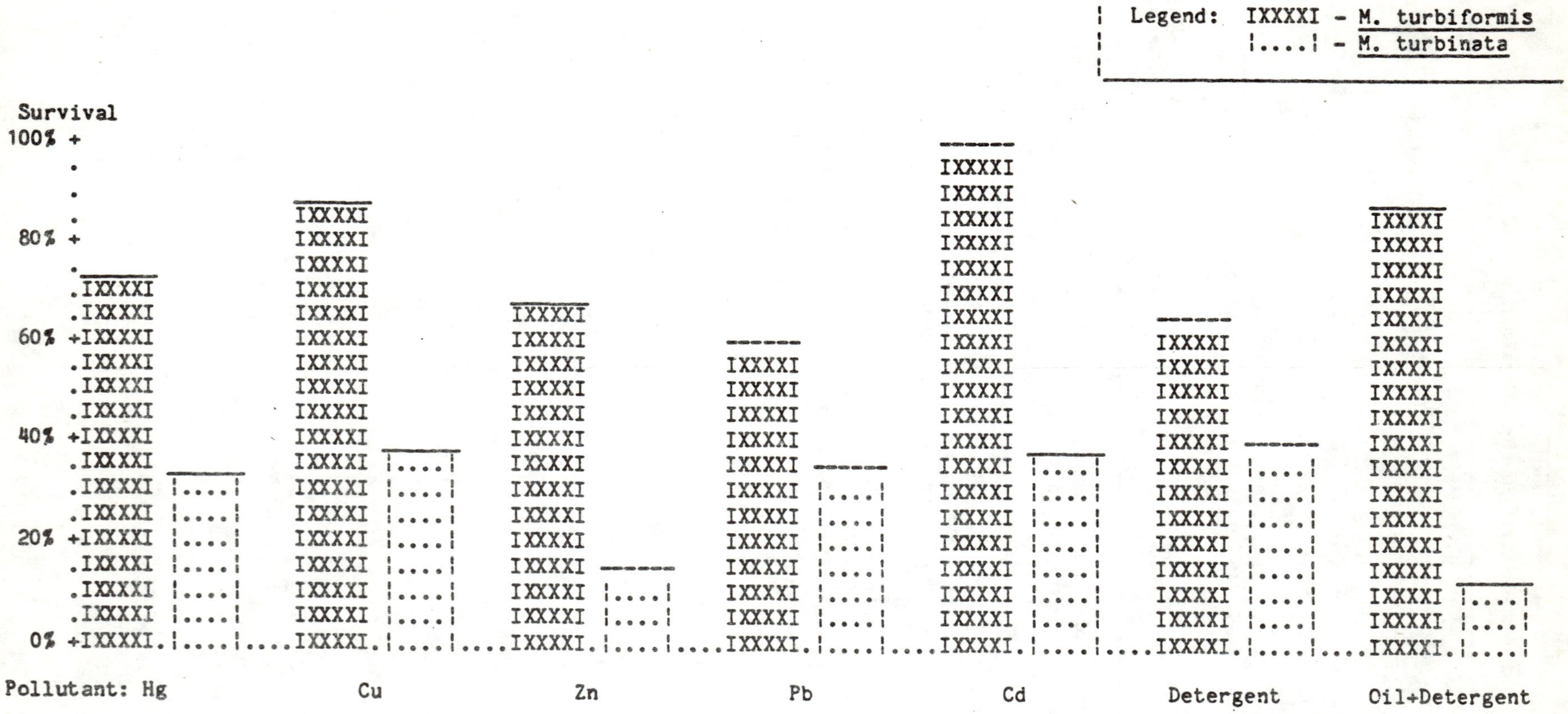

- 178 -

```
Survival                                    Legend:  IXXXXI - M. turbiformis
100% +                                               !....! - M. turbinata
      .
      .
      .
 80% +                                                 IXXXXI
      .                                                IXXXXI
      .                                                IXXXXI
      .                .IXXXXI                         IXXXXI
 60% +IXXXXI           .IXXXXI           IXXXXI        IXXXXI      IXXXXI
      .IXXXXI          .IXXXXI           IXXXXI        IXXXXI      IXXXXI
      .IXXXXI          .IXXXXI           IXXXXI        IXXXXI      IXXXXI
      .IXXXXI          .IXXXXI           IXXXXI        IXXXXI      IXXXXI
 40% +IXXXXI           .IXXXXI           IXXXXI        IXXXXI      IXXXXI
      .IXXXXI   :....: .IXXXXI           IXXXXI :....: IXXXXI      IXXXXI
      .IXXXXI   :....: .IXXXXI           IXXXXI :....: IXXXXI :...: IXXXXI
      .IXXXXI   :....: .IXXXXI           IXXXXI :....: IXXXXI :...: IXXXXI
 20% +IXXXXI   :....: .IXXXXI           IXXXXI :....: IXXXXI :...: IXXXXI
      .IXXXXI :....: .IXXXXI :...:      IXXXXI :....: IXXXXI :...: IXXXXI
      .IXXXXI :....: .IXXXXI :...:      IXXXXI :....: IXXXXI :...: IXXXXI
      .IXXXXI :....: .IXXXXI :...:      IXXXXI :....: IXXXXI :...: IXXXXI :...:
  0% +IXXXXI.:....:.IXXXXI.:...:...IXXXXI.:....:IXXXXI.:...:IXXXXI.:...:

Pollutant: Hg       Cu        Zn        Pb        Cd        Detergent  Oil+Detergent
```

Figure 1. Comparison of survivorship between the two species of Monodonta

Table IV presents the survivorship results for the two Littorina species. All 14 comparisons including both Akko and Shikhmona populations show higher survivorship for the more polymorphic species. The experiment was terminated after 4 days even in cases when no species reached LD50. For each pollutant the most impressive difference in survivorship was presented diagrammatically in Figure 2.

Table IV

Survivorship results for the two Littorina species

Pollutant	Concentration (mg/l)	Site	Days until LD50	Survival L. neritoides	L. punctata
Hg	1	Akko	3	.94	.24
		Shikmona	4	.82	.14
	10	Akko	3	.92	.11*
		Shikmona	3	.90	.47
Cd	5	Akko	2	.38	.22
		Shikmona	4	.84	.56
	20	Akko	1	.33	.0
		Shikmona	2	.94	.45*
Hg + Cd	Hg 1	Akko	4	.90	.84
	Cd 5	Shikmona	4	1.00	.42
	Hg 10	Akko	3	.78	.0*
	Cd 20	Shikmona	3	.72	.47
8.5% NaCl		Akko	3	.94	.16*
		Shikmona	2	.84	.20

* The results are presented diagrammatically in Fig.2

```
          Legend:  IXXXXI - L. neritoides
                   |....| - L. punctata

Survival
100% +
      .IXXXI         IXXXI                      IXXXI
      .IXXXI         IXXXI                      IXXXI
 80% +IXXXI          IXXXI                      IXXXI
      .IXXXI         IXXXI          IXXXI       IXXXI
      .IXXXI         IXXXI          IXXXI       IXXXI
      .IXXXI         IXXXI          IXXXI       IXXXI
 60% +IXXXI          IXXXI          IXXXI       IXXXI
      .IXXXI         IXXXI          IXXXI       IXXXI
      .IXXXI         IXXXI          IXXXI       IXXXI
      .IXXXI         IXXXI  ------  IXXXI       IXXXI
 40% +IXXXI          IXXXI |....|   IXXXI       IXXXI
      .IXXXI         IXXXI |....|   IXXXI       IXXXI
      .IXXXI         IXXXI |....|   IXXXI       IXXXI
 20% +IXXXI          IXXXI |....|   IXXXI       IXXXI
      .IXXXI         IXXXI |....|   IXXXI       IXXXI ------
      .IXXXI ------  IXXXI |....|   IXXXI       IXXXI |....|
      .IXXXI |....|  IXXXI |....|   IXXXI       IXXXI |....|
  0% +IXXXI.|....|...IXXXI.|....|...IXXXXI.|----|....IXXXI.|....|

Pollutant: Hg          Cd           Hg+Cd          NaCl 5%
```

Figure 2. Comparison of survivorship between the two species of Littorina.

Table V presents the survivorship results for the two _Cerithium_ species. All 10 comparisons show higher survivorship for the more polymorphic species.

Table V

Survivorship results for the two _Cerithium_ species

Pollutant	Concentration (mg/l)	Days until LD50	Survival _C. scabridum_	_C. rupestre_
Hg	3	7	.68	.31*
Cd	15	7	.54	.50
	25	6	.64	.46*
Hg + Cd	15 + 15	1	.35	.34
	20 + 25	1	.25	.20*
Nonionic detergent, Marlophen 89	65	1	.38	.34
	70	1	.42	.40
	75	2	.72	.48*
Crude oil + detergent 65mg/l	10	1	.84	.24*
	20	1	.45	.40

* The results are presented diagrammatically in Fig.3

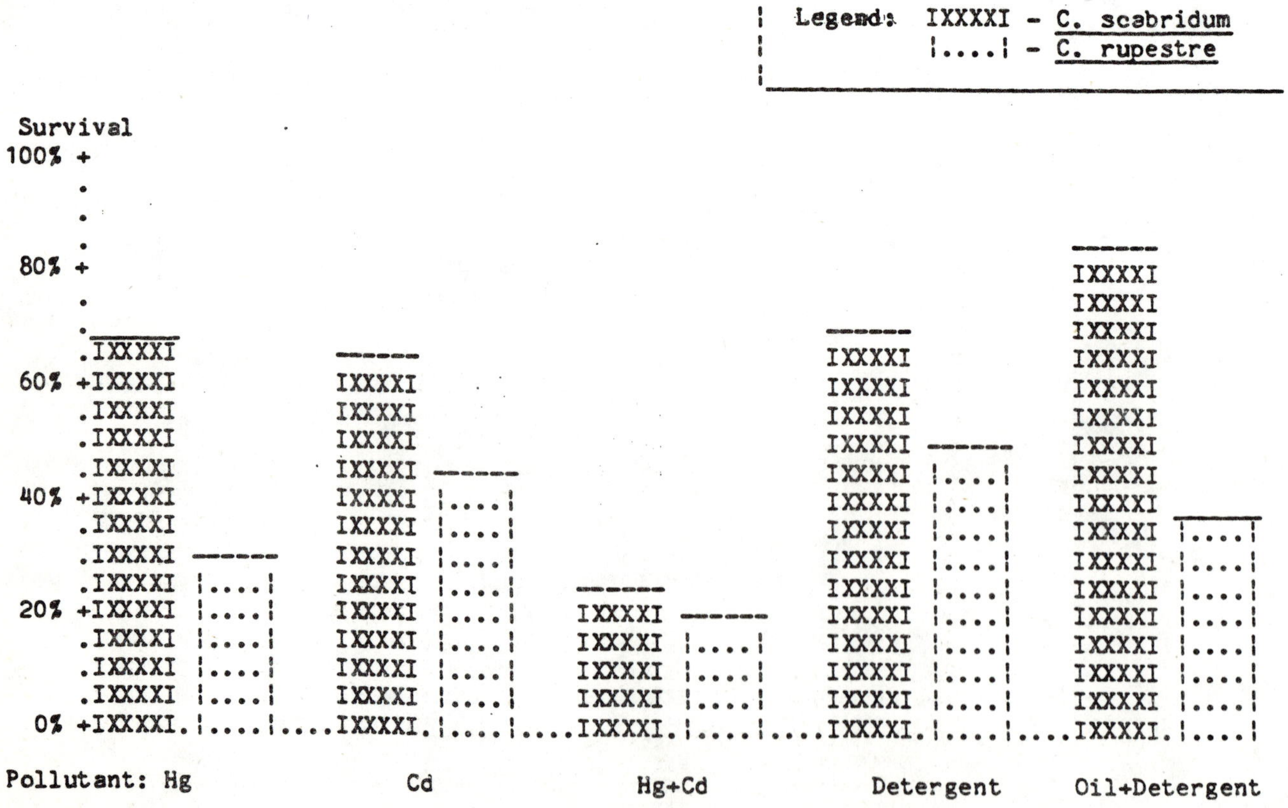

Legend: IXXXXI - _C. scabridum_
         I....I - _C. rupestre_

Figure 3. Comparison of survivorship between the two species of _Cerithium_.

4. DISCUSSION

The results presented here have both theoretical and practical implications. Theoretically, the species with higher levels of genetic diversity were more resistant to pollution <u>in the three pairs of species tested here</u>. Therefore, we hypothesize that this may represent a general pattern which is in accordance with the niche width variation hypothesis. The latter states that species exposed to broader niche widths harbor more genetic diversity enabling them to cope with the wider ecological spectra they experience. The idea of positive correlation between genetic and environmental variation, the niche-width-variation hypothesis, is widespread in evolutionary biology. It was tested critically and confirmed for karyotypic, electrophoretic and quantitative traits. It was also supported by allozyme studies comparing habitat specialist and generalist species and by life history analyses. Theoretically, the existence of a protected polymorphism is more likely in more heterogeneous environments. Likewise, models with spatial heterogeneity and limited migration may be more effective in maintaining genetic polymorphism than alternative models with less environmental heterogeneity and higher migration (see Nevo, 1983 and Nevo <u>et al.</u>, 1984 for references supporting all these theoretical claims).

The three species pairs tested here each comprises one narrow niche species characterized by a low level of genetic diversity and one broad niche species characterized by a higher level of genic diversity, corroborating the niche width variation hypothesis. It is noteworthy, that the broad niches are characterized by higher levels of ecological uncertainty.

The three species <u>M. turbiformis</u>, <u>L. neritoides</u> and <u>C. scabridum</u> are exposed to wider ranges of fluctuations of salinity, temperature and oxygen content. Their counterpart species experience narrower ranges of fluctuations of these and possible other parameters. The present study clearly demonstrates that the level of genetic diversity is also related to the range of pollution effects. The broad niche species were more resistant to diverse pollutants than the narrow niche species. Since pollution is part of the ecological background, the niche-width-variation hypothesis applies also to pollution ranges.

Practically, we have earlier shown (reviewed in Nevo <u>et al.</u> 1983) that a monitoring system based on changes in population genetic structure is not only theoretically commendable but practically feasible. This conclusion is supported by the rapid and precise genetic changes that follow the effect of pollution. Ideally, using in situ populations of different marine species, specific genetic loci, either singly or in combination, may prove sensitive to different pollutants and would be easily assayed by relatively quick and easy electrophoretic tests. Such direct genetic monitoring may become an indispensable tool, sometimes preferable (Luoma, 1977) to the chemical ones used routinely (Toribara <u>et al.</u>, 1977) as it alerts us to both the short and long-term genetic changes that populations undergo before their final extermination as a result of pollution.

We demonstrated here that the search for optimal marine species as promising potential genetic monitors should relate to their levels of genetic diversity. The higher the level of genetic diversity in a marine species, the higher the probability that it will provide a good genetic monitor for pollution. This is because it consists of diverse genotypic arrays from which <u>specific allozyme monitors</u> could be chosen for detecting the effects of specific pollutants.

5. ACKNOWLEDGEMENTS

We thank Shimeon Simson for field and laboratory assistance and wish to extend our deep gratitude to FAO/UNEP, to the Israel Discount Bank Chair of Evolutionary Biology and to the "Ancell-Teicher" Research Foundation for Genetics and Molecular Evolution, established by Florence and Theodore Baumritter of New York, for financial support.

6. REFERENCES

Lavie, B. and E. Nevo, Genetic diversity in marine molluscs: A test of the niche-width
1981   variation hypothesis. Mar.Ecol., 2:335-42

_____, Genetic diversity of marine gastropods: contrasting strategies of
1986   Cerithium rupestre and Cerithium scabridum. Mar.Ecol., 28:99-103

Luoma, S.N., Detection of trace contaminant effects in aquatic ecosystems
1977   J.Fish.Res.Board Can., 34(3):436-9

Nevo, E., Population genetics and ecology: The interface. In Evolution from molecules
1983   to men, edited by D.S. Bendall. Cambridge, U.K., Cambridge University Press, pp.287-321

Nevo, E., A. Beiles and R. Ben-Shlomo, The evolutionary significance of genetic
1984   diversity: ecological, demographic and life history correlates. In Evolutionary dynamics of genetic diversity, edited by G.S. Mani. Lect.Notes Biomath., 53:13-213

Nevo, E., B. Lavie and R. Ben-Shlomo, Selection of allelic isozyme polymorphisms in
1983   marine organisms: pattern, theory and application. In Isozymes: current topics in biological and medical research. Vol.10. Genetics and evolution, edited by M.C. Rattazi, J.G. Scandalias and G.S. Whitt, New York. Alan R. Liss, Inc., pp.69-92.

Noy, R., B. Lavie and E. Nevo, The niche-width-variation hypothesis revisted: Genetic diversity in the marine gastropods Littorina punctata and Littorina neritoides (in press)

Toribara, T.Y., et al., (eds). Environmental pollutants. New York, Plenum Press, 500 p.
1977

COMPARISON OF THE EFFECTS OF ORGANIC POLLUTION AND POLLUTION BY MINING
WASTE IN GREEK WATERS

by

A. NICOLAIDOU*, A. ZENETOS**, C. BOGDANOS** and D. PAPADOPOULOS*

*Zoological Laboratory, University of Athens, Panepistimiopolis,
GR-15784 Athens, Greece
**National Centre for Marine Research, GR-16604 Hellinikon, Greece

1. INTRODUCTION

There are various industrial solid wastes which are dumped into the sea as an economic method of disposal. These have received very little attention with regard to disturbance of marine benthic communities. The effects on the bottom fauna of china clay deposits have been studied in Cornwall by Howell and Shelton (1970) and Probert (1975). The impact of fly ash on benthos off the Northumberland coast was studied by Bamber (1983). Bourcier (1969), Bourcier and Zibrowius (1973) and Blackman and Wilson (1973) have studied the effects of "red mud" - residual waste from the extraction of aluminium from bauxite. Finally, brief mention of biological effects of mine waste disposal is made by Goyette and Nelson (1977) for an area in British Columbia and by Wong et al., (1978) for an area in Hong Kong.

Conversely, numerous papers have been published on the biological effects of organic enrichment in the marine environment. With reference to benthic communities, Pérès and Bellan (1972) summarise the impact of organic pollution on benthic populations and Pearson and Rosenberg (1978) examine the succession in macrobenthic communities in relation to organic enrichment of the marine environment.

The aim of this paper is to compare the little known effects of mine tailing with those produced by organic pollution. Comparison is made on the basis of the community structure and diversity and the similarities of the fauna. Data collected at two sites are considered. One off Larymna (Evoikos Gulf), the dumping site of metalliferous residue from the processing of laterite to extract nickel and the other in Elefsis Bay (Saronikos Gulf), close to the sewer outfall of Athens.

2. MATERIALS AND METHODS

The benthos off Larymna was studied as part of the assessment of the environmental impact of dumping in the area. In this paper six stations were taken into consideration. They were chosen to include two polluted stations (Station 5 in the centre of the spoil patch and Station 9 at the border of it), two clean stations (2 and 3) with similar sediment size composition as the polluted ones and two clean stations (7 and 8) with finer sediments. Two samples were taken at each station with a 0.1 m^2 van Veen grab.

In Elefsis Bay the samples were taken at four stations at increased distances from the outfall using a Ponar grab sampling 0.045 m^2 of the bottom.

In both cases the samples were sieved through a 1 mm sieve, stained with Rose Bengal and preserved in 4% formalin.

The sediment analysis was carried out by the geological laboratory of the National Centre for Marine Research, Athens; the pipette method was used and the nomenclature is after Folk (1954). The diversity of the fauna was calculated by the Shannon-Wiener diversity index (Shannon and Weaver, 1963). The ordinations were performed by the programs PRINC for Principal Components Analysis and CORRES for the Correpondence Analysis, from the computer library of the Marine Biological Association of the U.K., where the computing was carried out. In the analyses, species found only at one station and in numbers lower than three individuals were excluded from the calculations.

3. RESULTS

The depth and sediment characteristics of the sampling stations are shown in Table I which also summarises the number of species, number of individuals and diversity at each station. In Larymna, 159 species were found. The smaller number of species, 19, were found in the clean Station 7 with muddy sediment, and the highest in the clean sandy Stations 2 and 3. The last two stations also had the greatest number of individuals, 1595 and 1625 individuals/m^2 respectively. The smallest number of individuals, 295 individuals/m^2, was found at Station 8. The diversity was highest at the clean sandy stations, followed closely by Station 9 at the edge of the polluted patch. Lowest diversity was observed at the polluted sandy Station 5 and the clean but muddy Station 7.

Table I

Characteristics of sampling stations and communities

Stations	Depth (m)	Sediment type (after Folk)	% sand	No. of species*	No. of individuals per m^2	Diversity H'	Evenness
2	58	muddy sand	89	81	1,595	5.503	0.895
3	51	muddy sand	62	81	1,625	5.291	0.860
5	75	sand	95	25	970	2.461	0.546
9	80	muddy sand	61	33	370	4.056	0.829
7	84	mud	1	19	430	2.497	0.606
8	85	mud	3	28	295	4.169	0.894
E	36	sandy silt	33	6	2,926	0.926	0.358
F	31	sandy silt	47	24	12,342	2.551	0.557
M	20	sandy silt	27	61	6,127	3.124	0.727
N	90	sandy silt	27	56	1,430	5.172	0.887

* The unit area for this column is $0.2m^2$ for Larymna and $0.09m^2$ for Elefsis.

It is clear that in this area, the diversity does not decrease in relation to pollution but is also related to particle size of the sediment. This can be seen in Fig.1, where the diversity is plotted against the percentage sand (coarse fraction) in the sediment.

The abundances of individual species remained generally low, mostly below 50 individuals/m^2 over the whole sampling area of Larymna. The greatest abundance was that of a cirratulid polychaete at Station 7 (200 individuals/m^2) and the bivalve Kellyella miliaris (590 individuals/m^2) at Station 5. The last species, in the present survey, was only found in the polluted Stations 5 and 9 as was the bivalve Leptaxinus ferruginosus. No other striking differences were observed between the fauna of the polluted and clean stations. Some species were not found in the polluted stations but where they were present elsewhere they were not found in large numbers.

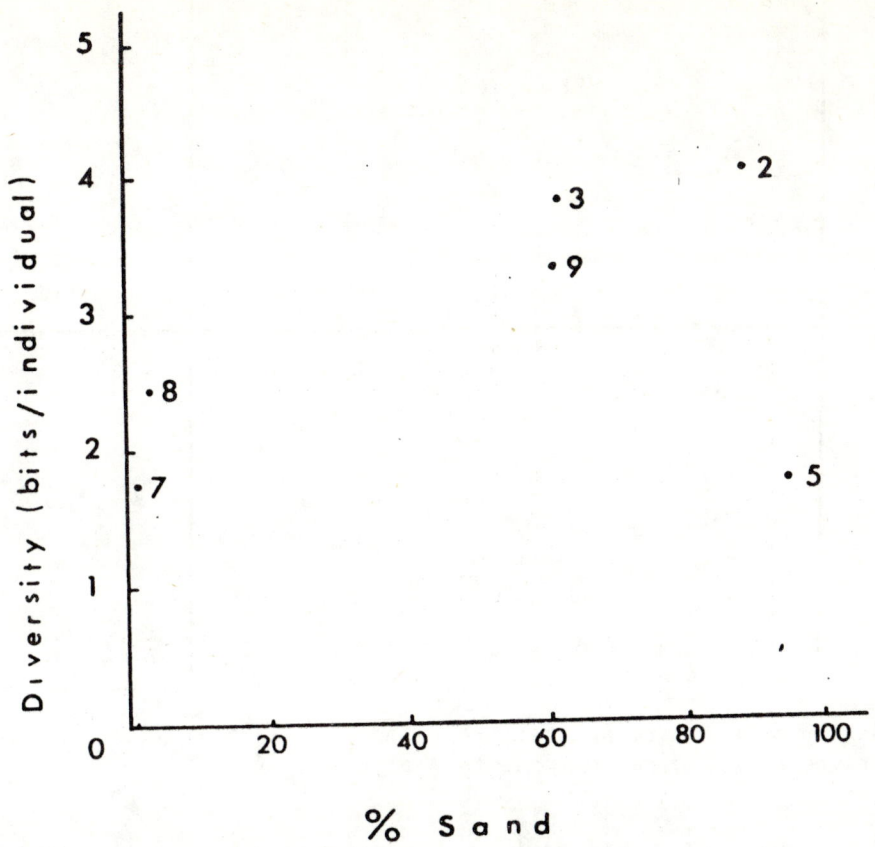

Figure 1. Relation between diversity and percentage of sand in the sediment for the Larymna stations.

In the area of Elefsis 105 species were found. The station closest to the outfall had only 6 species while the largest number, 61 species, was observed at the shallowest Station M. The diversity increased with increased distance from the source of pollution and ranged from 0.9 at Station E to 5.2 at the furthest Station N. The maximum number of individuals (12,342 individuals/m^2) was at Station F immediately after the highly polluted zone. A lot of species are missing from the polluted stations while others are found there either exclusively or in very large numbers. Such species are Capitella capitata with maximum density of 689 individuals/m^2, Cirratulus cirratus (4,344 individuals/m^2), Polydora antennata (2,266 individuals/m^2), Prionospio malmgreni (1,266 individuals/m^2), Scolelepis fuliginosa (2,378 individuals/m^2) and S. girardi (2,700 individuals/m^2). Most of the above species have been described by Bellan (1967) as indicators of organic pollution.

As diversity index was not sufficient to separate the polluted from the unpolluted stations in Larymna some ordination techniques were employed. The principal component analysis - R mode - was used for the stations of Larymna. The position of the stations on Axes 1 and 2 and 1 and 3 are shown in Fig.2a and 2b respectively. As the three first axes included 83% of the variability, further axes were not considered. When the first two axes are used the polluted coarse Stations 5 and 9 are grouped together with the clean but fine Stations 7 and 8, probably due to the much larger number of species of the coarse clean Stations 2 and 3 (Fig.2a). However, when the first and third axes are plotted there is a definite distinction between coarse, fine and polluted stations (Fig.2b).

When the same analysis was applied to both sets of data from Larymna and Elefsis, the stations were ordinated according to their species richness whichever combination of the three axes was used. (Plot of stations on Axes 1 and 2 is shown in Fig.3).

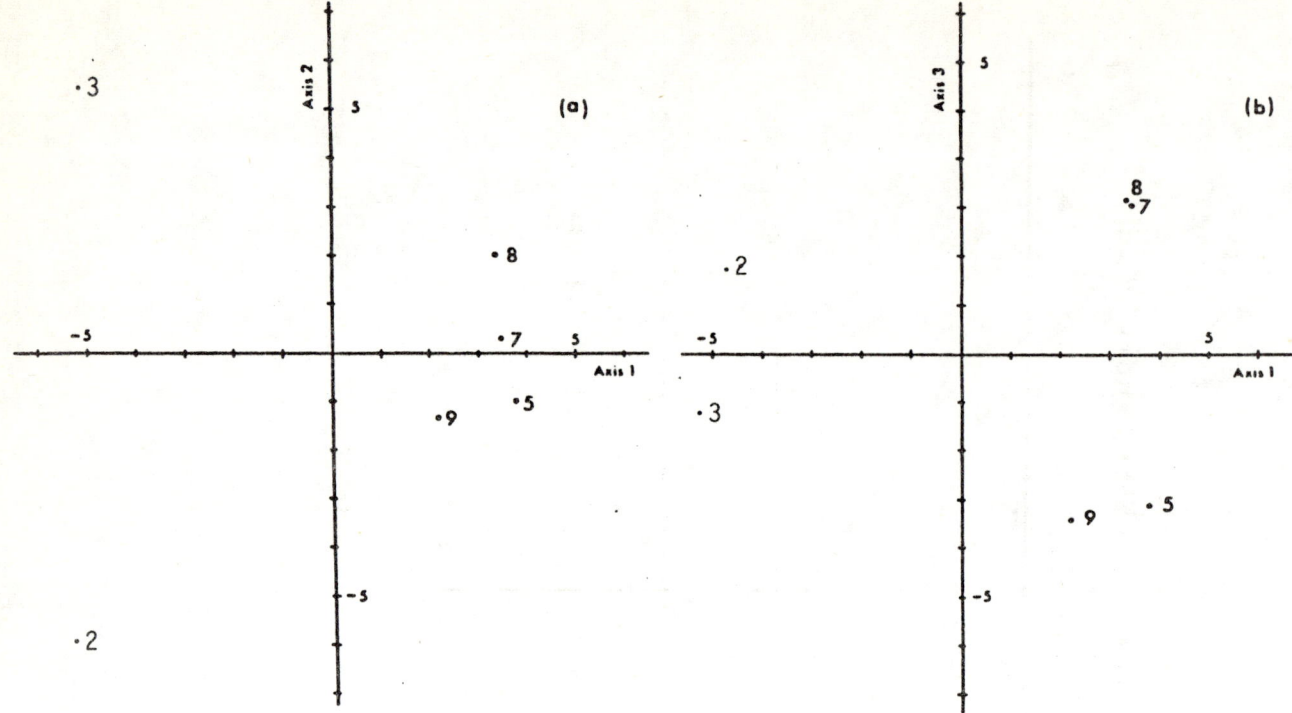

Figure 2. Position of the Larymna stations on Axes 1 and 2(a) and 1 and 3(b) derived by Principal Components Analysis.

Correspondence analysis applied to the same data separated clearly the highly polluted stations of Elefsis Bay along Axis 1. Although on Axis 2 there appears no clear distinction between the Larymna stations, on Axis 3 there is a sequence of coarse and polluted stations at lower values on the axes, followed by coarse and clean stations, with the fine clean stations having highest scores on this axis.

Fig.4a and b shows the position of stations along Axes 1 and 2 and Axes 1 and 3 respectively. The high scores of Stations E and F on Axis 1 are due to the high densities of some species occurring exclusively in high abundance in those stations. This can be seen from the plotting of the species scores on the same axes 1 and 3 in Fig.5. For example the species Scolelepis fuliginosa (No.57) has a density of 2,378 individuals/m^2 at Station 5 and 78 individuals/m^2 at Station F. Scolelepis girardi (No.56) with 2,700 individuals/m^2 appears only at Station F. Capitella capitata (No.14) appears only at Stations E and F with respective densities of 444 and 689 individuals/m^2.

4. DISCUSSION

Although there were differences in the physical environment as well as in the samplers used at each site, the data obtained seemed sufficient for at least a preliminary comparison between the effects of the two types of pollution. It appears that the effects of organic pollution on the benthic fauna are the most dramatic. The site of Elefsis showed more or less the characteristic zonation described by Bellan (1967) with the highly polluted, nearly azoic, area close to the pollution source (Station E), a polluted zone with high numbers of individuals and few species (Station F) and the cleaner zone at Stations M and N. The pollution effects in the area may become more catastrophic later in the summer when anoxic conditions occur in the Bay (Zarkanellas, 1979).

It is believed that, unlike organic pollution, dumping of mine residue has a mechanical effect on the benthic animals, either directly killing them or making the environment so unstable that few species are able to survive (Carter, 1975).

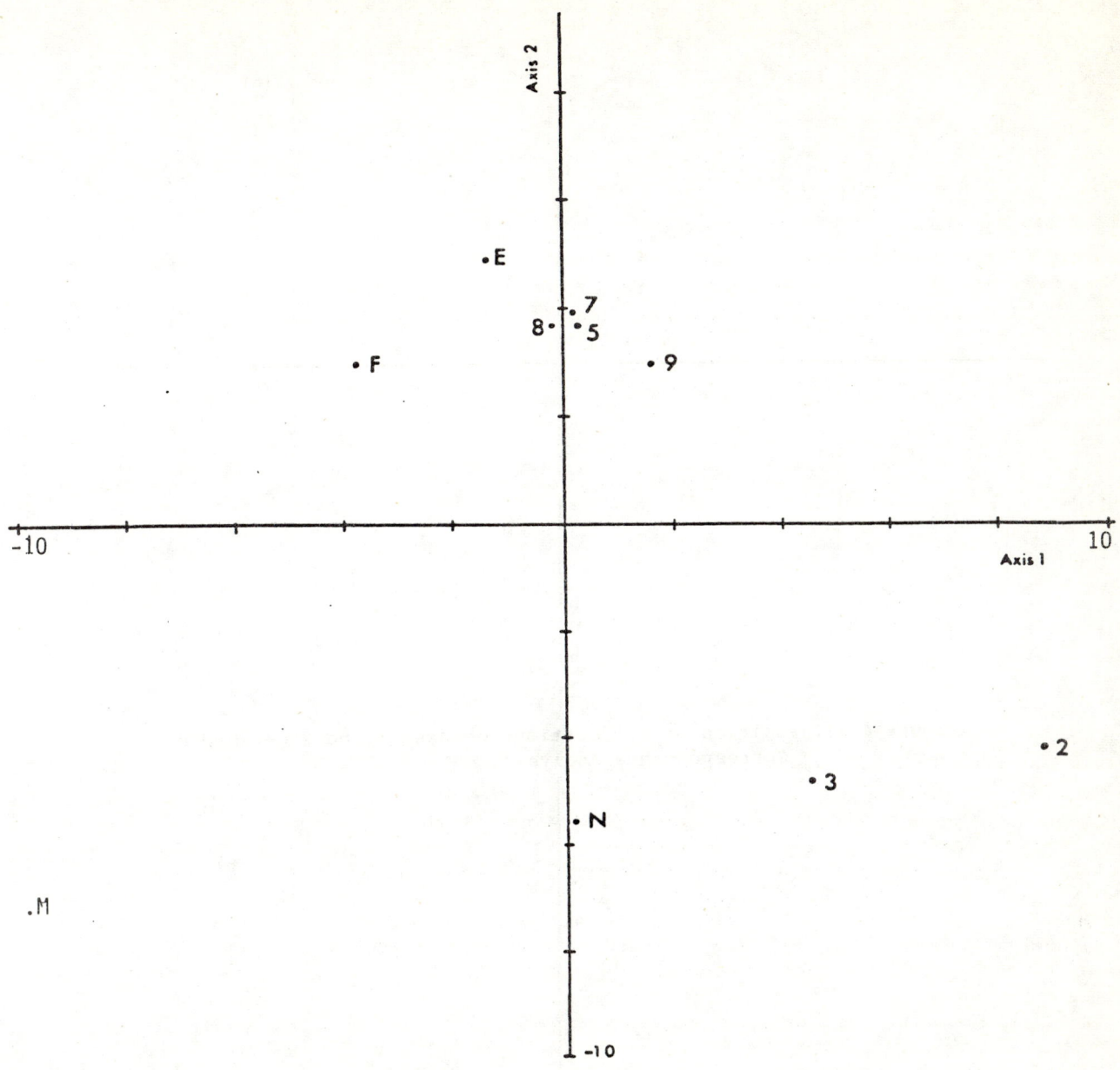

Figure 3. Position of all stations on Axes 1 and 2, derived by Principal Components Analysis.

No completely azoic area was found here, not even at Station 5 at the centre of the spoil ground. At the time of sampling a thin layer of fine sediment was observed over the dumped material and this seemed sufficient to support some animal life. Evans et al., (1973) mention that in Rupert Inlet, Canada, the benthos was completely obliterated, only where the mine waste amounted to 50cm or more. Besides, it is known that some species have the ability to return to the surface after burial and polychaetes in general (Goyette and Nelson, 1977) are more resistant to burial than other benthic organisms. Presumably an azoic area would be found here too if the sampling happened to take place immediately or very soon after dumping on the spot.

In the polluted stations of Elefsis there is an increase in the abundance of opportunist species such as Capitella, Polydora and Scolelepis, which are typical of organically enriched areas (Bellan, 1967; Pearson and Rosenberg, 1978). However, their presence, at least of Capitella, in most polluted areas is not due to their better tolerance but to their better adaptive strategy (r-selection) (Gray, 1981). On the

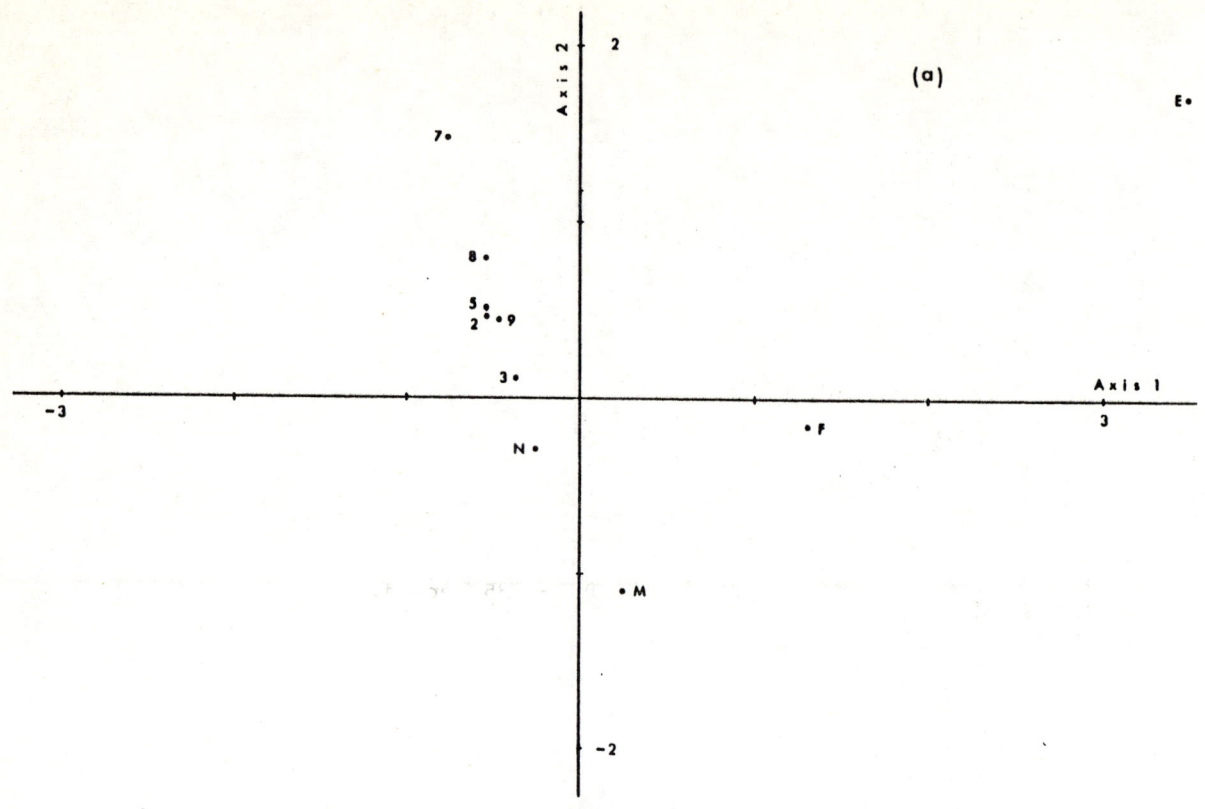

Figure 4(a). Position of all stations on Axes 1 and 2 derived by Correspondance Analysis.

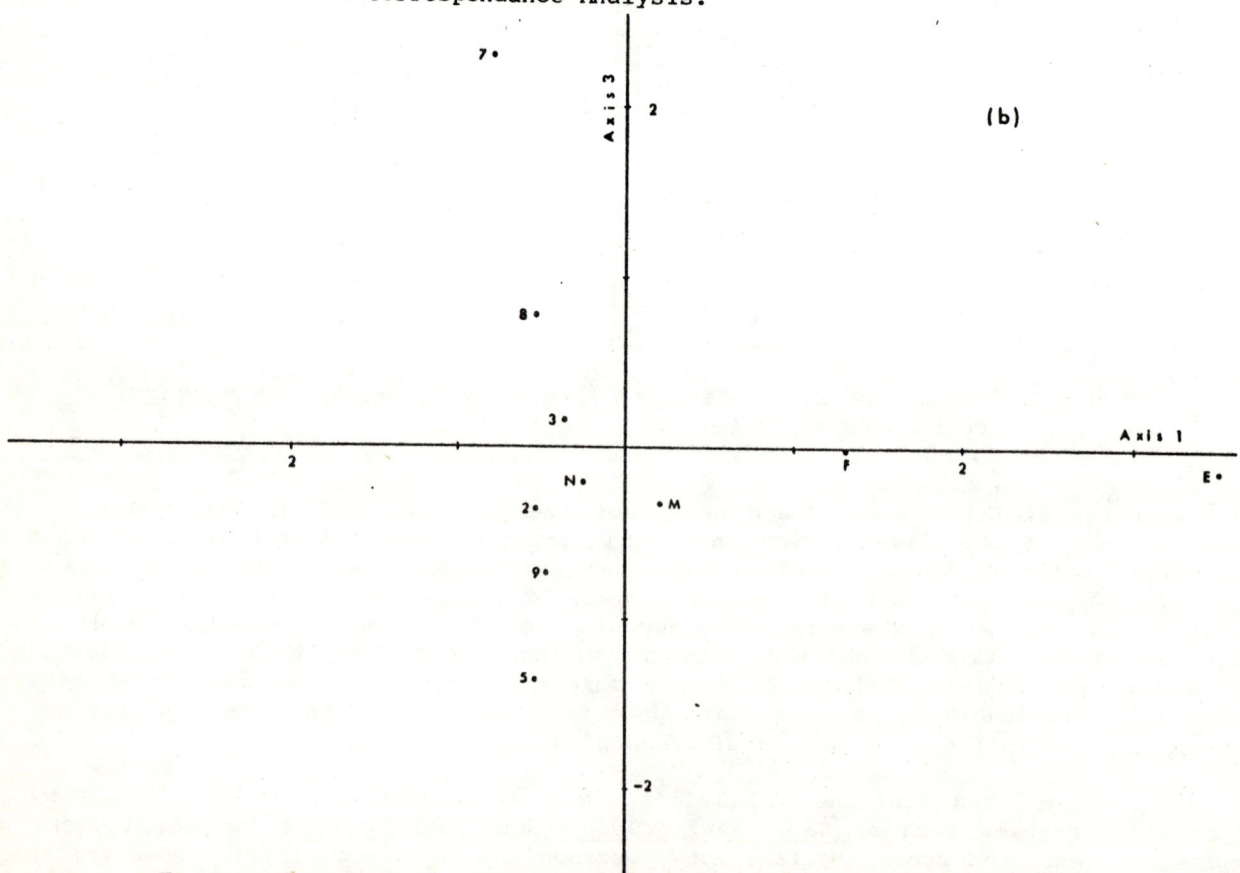

Figure 4(b). Position of all stations on Axes 1 and 3 derived by Correspondence Analysis.

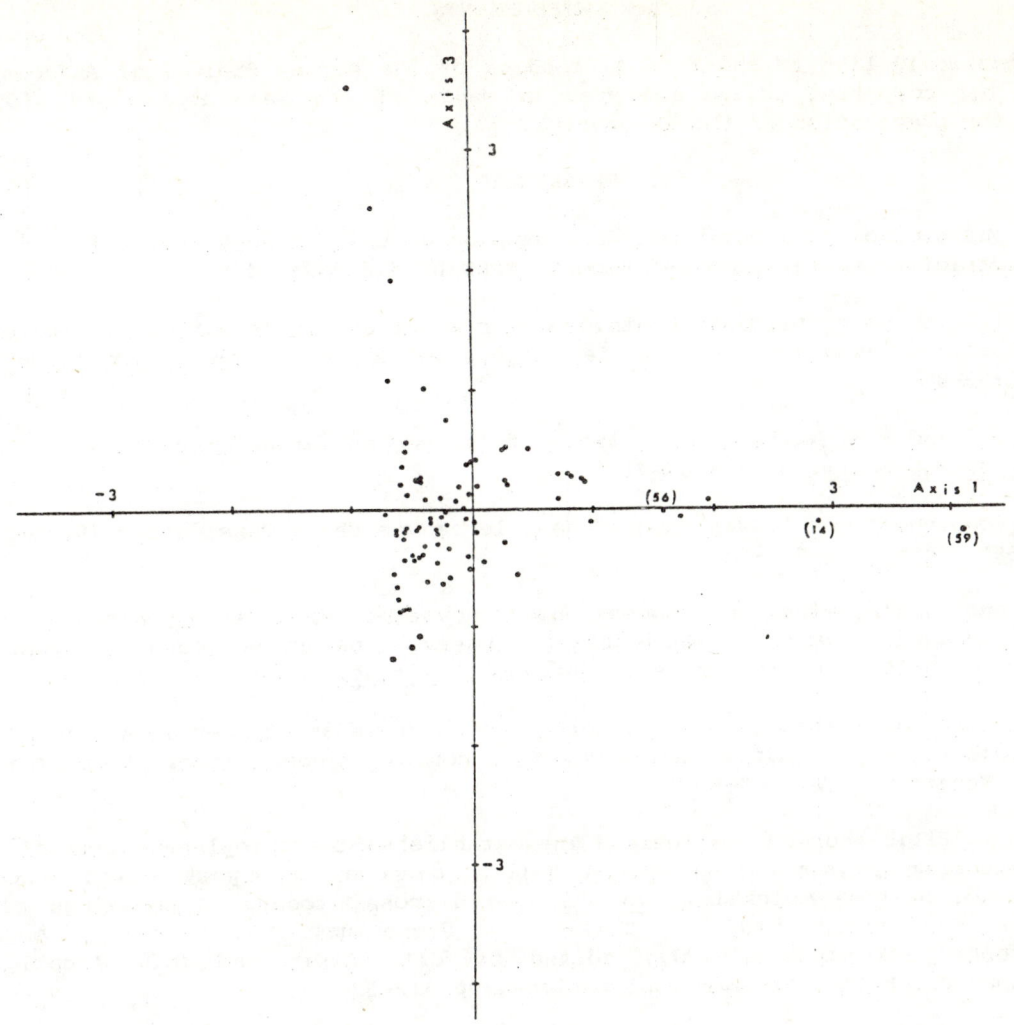

Figure 5. Position of species along Axes 1 and 3 derived by Correspondence Analysis - (explanation of numbers in text).

contrary it is thought that the otherwise rare species *Leptaxinus ferruginosus* and *Kellyella miliaris* of the order *Lucinacea* (van Aartsen, pers.comm.), found in increased numbers in the polluted stations in Larymna, may actually be favoured by the iron in the dumped material. It is known from the literature (Tebble, 1966) that *L. ferruginosus* accumulates grains of iron on its shell. There is no specific mention for *K. miliaris* but as it belongs to the same order of which other members are known to collect iron grains, it may be inferred that its high abundance in the area is not accidental.

Diversity indices have been recommended by Stirn (1981) as a good measure for the detection and assessment of all forms of marine pollution. The same author stresses, however, that comparisons should be made over a uniform environment. In principle then, diversity indices should not be applied to detect pollution by solids, as dumping of particles of size and composition different from the native sediment, would change the environmental conditions. As in the present case, pollution may even cause an increase in diversity if coarse material is dumped over fine sediments. (The importance of sediment size in community structure and diversity is reviewed by Gray, 1974). This make the very broad definition of pollution by Patrick (see Stirn, 1981) as "anything which brings about a reduction in the diversity of aquatic life and eventually destroys its balance", inaccurate. Multivariate analysis has been often used in community ecology (Gauch, 1982) to delimit communities and to describe trends in them, but it is not used very much in pollution studies. Here, ordination has proved a useful tool, supplementing diversity, in separating polluted and unpolluted stations, as well as the two different types of pollution.

5. ACKNOWLEDGEMENTS

The authors would like to thank Dr L. Maddock of the Marine Biological Association, Plymouth, for her computing advice and Mrs. L. Mavin of the same laboratory, for her assistance in the preparation of the manuscript.

6. REFERENCES

Bamber, R.N., The benthos of a marine fly ash dumping ground. London, Central
1983   Electricity Generating Board Report (TPRD/L/2414/R82):14 p.

Bellan, G., Pollution et peuplements benthiques sur substrat meuble dans la region de
1967   Marseille. Première partie. Le secteur de Cortiou. Rev.Int.Océanogr.Méd., 6-7:53-87

Blackman, R.A.A., and K.W. Wilson, The effects of red mud on marine animals.
1973   Mar.Pollut.Bull., 4(11):169-71

Bourcier, M., Ecoulement des "boues rouges" dans le canyon de la Cassidaigne (Décembre
1969   1968). Téthys, 1:779-82

Bourcier, M., and H. Zibrowius, Les "boues rouges" déversées dans le canyon de la
1973   Cassidaigne (region de Marseille). Observations en soucoupe plongeante SP 350 (juin 1971) et résultats de dragages. Téthys, 4 (4):811-40

Carter, L., Disposal of wastes to tidal waters. In Minerals and the environment:
1975   Conference proceedings, edited by M.J. Jones. London, Institution of Mining and Metallurgy, pp.149-57

Evans, J.B., D.V. Ellis and C.A. Pelletier, The establishment and implementation of a
1973   monitoring program for underwater tailing disposal in Rupert Inlet, Vancouver Island, British Colombia. In Tailing disposal today. Proceedings of the First International Tailing Symposium, Tucson, Arizona, October 31-November 3, 1971, edited by C.L. Alpin and G.O. Argall, San Francisco, Miller Freeman Publications, pp.512-52

Folk, R.L., Distinction between grain size and mineral composition in sedimentary rock
1954   nomenclature. J.Geol., 62:334-59

Gauch, H.G. Jr., Multivariate analysis in community ecology. Cambridge, U.K., Cambridge
1982   University Press, Cambridge studies in ecology, 298 p.

Goyette, D., and H. Nelson, Marine environmental assessment of mine waste disposal into
1977   Rupert Inlet, British Colombia. Fisheries and Environment Canada. Environmental Protection Service, Surveillance Report, Report EPS PR-77-11

Gray, J.S., Animal-sediment relationships. Oceanogr.Mar.Biol., 12:223-61
1974

_____, The ecology of marine sediments. An introduction to the structure and
1981   function of benthic communities. Cambridge, U.K., Cambridge University Press, Cambridge studies in modern biology, 2:185 p.

Howell, B.R. and R.G.J. Shelton, The effect of china clay on the bottom fauna of
1970   St. Austell and Mevagissey Bays. J.Mar.Biol.Assoc.U.K., 50:593-607

Pearson, T.H. and R. Rosenberg, Macrobenthic succession in relation to organic
1978   enrichment and pollution of the marine environment. Oceanogr.Mar.Biol. 16:229-311

Pérès, J.M., and G. Bellan, Apercu sur l'influence des pollutions sur les peuplements
1972   benthiques. In Marine pollution and sea life, edited by M. Ruivo. West Byfleet, Surrey, U.K., Fishing News Books for FAO, pp.375-86

Probert, P.K., The bottom fauna of china clay waste deposits in Mevagissey Bay.
1975   J.Mar.Biol.Assoc.U.K., 55(1):19-44

Shannon, C.E., and W. Weaver, The mathematical theory of communication. Urbana,
1963   Illinois, University of Illinois Press, 117 p.

Stirn, J., Manual of methods in aquatic environmental research. Part 8. Ecological
1981   assessment of pollution effects. (Guidelines for the FAO(GFCM)/UNEP Joint Coordinated Project on Pollution in the Mediterranean). FAO Fish.Tech.Pap.,(209):70 p.   Issued also in French

Tebble, N., British bivalve sea-shells, a handbook for identification. London Trustees
1966   of the British Museum (Natural History),

Wong, M.H., K.C. Chan and C.K. Choy, The effect of the iron ore tailings on the coastal
1978   environment of Tolo Harbour, Hong Kong. Environ.Res., 15:342-56

Zarkanellas, A.J., The effects of pollution induced oxygen deficiency on the benthos in
1979   Elefsis Bay, Greece. Mar.Environ.Res., 2(3):191-207

ETUDE DE LA DISTRIBUTION ET DE LA PHENOLOGIE DE
L'ANGIOSPERME MARINE Posidonia oceanica

par

P. PANAYOTIDIS et A.V. CATSIKI
Centre National des Recherches Marines
Aghios Kosmas
166 04 Hellinikon
Athènes, Grèce

1. INTRODUCTION

L'étude de l'impact de la pollution sur les herbiers de Posidonia oceanica (L.) Delile dans le golfe Saronikos (Mer Egée, Grèce) a commencé recemment par le Centre National des Recherches Marines dans le cadre de l'activité "I" de la composante recherche de MED POL - Phase II.

Cette recherche comporte l'étude de la distribution des herbiers et de la phénologie de la phanérogame marine P. oceanica à l'aide d'une méthodologie élaborée dans les années 1975-1985 par les chercheurs des laboratoires méditerranéens de biologie marine et surtout du laboratoire de Biologie Vegétale de Marseille-Luminy, de la Stazione Zoologica di Napoli et du Centre National des Recherches Marines, Athènes.

2. OBJECTIF DE LA RECHERCHE

Nos observations sont étendues dans l'ensemble du golfe, mais les comptages et mesures s'effectuent sur les herbiers peu profonds de 4 à 6 mètres, ce qui correspond à la frange supérieure de l'herbier. Le choix de la frange supérieure comme objet principal d'étude est imposé par le fait que cette portion de l'herbier est soumise directement aux activités humaines et semble la plus fragile du point de vue écologique (Pérès et Picard, 1975). La pêche au "gangui", l'ancrage, les modifications du rivage par les constructions humaines et la pollution urbaine et industrielle sont les causes principales de la dégradation des herbiers superficiels, (Wittman, 1984). A ces phénomenes se superposent des procedés naturels tels que l'érosion, la compétition entre biocénoses et des phénomènes du cosmos (tâches solaires, etc.) qui à notre avis, influencent l'avancement ou le recul de P. oceanica probablement par l'intermédiaire de la fréquence de la floraison.

L'objectif de notre travail est de décrire l'état actuel de la frange supérieure de l'herbier dans le golfe et tenter par la suite de distinguer la part de l'impact dûe aux moyennes mécaniques (pêche, ancrage, erosion) et la part causée par la pollution proprement dite (urbaine, chimique, etc.). Etant donné que la densité des faisceaux au mètre carré et la surface foliaire est affectée par les deux catégories de nuisances, nous avons appliqué une méthodologie basée sur des paramètres morphologiques et physiologiques de la plante afin d'individualiser la deuxième catégorie de nuisances sur l'herbier. Par ailleurs l'étude des herbiers dans des secteurs qui ne sont pas touchés par les activités humaines offrent une estimation de la variabilité naturelle du milieu marin.

3. DEFINITIONS ET METHODOLOGIE PROPOSEE

L'estimation de la densité de l'herbier comporte des comptages en plongée, l'unité de comptage étant le faisceau (Figure 1). Les comptages s'effectuent sur une surface de 400 cm^2 définie par un cadre de 20 cm x 20 cm posé de place en place dans l'herbier étudié.

Parallèlement aux comptages en plongée, l'opérateur note la position des faisceaux (verticale ou horizontale) et la présence ou absence de matte (1). Ces critères permettent de définir trois types d'herbier (Meinesz et al., 1981):

---

(1) matte: enchévauchement de racines et de tiges de P. oceanica pouvant atteindre plusieurs mètres d'épaisseur, issus de la croissance verticale de la plante et dont seules quelques disaines de centimètres superficielles sont vivantes.

- le type 1 correspond à un herbier de faible densité (recouvrement inférieur à 50%, dont les rhizomes sont en position horizontale).

- le type 2 correspond à un herbier de densité élevée (recouvrement supérieur à 50%), dont les rhizomes, en position verticale ne constituent pas encore une matte importante.

- le type 3 correspond à un herbier de densité très élevée (recouvrement supérieur à 75%) dont les rhizomes en position verticale, constituent une matte épaisse (supérieure à 20 cm).

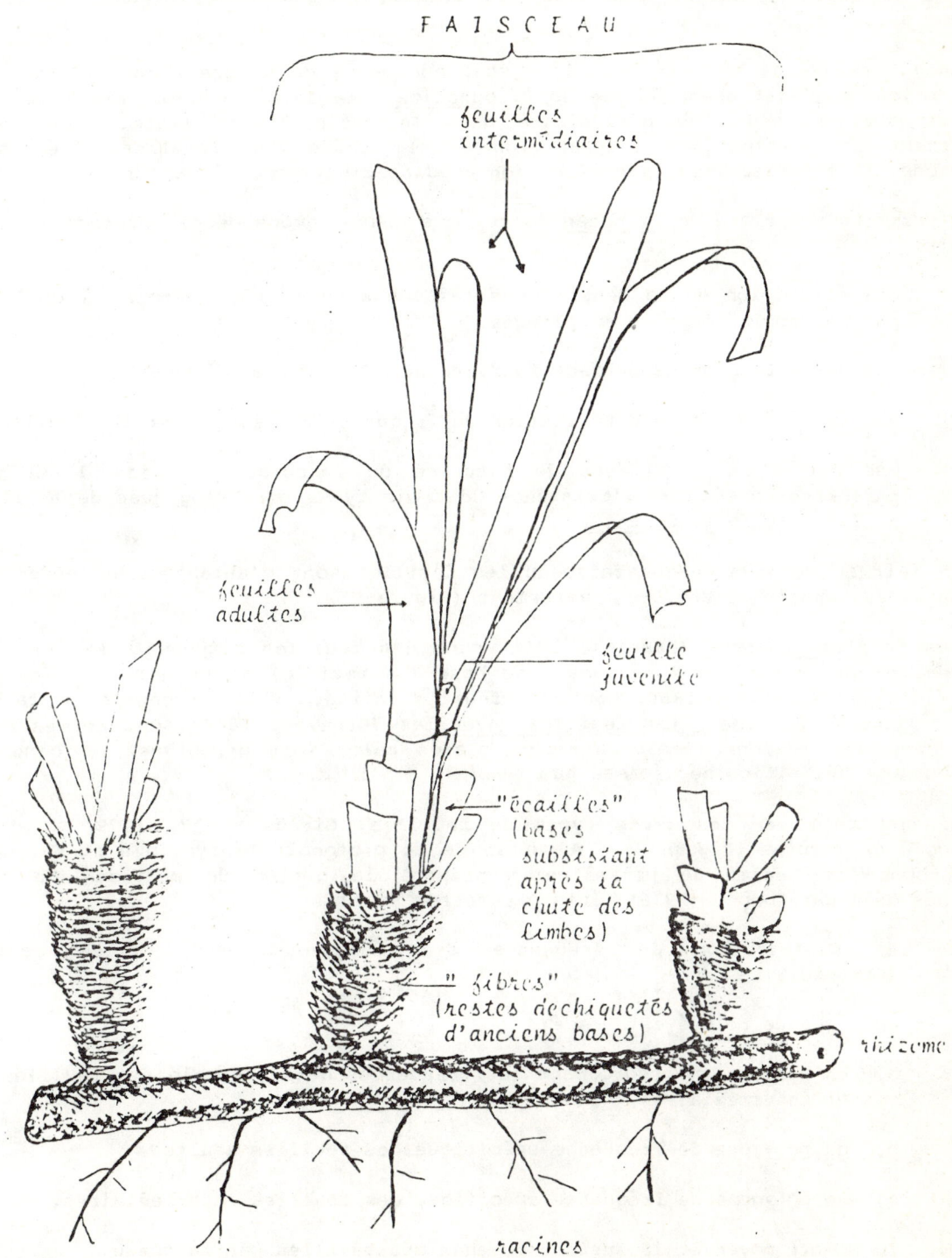

Figure 1. Représentation schématique d'un faisceau de Posidonia oceanica x 3/4 d'après Panayotidis, 1980.

Les prélèvements des faisceaux s'effectuent également en plongée, à l'aide du même cadre de 20 cm x 20 cm. Les faisceaux sont prélevés avec rhizomes et racines.

Un élément important dont il ne faut pas sousestimer lors du prélèvement des faisceaux de P. oceanica, est la microstructure de l'herbier. En effet les études de Panayotidis et al. (1981) sur les problèmes d'échantillonage des faisceaux ont montré l'existence au sein d'un herbier donné, sur de très courtes distances des "tâches" (patchiness) de faisceaux, présentant des différences phénologiques, au moins dans certaines phases de leur cycle. Il est donc nécessaire, pour tenir compte de la microstructure de l'herbier, de prélever plusieurs lots (au moins quatre) de faisceaux espacés de quelques mètres les unes des autres, chaque lot provenant d'une surface 400 cm^2 sur le fond.

Chaque relevé de 400 cm^2 fait l'objet d'une série de mesures, qui ont comme but la déscription de l'état phénologique de P. oceanica. La terme phénologie est défini comme l'étude des phénomènes de périodicité dans le cycle d'une plante. Ces phénomènes concernent la germination, l'apparition et la chûte des feuilles, l'éclosion des bourgeons, la floraison, la fructification et dissémination des diaspores.

A partir d'un relevé de P. oceanica tel que nous l'avons décrit précédemment on peut obtenir :

a) une estimation de la densité des faisceaux au m^2 plus précise à celle obtenue par un simple comptage en plongée.

b) une estimation de la surface foliaire au m^2 et de la biomasse.

c) une estimation de la distribution de fréquence des longueurs des feuilles.

d) une estimation de l'état de maturité du faisceau à l'aide d'une série de paramètres basés sur l'existence de trois types morphologiques de feuilles dans un faisceau de P. oceanica.

En fait au sein de chaque faisceau les feuilles sont d'autant plus jeunes qu'elles sont situées vers le centre, et inversement (Figure 2).

Les feuilles adultes (Grenier, 1860) sont des feuilles situées à la périphérie du faisceau et présentent deux régions: une base (ou petiole) et un limbre. Les feuilles intermédiaires (Grenier, 1860) sont des feuilles situées vers le centre du faisceau et sont dépourvues de base. Les feuilles juvéniles (Grenier, 1860) sont celles du centre même du faisceau, sont incolores ou vert très pâle, sont dépourvues de base et leur longueur par définition ne dépasse pas 50 mm.

Tenant compte de ces trois types de feuilles, Giraud (1977, 1979) a proposé un protocole de mesures (Figure 3). A partir de ce protocole Panayotidis et Giraud (1981) ont proposé une série de paramètres exprimant la notion de maturité du faisceau, utilisée déjà par Cooper (1976). Ces paramètres sont:

1. La distribution de fréquence des longueurs de feuilles adultes et intermédiaires.

   Pour chaque relevé est tracé:

   a) un histogramme regroupant les fréquences de l'ensemble des feuilles adultes et intermédiaires,

   b) un polygone de fréquence spécifique des feuilles adultes,

   c) un polygone de fréquence spécifique des feuilles intermédiaires.

2. Le nombre moyen et la surface moyenne des feuilles par faisceau.

3. Les indices R_n et R_s.

L'indice Rn est défini comme la moyenne de la somme des rapports du nombre des feuilles adultes sur le nombre des feuilles adultes plus celui des intermédiaires:

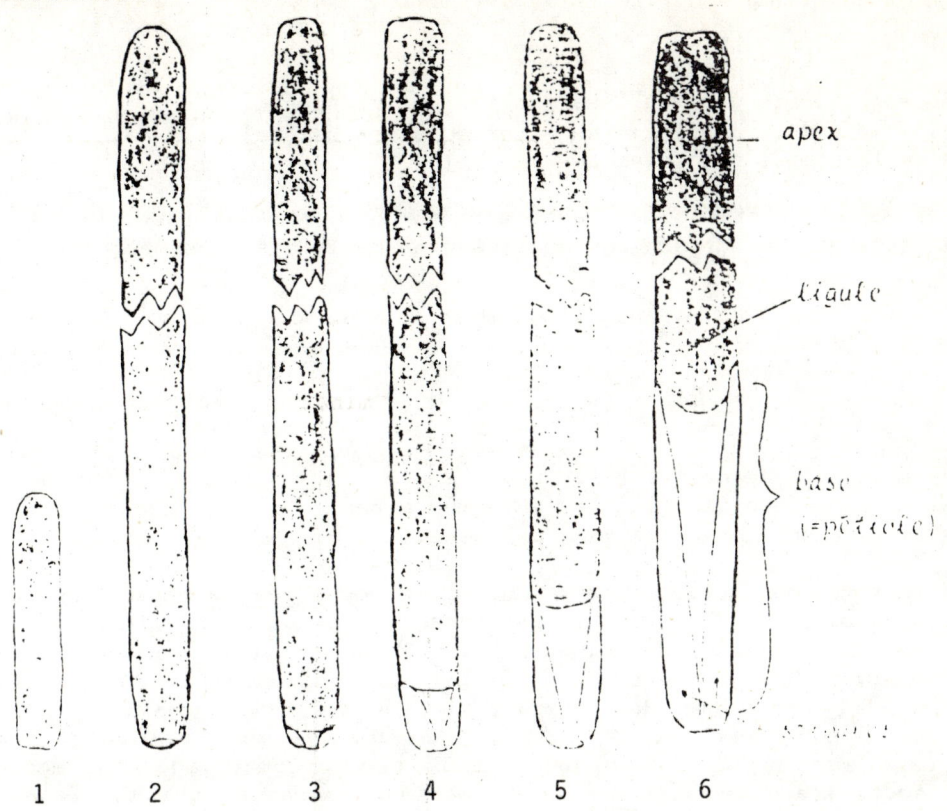

Figure 2. Passage d'une feuille du stade juvénile (1) au stade intermédiaire (2, 3) et adulte (4, 5, 6). Dimensions naturelles d'après Giraud, 1977

Date : Lieu : Profondeur : Faisceau N° :	Dimensions en mm Entre paranthèses : Feuilles entières			Remarques particulières
TYRE de feuille :	Longueur de la base	Longueur totale	Longueur moyenne	
ADULTES				
INTERMEDIAIRES				
JUVENILES				

Figure 3. Fiche de mesures sur un faisceau. D'après Panayotides, 1980

$$R_n = \frac{\sum \frac{na}{na+ni}}{N}$$

ou na = le nombre des feuilles adultes par faisceau
    ni = le nombre des feuilles intermédiaires par faisceau
    N = le nombre des faisceaux par relevé

L'indice Rs est défini comme la moyenne de la somme des rapports de la surface des feuilles adultes sur la surface des adultes plus celle des intermédiaires:

$$R_s = \frac{\sum \frac{Sa}{Sa+Si}}{N}$$

ou Sa = la surface des feuilles adultes
    Si = la surface des feuilles intermédiaires
    N = le nombre des faisceaux par relevé

Les différentes catégories de feuilles de P. oceanica expriment un niveau différent d'activité métabolique. Il est donc intéressant d'étudier parallèlement aux phénomènes de croissance, le phénomène de bioaccumulation de certains polluants par les feuilles adultes et intermédiaires aussi bien que par les rhizomes et racines de P. oceanica, pour éclaircir le problème de l'impact de la pollution chimique (par exemple la pollution par les métaux lourds) sur la physiologie de la plante. Un point très important dans le cas de dosage des métaux est le nettoyage des échantillons afin d'éliminer toute trace de sédiment et de matière étrangère qui pourraient faucher les dosages. Les écailles des feuilles mortes sont éloignées des tiges; tous les organes et spécialement les feuilles sont dispensées des épiphytes à l'aide d'une spatule de P.V.C.

Les tissus ainsi recuperés sont lyophilisés et homogéneisés (Bernhard, 1976). Après la digestion de 0,5 gr. de tissus par 5 ml d'acide nitrique dans des bombes teflon pendant deux heures aux 150° C on réalise le dosage des métaux par spéctrophotométrie d'absorption atomique sans flamme.

4. RESULTATS ET DISCUSSION

Les résultats obtenus durant les premiers mois d'étude des herbiers dans 5 stations du golfe Saronikos (Figure 4) n'ont qu'un caractère préliminaire. Néanmoins ces résultats nous permettent de comparer les herbiers du golfe Saronikos à ceux du golfe de Marseille et de Napoli qui ont été étudiés suivant une méthodologie equivalente.

4.1 Répartition des herbiers dans le golfe Saronikos et densité des faisceaux au m^2

D'après nos observations les herbiers couvrent de grandes surfaces de la côte du golfe Saronikos, surtout dans sa partie extérieure, limitée par la presqu'île de Méthana, l'île Aegina et les parages de l'Attique, de Pirée au Cap Sounio (secteur 1 sur la carte). Plus à l'ouest entre l'île Aegina et les îles Angistri et Métopi, une grande surface du fond est couverte de matte morte et de larges intermattes couvertes d'un peuplement de remplacement (Giaccone, 1974) à Caulerpa prolifera, Gracilaria verrucosa et Ulvacés accompagnés par Cymodocea nodosa et Halophila stipulacea, cette dernière étant beaucoup plus commune que l'on croyait recémment (Hartog, 1970; Lipkin, 1975). Sur cette partie du golfe (secteur 2 sur la carte) existent encore par localités des herbiers très denses. Enfin tout le long de la côte ouest du golfe, de l'île Salamis à l'isthme de Corinthe et delà au port d'Epidaure (secteur 3 sur la carte), P. oceanica est rare. Sur cette partie de la côte du golfe Saronikos la végétation dominante sont des herbiers mixtes à Cymodocea nodosa, Caulerpa prolifera et Halophila stipulacea.

Sur les côtes du secteur 1 et le long d'un axe nord-sud nous avons choisi cinq herbiers de référence sur lesquels nous avons l'intention de faire des mesures à long terme. Ces herbiers sont du stade 1 de Meinesz et al. (stations Ag. Kosmas et Vayia) ou stade 3 (stations Méthana et Métopi) et du stade intermédiaire 2 (stations Perdica et Ag. Marina).

La présence dans les herbiers de Perdica, d'Ag. Marina, de Métopi et de Méthana (herbiers des stades 2 et 3) d'un grand nombre de faisceaux à croissance horizontale, coexistant avec une matte d'épaisseur supérieure à 20 cm montre que les herbiers subissent les conséquences des activités humaines, mais sont encore à mesure de régénerer.

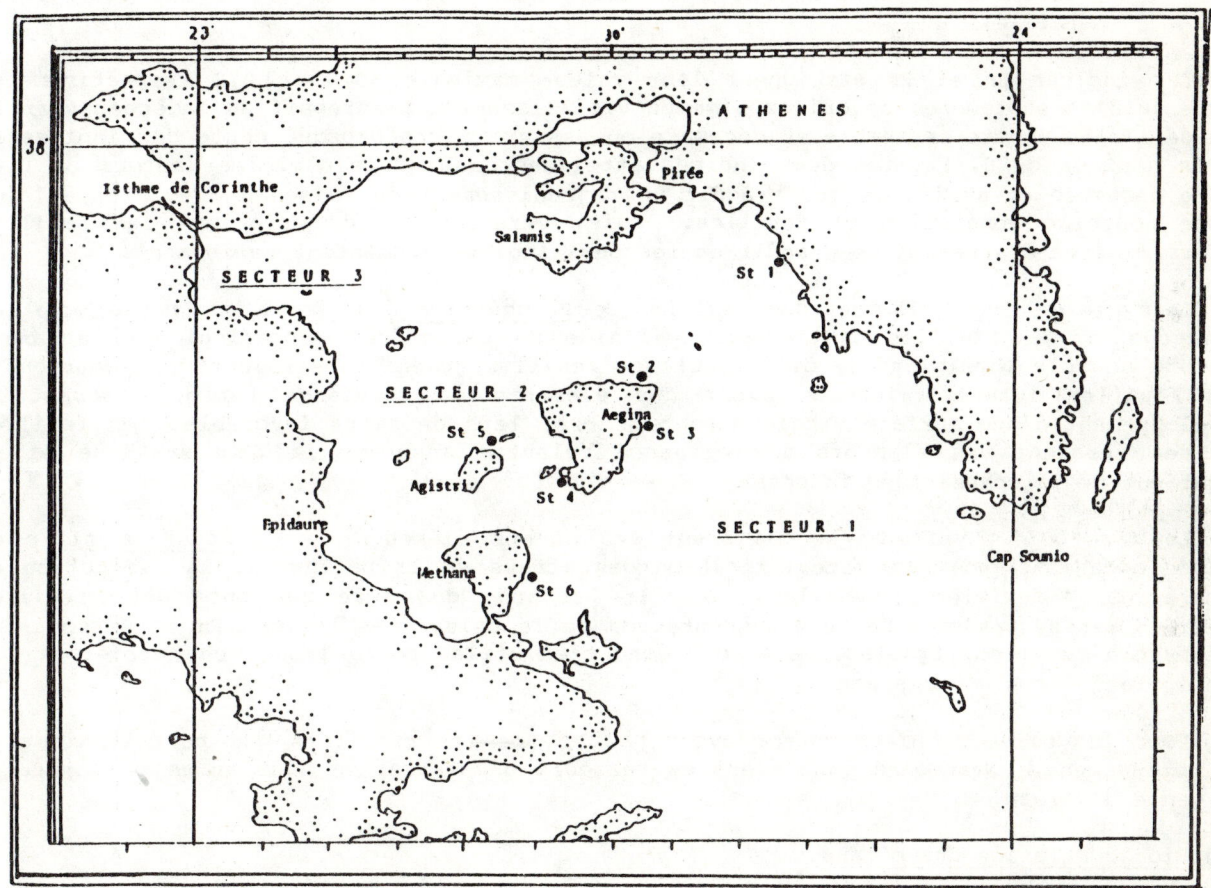

Figure 4. Stations étudiées: St. 1 = Ag. Kosmas, St. 2 = Vayia, St. 3 = Ag. Marina, St. 4 = Perdica, St. 5 = Metopi, St. 6 = Methana.

4.2 Surface foliaire et biomasse

Les mesures de la surface foliaire de P. oceanica en m^2/m^2 (indice foliaire) dans les cinq herbiers de référence du golfe varient de 19 (Méthana) à 10 (Ag. Kosmas) avec une médiane de 12 pour les stations de l'île Aegina. Ces mesures qui correspondent à l'aspect printanier de l'herbier sont nettement inférieures du max de 59 m^2/m^2, mesuré par Giraud, 1977; par contre elles sont proches aux valeurs données par Ott (1980) à Ischia.

En ce qui concerne la biomasse qui correspond à nos mesures, elle est de l'ordre de 600-700 gr dw/m^2.

4.3 Distribution des fréquences des longueurs des feuilles. Indices Rn et Rs

L'étude des histogrammes de fréquence des différentes longueurs des feuilles de P. oceanica dans le golfe Saronikos a confirmé le modèle du cycle de renouvellement des feuilles proposé par Panayotidis et Giraud (1981) (Figure 5). La réproductivilité des phénomènes saisoniers nous permet de faire l'analyse suivante. En été (septembre) le polygone des fréquences des feuilles intermédiaires se trouve toujours à gauche (classes de faible taille). En automne (décembre) la disposition des deux polygones est la même, mais l'écart entre eux est plus faible.

- Au début du printemps (avril), la superposition des deux polygones amorcée en hiver est maintenant complète.

- Vers la fin du printemps (juin), le polygone des feuilles intermédiaires se trouve maintenant à droite de celui des feuilles adultes (classes de forte taille).

Les indices Rn et Rs atteignent leur valeur maximale (0,6 et 0,8 respectivement) entre juin et septembre, ce qui exprime un vieillissement progressif du faisceau vers la fin de l'été. Entre septembre et décembre on assiste à une brusque chute des indices Rn et Rs (autour de 0,5). Les deux indices atteignent leur valeur minime (autour de 0,4) entre décembre et avril, ce qui traduit un rajeunissement du faisceau avec l'apparition d'une nouvelle génération de feuilles. Puis, vers la fin du printemps les feuilles intermédiaires s'arrêtent de grandir et les deux indices augmentent rapidement.

Le cycle de renouvellement des feuilles de P. oceanica nous permet de penser que les 3 catégories de feuilles distinguées à l'aide de caractères morphologiques, auraient chacune un rôle physiologique différent: les feuilles juvéniles seraient des "embryons", les feuilles intermédiaires seraient des éléments à croissance rapide assurant le développement d'une surface vitale durant un bref laps de temps favorable; les feuilles adultes seraient des éléments à croissance ralentie assurant la photosynthèse et le maintient de la cohésion du faisceau.

Le caractère physiologique différent des divers éléments de P. oceanica offre de bonnes occasions pour une matérialisation des effets de la pollution: la diminution du nombre des feuilles juvéniles, ou le retard du développement des feuilles intermédiaires, traduits par une augmentation de la valeur des indices Rn et Rs sont des effets clairs et chiffrables. Ils sont donc très utiles pour classer un herbier et par extrapolation son environnement.

Pour le golfe Saronikos nous n'avons pas encore les résultats qui permettraient un tel classement. Néanmoins dans d'autres secteurs d'observation nous avons pu constater ce genre d'effets.

4.4 Bioaccumulation des métaux

Les effets de la pollution sur le métabolisme de P. oceanica ont été constatés dans les herbiers qui se trouvent souvent dans la nappe polluée de la zone industrielle et de l'émissaire central des égouts d'Athènes, par le jeu des facteurs hydrologiques (carte 1 - secteur 1, parages de l'Attique).

Dans ce secteur la concentration des ions mercuriques dans les tissus était 0,642 ppm (feuilles intermédiairaes), 0,289 (feuilles adultes), 0,157 (rhizomes) et 0,165 (racines). Augier et al. (1977), trouve également dans des secteurs similaires une concentration de mercure dans les feuilles quatre fois plus forte que dans les racines rhizomes, tandis que dans les secteurs propres les feuilles accumulent le mercure au même niveau que les rhizomes et racines.

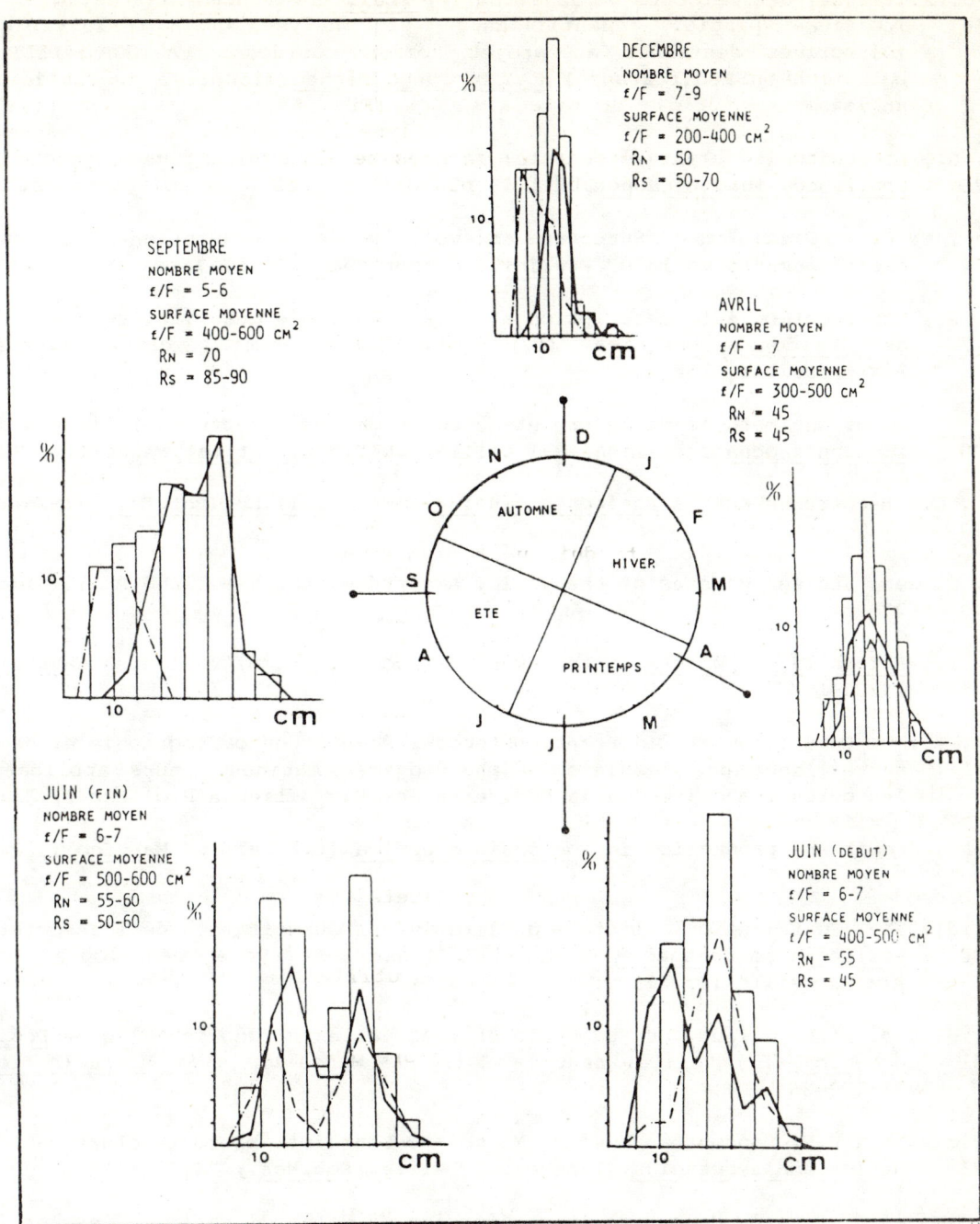

Figure 5. Cycle annuel de renouvelement de feuilles de P. oceanica proposée par Panayotidis et Giraud, 1981. L'histogramme des frequences correspond à l'ensemble des feuilles. Le polygone de frequence en trait discountinu correspond aux feuilles intermediaires. Le polygone de frequence en trait continu correspond aux feuilles adultes. f/F signifie "feuilles par faisceau".

5. BIBLIOGRAPHIE

Augier, H., G. Gilles et G. Ramonda, Recherche sur la pollution mercurielle en rade
1977 d'Hyères et dans l'Archipel des Stoechades (Méditerranée, France). 3. Teneur en mercure de la phanérogame marine Posidonia oceanica en fonction de la profondeur et de la pollution dans l'île de Port-Cros. Comparaison avec d'autres régions du littoral méditerranéen français. Trav.Sci.Parc Natl.Port-Cros., (3):27-38

Bernhard, M., Manuel des méthodes de la recherche sur l'environnement aquatique
1976 Troisième partie. Echantillonnage et analyse du matériel biologique. (Directives destinées au projet commun coordonné FAO(CGPM)/PNUE sur la pollution en Méditerranée) FAO Doc.Tech.Pêches,(158):132 p. Publié aussi en anglais

Cooper, G., La Posidonie, plante étonnante; la pêche ou la mariculture.
1976 Coh.Assoc.-Fond.G.Cooper, (1):55 p.

Giaccone, G. et L. Rizzi-Longo, Structure et évolution de la végétation marine dans les
1974 environnements pollués. Rev.Int.Océanogr.Méd., 34:67-72

Giraud, G., Contribution à la description et à la phénologie quantitative des herbiers
1977 de Posidonia oceanica (L.) Del. Thèse 3ème Cycle, Université de Aix-Marseille, 150 p.

_____, Sur une méthode de mesure et de comptage des structures foliaires de
1979 Posidonia oceanica (Linnaeus) Delile. Bull.Mus.Hist.Nat.Marseille, (39):33-9

Grenier, C., Recherches sur la Posidonia caulini König. Bull.Soc.Bot.Fr., 7(5/6):362-7
1860

Hartog, C. den, The sea grasses of the world. Amsterdam, North-Holland Publishing Co.,
1970 275 p.

Lipkin, Y., Halophila stipulacea, a review of a successful immigration. Aquat.Bot.,
1975 1:203-15

Meinesz, A., M. Cuvelier et R. Laureant, Méthodes récentes de cartographie et de
1981 surveillance des herbiers de phanérogames marines. Leurs applications sur les côtes françaises de la Méditerranée. Vie Milieu(A Biol.Mar.), 31:27-34

Ott, J.A., Growth and production in Posidonia oceanica (L.) Delile. Mar.Ecol., 1:47-64
1980

Panayotidis, P., Contribution à l'étude qualitative et quantitative de l'association
1980 Posidonietum oceanicae FUNK, 1927. Thèse 3ème Cycle Océanologie, Université Aix-Marseille II, 213 p.

Panayotidis, P. et G. Giraud, Sur un cycle de renouvellement des feuilles de Posidonia
1981 oceanica (L.) Delile, dans le Golfe de Marseille. Vie Milieu(A Biol.Mar.), 31:129-36

Panayotidis, P., C.F. Boudouresque et J. Marcot-Coqueugniot, Microstructure de l'herbier
1981 de Posidonia oceanica (Linnaeaus) Delile. Bot.Mar., 24:115-24

Pérés, J.M. et J. Picard, Causes de la raréfaction et de la disparition des herbiers de
1975 Posidonia oceanica sur les côtes françaises de la Méditerranée. Aquat.Bot., 1:133-9

Wittmann, K.J., Temporal and morphological variations of growth in a natural stand of
1984 Posidonia oceanica (L.) Delile. Mar.Ecol., 5(4):301-16

THE IMPACT OF POLLUTION ON THE COPEPOD COMMUNITY OF THE KASTELA BAY

by

D. REGNER
Institute of Oceanography and Fisheries,
Split, Yugoslavia

1. INTRODUCTION

This paper presents some results from the most recent studies on the copepod plankton community from the area of Kastela Bay (middle Adriatic). This area has recently been strongly affected by pollution of agricultural, domestic and industrial effluents.

Copepods are the best represented group of net zooplankton. This study of copepods was undertaken, on the one hand, due to the fact that they contribute 70 to 99% of the total zooplankton in the Kastela Bay and, on the other hand, to their significance in the nutrition chains in the sea (first level of consumers of organic matter produced in the sea and at the same time food of some commercially important fish species). Since copepods respond very quickly, both by composition and quantity, to the changes in the environment they inhabit, the most recent results obtained for the 1982-1985 period within the framework of the FAO/UNEP project (YUG/21) are brought out and compared with the results of some earlier studies.

2. MATERIALS AND METHODS

Zooplankton samples for qualitative and quantitative analyses were collected in vertical hauls from bottom to surface using the "Hensen" net (73/100, silk No.3).

Material was fixed in 2% neutralized formaline. One twentieth of the sample was counted and the whole catch examined for rare species.

Diversity index $\underline{d}$ as quantitative indicator of copepod group structure was estimated from the formula developed by Margalef (1951).

$$d = \frac{S-1}{\log_n N}$$

where S is the number of species, N the number of individuals per cubic metre and $\log_n$ the natural logarithm of the number of individuals.

3. STUDY AREA

Material was collected from five stations in the Kastela Bay (Fig.1). Station 1 (10m depth) is under the strongest impact of agricultural wastes, Station 2 (10m depth) is directly affected by industrial wastes-chemical industry, Station 3 (20m depth) is under the common influence of different industrial plants (food, shipyard, breaker's yard, etc.) and Station 4 (40m depth) is at the entrance to the Kastela Bay and Station 5 (35m depth) in the middle of the Bay.

Kastela Bay is one of the largest bays in the middle Adriatic (61 km^2), located in a markedly shallow and enclosed area under the direct influence of the land. It is connected with the adjacent sea through a small strait near the town of Trogir and a larger one opening into the Brac Channel.

The bottom is mostly muddy. The mean depth is about 23 m with a maximum of 47 m.

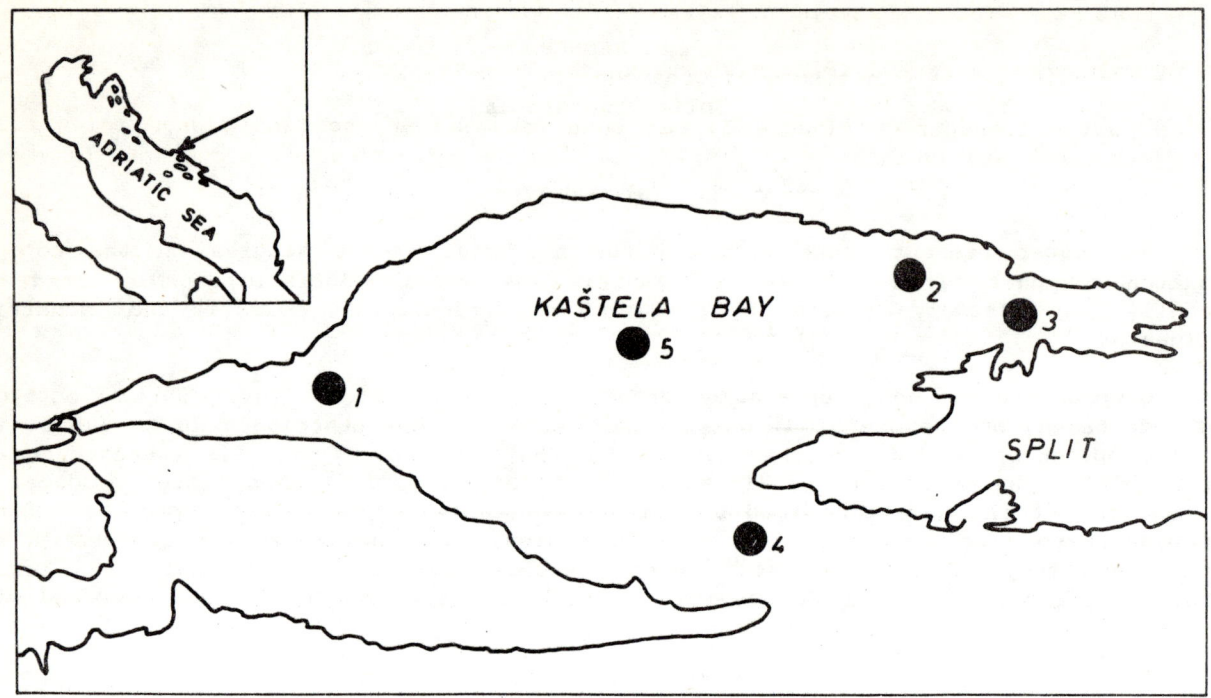

Figure 1. The study area

The freshwater runoffs come from the Jadro River, Pantan stream and the springs along the coast, of which some are submarine.

Mean current speed in the bay is up to 13cm/sec in the surface layer, and up to 6cm/sec at 15 and 30m depths. From the viewpoint of dynamics the eastern part of the bay is somewhat isolated (Zore-Amanda et al., 1976).

Water enters the bay predominantly in the surface layer and leaves it in the intermediate and bottom layers. The W direction is prevalent in winter, NW in spring and early summer E in summer and S in autumn (Zore-Armanda et al., 1974). From current roses constructed by seasons Zore-Armanda et al., (1976) calculated that the complete exchange of water with the Brac Channel takes place twice in a month's period.

Studies of basic hydrographic parameters of the bay such as temperature, salinity, density, transparency, oxygen content (O_2), oxygen saturation ($O_2\%$), pH have shown that the bay is particularly strongly affected by the land factors, that is freshwater and waste water discharges (industrial and agricultural wastes) which are reflected upon chemical and biological properties of the bay, particularly in its easternmost part (Zore-Armanda et al., 1976).

Thus the surface temperature ranges from 7 to 25°C, and exceptionally may even exceed 28°C.

Maximum salinity recorded in August reaches as high a value as 38.19°/oo, and the lowest is diminished to 28.17°/oo, which is further indicative of the strong land and freshwater impacts.

Phosphate levels usually lie in the range of 0-0.13 µmol/l but they reach 0.39 µmol/l in the surface layers of the Jadro River.

Silica levels vary from 1.78-13.57 µmol/l in the middle of the bay whereas the freshwaters discharged into the bay contain up to 50 µmol/l Si.

Annual mean of $CaCO_3$ levels is rather high, somewhere about 120mg/l.

pH values range from 8.18 to 8.28 during the year.

Sea water transparency (Table I) has been reduced for the last decade being half that of 10 years ago (Vukadin et al., 1984). This is indicative of strong land effects, that is freshwater and wastewater runoffs (Buljan and Zore-Armanda, 1971).

Table I

Seawater transparency in the middle of the Kastela Bay (Station 5) and in the eastern part (Station 3).

Year Station no.	1972/73	1975/76	1977/78	1982/83
5	12.4 m	10.4 m	8.5 m	9.0 m
3	7.8 m	6.0 m	-	4.2 m

Oxygen content varies from 4.51 to 6.16 ml/l O_2, and oxygen saturation from 87 to 109%. Greater positive and negative deviations occur predominantly in the eastern part of the bay. Due to the reduced circulation of water masses, particularly in summer, from time to time intensive blooms of individual phytoplankton red tide organisms occur in this area to which new quantities of organic matter and nutrients are incessantly carried by freshwaters so that the capacity of autopurification is considerably reduced. These blooms may even be accompanied by mass fish mortality as in 1980 and 1982.

As to productivity, Kastela Bay belongs to the most productive areas of the Adriatic. This is because, thanks to photosynthesis, the phytoplankton fixes 115-240 g/C/year per m^2 of illuminated water column (Pucher-Petkovic et al., 1980). According to world seas classification the Kastela Bay is placed into the fourth category of 5 categories with a mean assimilation of 430 mg C/m^2/day.

The increase in the number of heterotrophic bacteria, as well as of other biological parameters in the easternmost part of the bay, are also indicative of the large quantities of organic matter in that area.

Organic pollutants (phenols, mineral oils) has not yet reached the level of toxicity for marine organisms (Zore-Armanda et al., 1976).

The content of heavy metals in sediments confirms the fact that the eastern part of the bay is strongly affected by industrial wastes since the values of mercury, cadmium, zinc, copper and lead are considerably increased in the surface sediment layers.

The next chapter describes the effects of these changes caused by land pollution on qualitative composition, and quantity of copepod groups in the Kastela Bay and particularly in its eastern part.

4. RESULTS AND DISCUSSION

Studies of seasonal and long-term variations of the composition and quantity of copepod groups have been intensively (on a monthly basis) carried out since 1960 on the material collected from Station 5 in the middle of the bay (Regner, 1973, 1976, 1977, 1978, 1979, 1981, 1982, 1984, 1985; Regner and Vucetic, 1980).

The financial assistance given by UNEP for the YUG 21 project made it possible to explore the most threatened shallowest marginal parts of the bay, strongly affected by the land. Even though some earlier data (Zore-Armanda et al., 1976) have already shown the changes in composition and quantity of copepods, they cannot be compared with the results of these recent researches since the samples were not collected from the same stations and in the same months. Therefore, this paper presents the detailed results of the most recent studies carried out in the 1983-1985 period.

Station 1 - located in front of Trogir is under the strongest influence of agricultural wastes from the agricultural plantations stretching along the coast (Table II).

A total of 20 species and 2 genera, the determination of which was not possible, were recorded from this station (Table II). Considering the fact that the depth at this station does not exceed 10m, the number of recorded species is quite usual for coastal areas. The occurrence of species such as Calanus halgolandicus, C. tenuicornis, Microsetella norvegica and Euterpina acutifrons is indicative of some open sea effects since all these are pelagic species.

The highest number of species was recorded in autumn. Acartia clausi was best represented and constituted 50% of the total copepod counts. Its presence varied between 15-50%.

The number of individuals was highest in summer and lowest in winter. However, the number of copepods was considerably high all the year round (particularly if the shallowness of the station is taken into account), since they probably had enough food available due to favourable environmental conditions.

All these data on copepod density and composition show a strong influence of land runoffs at this station.

This is evident from disturbed relations between individual species, that is, marked dominance of Acartia clausi, as well as from the changes in the rythm of seasonal density oscillations which showed lower copepod numbers only in winter. In addition two annual maxima did not occur since the number of copepods was high throughout the year.

Station 2 - in front of the small town of Kastel Sucurac is directly and particularly strongly affected by industrial wastes (chemical industry) (Table III).

As at the previous station, the depth does not exceed ten metres and therefore the number of determined copepod species was almost the same (Table III). A total of 21 copepod species and two genera were recorded. Neritic species, which usually occur in greater numbers, dominated. However, the copepods characteristic for the open sea were also recorded: Calanus helgolandicus, C. tenuicornis, Clausocalanus parapergens, Euterpina acutifrons and some others. The largest number of species was recorded in summer 1983, and the smallest in spring 1985. Acartia clausi was markedly dominant. It contributed from 17-93%, which had not been earlier recorded from Kastela Bay. These high percentages of Acartia clausi were found in spring 1982.

The number of copepods per cubic metre showed greater variations than at the previous station. Maximum values were ten times the minimum ones, which was evidently due to the land influence. The values recorded in spring-summer were much higher. The autumn maximum otherwise usually recorded was not recorded at all. The number of copepods was very low, in autumn and winter particularly in 1984.

Accordingly, the situation at Station 2 is very similar to that at Station 1. High values of Acartia clausi may even be held as a monoculture.

Rythm of seasonal density oscillations has also changed, since no marked maxima in spring and at the beginning of autumn were recorded, which in fact was usual, owing to the position of the Adriatic Sea in the temperature climatic zone. All these phenomena are another proof of disturbed relations in the copepod community at this station.

Table II

The qualitative and quantitative composition of copepods at Station 1.

	Feb 82	May 82	Aug. 82	Oct 82	Jun 83	Mar 84	Aug 84	Dec 84	May 85
Calanus helgolandicus	0	20	0	0	5	0	20	0	0
Calanus tenuicornis	0	0	0	0	0	20	20	0	20
Calocalanus contractus	0	20	0	0	20	0	0	0	0
Paracalanus parvus	260	140	260	240	280	40	240	40	560
Ctenocalanus vanus	200	20	0	40	20	0	40	0	40
Clausocalanus jobei	0	0	0	20	0	0	0	0	0
Clausocalanus arcuicornis	20	0	0	0	0	0	160	0	0
Clausocalanus pergens	20	60	0	60	20	0	40	0	0
Clausocalanus furcatus	0	0	60	20	20	0	0	0	0
Centropages typicus	80	260	0	180	40	180	80	0	180
Centropages kröyeri	0	0	240	0	60	0	20	20	0
Temora stylifera	0	0	460	440	320	0	780	20	0
Temora longicornis	0	680	40	0	80	60	20	40	0
Acartia clausi	560	1280	1800	480	1040	100	440	60	460
Diaixis pygmaea	20	0	0	20	5	0	0	0	0
Oithona sp.	60	160	60	160	80	280	240	120	160
Microsetella norvegica	0	20	0	0	0	0	0	0	0
Euterpina acutifrons	0	0	0	20	0	5	20	0	0
Oncaea sp.	0	0	0	0	0	0	0	0	0
Corycaeus typicus	0	0	0	20	0	0	20	20	0
Corycaeus brehmi	0	0	0	0	20	0	0	0	0
Corycella rostrata	0	0	40	20	5	0	0	0	0
Copepodits and other copepods	240	1180	600	1200	1040	300	600	160	400
Total	1460	3220	3560	2920	3060	985	2740	480	1820
Total/m^3	487	1073	1188	973	1020	328	913	160	606

Table III

The qualitative and quantitative composition of copepods at Station 2.

	Feb 82	May 82	Aug 82	Oct 82	Jun 83	Mar 84	Aug 84	Dec 84	May 85
Calanus helgolandicus	0	0	0	0	20	0	0	0	0
Calanus tenuicornis	80	0	0	0	40	20	5	0	0
Calocalanus contractus	0	0	0	0	0	0	0	0	0
Paracalanus parvus	0	20	0	20	220	40	60	160	0
Paracalanus nanus	0	0	0	40	0	0	0	0	0
Ctenocalanus vanus	140	0	0	0	80	0	0	120	0
Clausocalanus jobei	0	0	0	0	80	0	40	0	0
Clausocalanus pergens	40	0	0	20	40	0	20	0	0
Clausocalanus parapergens	0	0	0	0	160	0	0	0	0
Clausocalanus furcatus	0	0	20	0	0	0	20	20	0
Centropages typicus	60	0	60	20	40	40	0	0	0
Centropages kröyeri	80	40	380	40	320	0	160	0	0
Temora stylifera	40	0	680	200	260	0	360	40	0
Temora longicornis	0	20	0	0	20	0	0	0	60
Isias clavipes	0	0	0	0	40	0	0	0	0
Diaixis pygmaea	20	0	0	0	0	0	0	0	0
Acartia clausi	1880	4060	3360	180	1600	60	860	60	1260
Oithona sp.	260	40	80	180	80	0	40	20	0
Oncaea sp.	0	0	0	0	0	0	40	20	0
Euterpina acutifrons	20	0	0	0	0	0	20	0	20
Corycaeus typicus	20	0	0	20	0	0	20	40	20
Corycaeus brehmi	0	0	0	20	40	20	0	40	0
Corycella rostrata	0	0	0	20	0	0	5	0	0
Copepodits and other copepods	300	180	180	260	380	280	400	140	60
Total	2140	4360	4760	1020	3440	460	1990	680	1420
Total/m^3	713	1453	1588	340	1146	153	663	113	473

Station 3 - of 30m depth is in the eastern part of the bay. It is affected by a variety of land pollution sources: food industry, breaker's yard, shipyard, etc.

A total of 22 copepod species and two genera with almost identical species to those from two preceding stations (Table IV) were found. *Acartia clausi* was also markedly dominant contributing some 60% (recorded in summer 1983).

The number of individuals was considerably lower than at two preceding stations. However, seasonal density oscillations showed almost the same trend like those at Stations 1 and 2, with increased values in spring-summer.

With respect to the fact that the depth is almost twice that at Stations 1 and 2, already mentioned changes in the plankton community are still better felt here.

Station 4 - at the entrance to the bay is of 40m depth. It is not a typical bay station due to its location at the entrance.

All the water mass dynamic studies carried out in this area so far (Table V) show that under certain conditions (south wind "jugo") water enters the bay from the Split town port and under some others (north wind "bura") from the Brac Channel, as a compensatory current to the outgoing component into the Brac Channel.

The Station's depth as well as its peculiar position are reflected upon the species composition. Thus 28 species and 2 genera were recorded (Table V). Many of the species recorded belong to the species more frequent in the open sea: *Calanus halgolandicus*, *C. tenuicornis*, *Calocalanus contractus*, *Calocalanus plumulosus*, *Clausocalanus parapergens*, *Corycaeus furcifer* and some others which is evidence of a good water exchange with the open sea.

The highest number of species encountered in the colder part of the year confirms (or "agrees with") the results of our earlier studies of seasonal changes of qualitative composition of copepod group (Regner, 1973, 1979).

Species *Temora stylifera*, *Acartia clausi*, *Centropages typicus*, *Paracalanus parvus* and *Ctenocalanus vanus* were quantitatively best represented. Other species, however, occurred in considerably smaller numbers.

Total number of individuals varied during the year with a peak in spring and rather low values in summer. Minimum numbers were recorded in autumn 1982.

Species composition and number and dominant species, characteristic for the overall east Adriatic coast, as well as the annual variation rythm usual in the coastal sea, were at this station quite different from those at the previous stations located in the shallowest and most threatened part of the bay.

Station 5 - is in the middle of Kastela Bay. The best represented species of 24 species and 2 two genera recorded were: *Acartia clausi*, *Temora stylifera*, *Centropages typicus*, *Paracalanus parvus* and *Ctenocalanus vanus* (Table VI). The majority of species were recorded in the colder part of the year like at Station 4. Besides neritic species the species more frequent in the open sea were also recorded from this station.

The seasonal variation rhythm is similar to that at the previous station with the highest number of individuals in winter-spring.

The composition and character of species, higher number of dominant species as well as the rhythm of seasonal variations have not recently significantly changed. Therefore it may be concluded that this station has not yet become strongly influenced by the land.

Somewhat different properties from those at other stations may be partly due to some topographic and hydrographic features. Thus the depth is greatest at this station, it is most apart from the land and direct pollution sources and the dynamics is more intensive that in the marginal parts of the bay.

Table IV

The qualitative and quantitative composition of copepods at Station 3.

	Feb 82	May 82	Aug 82	Oct 82	Jun 83	Mar 84	Aug 84	Dec 84	May 85
Calanus helgolandicus	0	0	0	0	0	0	0	0	0
Calanus tenuicornis	160	0	0	0	0	0	0	0	160
Calanus contractus	0	0	0	0	20	0	0	0	0
Paracalanus parvus	320	0	60	80	60	80	80	280	260
Paracalanus pygmaeus	60	0	0	0	0	0	0	0	0
Ctenocalanus vanus	200	0	0	20	0	20	320	200	420
Clausocalanus jobei	0	0	0	40	0	0	0	20	40
Clausocalanus arcuicornis	0	0	0	20	0	0	60	0	60
Clausocalanus pergens	20	20	20	40	0	20	0	0	60
Clausocalanus parapergens	0	0	0	0	0	0	0	20	20
Clausocalanus furcatus	0	0	0	40	0	0	0	0	0
Centropages typicus	400	60	40	0	80	300	280	0	420
Centropages kröyeri	120	40	400	40	240	80	80	40	0
Candacia armata	0	0	0	0	0	0	0	0	60
Temora stylifera	0	0	580	560	40	0	360	0	0
Temora longicornis	0	100	0	0	0	40	0	0	500
Acartia clausi	2500	240	1600	480	1100	120	1600	20	1200
Diaixis pygmaea	20	0	0	0	0	20	20	20	40
Oithona sp.	100	20	80	140	60	20	80	80	180
Oncaea sp.	0	0	0	0	0	0	0	0	20
Euterpina acutifrons	60	0	0	5	40	0	0	0	20
Corycaeus typicus	0	0	0	0	0	20	80	20	0
Corycaeus brehmi	20	0	5	0	0	20	20	0	20
Corycella rostrata	0	0	0	20	0	0	20	0	20
Copepodits and other copepods	1600	380	500	760	180	760	500	280	560
Total	5580	860	3285	2245	1820	1500	3480	980	4080
Total/m³	930	143	547	374	303	250	580	163	680

Table V

The qualitative and quantitative composition of copepods at Station 4.

	Feb 82	May 82	Aug 82	Oct 82	Jun 83	Mar 84	Aug 84	Dec 84	May 85
Calanus helgolandicus	0	0	0	5	0	0	0	0	0
Calanus tenuicornis	380	60	20	5	0	0	0	20	40
Calocalanus pavo	0	40	0	0	0	0	0	20	0
Calocalanus contractus	60	0	0	0	0	0	0	0	0
Calocalanus plumulosus	20	0	0	0	0	0	20	0	0
Paracalanus parvus	380	980	140	100	20	320	300	360	580
Macrosetella gracilis	0	0	0	0	0	0	0	0	60
Ctenocalanus vanus	480	360	240	0	0	140	360	220	300
Clausocalanus arcuicornis	0	0	0	0	0	0	40	0	80
Clausocalanus jobei	0	60	140	40	20	160	160	40	0
Clausocalanus pergens	160	140	0	60	0	0	0	0	120
Clausocalanus parapergens	0	60	0	0	0	0	0	0	180
Clausocalanus furcatus	80	60	0	120	0	0	0	0	0
Centropages typicus	560	1360	580	0	140	460	140	40	600
Centropages kröyeri	0	80	280	0	50	120	0	0	0
Temora stylifera	60	140	1210	420	600	2420	0	140	420
Temora longicornis	0	80	80	0	0	0	460	0	900
Acartia clausi	360	1200	1040	60	560	1360	340	1860	300
Diaixis pygmaea	20	60	20	0	20	0	0	0	0
Candacia armata	0	0	0	5	0	0	0	0	0
Lucicutia flavicornis	0	0	0	0	0	0	5	0	0
Oithona sp.	720	580	380	140	160	320	40	460	120
Euterpina acutifrons	0	20	0	5	0	0	0	0	100
Labidocera wollastoni	0	0	0	55	0	0	0	20	0
Oncaea sp.	0	0	0	60	0	40	0	0	0
Corycaeus typicus	60	0	0	0	0	40	60	80	180
Corycaeus brehmi	60	0	0	20	40	0	0	140	0
Corycella rostrata	20	0	0	40	0	0	0	0	0
Corycaeus latus	40	0	0	0	0	0	0	0	0
Corycaeus furcifer	0	0	0	0	0	0	0	40	0
Copepodits and other copepods	1080	1280	240	80	440	1120	1060	380	1300
Total	4540	6560	4370	1165	2060	6500	2985	3880	5280
Total/m^3	378	547	354	97	171	542	249	323	440

Table VI

The qualitative and quantitative composition of copepods at Station 5.

	Feb 82	May 82	Aug 82	Oct 82	Jun 83	Mar 84	Aug 84	Dec 84	May 85
Calanus helgolandicus	20	20	0	0	0	20	0	0	0
Calanus tenuicornis	100	20	5	20	0	40	0	0	280
Calocalanus plumulosus	0	40	0	0	0	20	20	0	0
Paracalanus parvus	1340	640	20	240	120	220	200	180	460
Ctenocalanus vanus	780	500	80	420	160	120	380	180	160
Clausocalanus arcuicornis	0	0	40	80	120	200	0	0	140
Clausocalanus jobei	0	160	20	130	80	180	260	0	140
Clausocalanus pergens	60	80	60	0	20	80	0	80	40
Clausocalanus parapergens	20	80	0	60	0	0	0	20	40
Clausocalanus lividus	0	20	0	0	0	0	0	0	0
Centropages typicus	180	805	100	420	300	220	600	280	280
Centropages kröyeri	140	0	40	0	60	80	200	20	0
Temora stylifera	60	0	740	540	420	60	200	360	0
Temora longicornis	0	20	20	40	0	0	0	80	780
Acartia clausi	2280	780	700	395	1245	700	1020	520	2300
Diaixis pygmaea	60	60	0	0	0	0	0	80	5
Candacia armata	20	0	0	5	0	0	0	0	0
Mecynocea clausi	20	0	0	40	0	0	0	0	0
Euterpina acutifrons	0	0	5	0	0	0	0	80	0
Corycaeus typicus	0	0	0	0	0	20	0	0	0
Corycaeus brehmi	20	0	0	0	20	0	0	40	0
Corycella rostrata	0	0	0	0	0	10	0	0	40
Oithona sp.	140	220	440	160	280	200	340	380	260
Copepodits and other copepods	1300	1280	400	820	850	600	400	760	1500
Total	6540	4725	2670	3360	3675	3670	3620	3496	7560
Total/m^3	623	450	254	320	350	350	345	333	720

Summing up the results presented here, the kinds of changes that have occurred in the plankton copepod community due to the land pollution may be established. These changes differ from one station to another since, as shown earlier, the land pollution sources differ as well as the pollution intensity.

Firstly, the changes of species composition of copepod group did not show any great variations in numbers and were rather uniform at all stations (Table VII).

Table VII

The oscillations of the number of species at the investigated area.

Station	Winter	Spring	Summer	Autumn	Summer	Spring	Summer	Winter	Spring
1	7+1	9+1	7+1	12+1	14+1	6+1	13+1	6+1	5+1
2	10+1	5+1	5+1	10+1	15+1	5	10+2	8+1	3+1
3	11+1	5+1	7+1	12+1	7+1	10+1	10+1	8+1	16+2
4	15+1	16+1	10+1	13+2	8+1	10	10+1	12+1	13+1
5	14+1	13+1	12+1	12+1	9+1	14+1	13+1	11+1	12+1

Somewhat higher numbers of species were mainly recorded during the colder part of the year. It is suggested that they may be connected with the intensified circulation in winter-spring and more intensive surface water inflow, particularly marked at Stations 4, 5, 3 and 2 and at Station 1 only in 1982.

However, the percentage proportions of individual species have been seriously disturbed, which is particularly shown by the dominance of *Acartia clausi*.

The following table shows that there was a very large number of individuals of this species at all the stations. It constituted from 21 to 66% of the total number of individuals.

Table VIII

Variations of percentages of *Acartia clausi* at study area stations.

	Total copepod numbers	Total number of *Acartia clausi* individuals	Proportion of *Acartia clausi* (in %)	Maximum proportion of *Acartia clausi* (in %)
1	20245	6220	30.7%	50%
2	20270	13320	65.7%	93%
3	23830	8860	37.2%	60%
4	37340	7680	20.6%	48%
5	32026	8220	25.7%	35%

It is also evident from the table that its percentages were lowest at Station 4. This is quite normal since this station is under the poorest influence of the land. Station 5, which is in the middle of the bay, followed, with respect to these changes which were less marked than at the shallow stations most exposed to land pollution. Acartia clausi constituted 66% of the total copepod counts at Station 2, (with maximum presence of as much as 93%). This is an apparent evidence of markedly disturbed normal relations in the plankton copepod community. At Stations 1 and 3 Acartia clausi formed 31 and 37% of the total counts respectively. These are, as well, rather high values.

These most recent results were compared with the results of studies carried out in 1960-1964 and 1970-1974 at Station 5 at which land effects are not so strongly felt as at markedly inshore stations along the margins of the bay (Table IX) (Regner, 1970, 1979).

Table IX

Variations of percentage proportions of Acartia clausi at Station 5 in 1960-1964 and 1970-1974 periods and in 1982-1985 period.

1960-64	1970-74	1982-85
20%	23%	26%

A trend of increase of percentage proportion of Acartia clausi was encountered at this station, as well. This is another confirmation that the eutrophication affects the copepod species composition evident by the disturbed relations between them.

Changes in the quantities of copepods at studied stations, by seasons, are presented in Table X.

Table X

Variations of the number of copepods per cubic metre at study area stations

Month Station	Feb 82	May 82	Aug 82	Oct 82	Jun 83	Mar 84	Aug 84	Dec 84	May 85
1	487	1073	1188	973	1100	328	913	160	606
2	713	1453	1588	340	1146	153	663	113	473
3	930	143	547	374	303	250	580	509	680
4	378	547	354	90	171	542	249	323	440
5	623	450	254	320	350	350	345	333	720

It is evident from Table X that usual spring maxima occurred whereas the values were considerably lower at Station 4 in summer. The number of copepods was considerably high throughout the warmer part of the year. Stations 1 and 2 showed summer maximum with the highest number of individuals.

Accordingly, variations in the number of individuals showed a changed seasonal rhythm otherwise characteristic for the temperate climatic belt to which the Adriatic belongs. Thus instead of the marked spring maximum and slightly increased values in autumn usually recorded. This is particularly applicable to Stations 1, 2 and 3 with respect to the fact that copepods have enough phytoplankton throughout the year and particularly in summer.

Similar changes of annual copepod cycle with increased numbers from July to September were recorded by Moraitou-Apostolopoulou (1981) from the polluted Elefsis Bay (Greece) the characteristics of which are very similar to those of the Kastela Bay.

Our earlier studies (Regner, 1984) established a summer maximum at Station 5 in July. However, here we do not have the data for this month. If the variations in the number of individuals per cubic metre at Station 5, obtained during the present study, are compared with the mean values for the 1970-1974 period (Table XI) it will become obvious that all recent measurements show increased numbers in relation to those 10 years ago.

Table XI

Variations of the number of individuals per cubic metre of the sea at Station 5 in 1970-1974 and 1982-1985 period.

Year	February	March	May	June	August	October	December
1970-74	227	156	263	188	330	295	179
1982-83	623		509	350	254	320	
1984		350			345		333
1985			720				

This information is proof that the effects of land runoffs are felt at Station 5 as well, even though much less than at Stations 2, 3 and 1.

The values of the diversity index obtained from the number of species and the number of individuals (Margalef, 1951), which is in fact an expression of the copepod community structure (Table XII), may help us to establish the characteristics of the copepod community in the study area. Diversity index values were considerably higher at Stations 4 and 5 than at Stations 1, 2 and 3.

Table XII

Variations of the diversity index (d) at the investigated stations.

Month Station	Feb 82	May 82	Aug 82	Oct 82	Jun 83	Mar 84	Aug 84	Dec 84	May 85
1	1.13	1.29	0.99	1.74	2.02	1.04	1.91	1.18	0.78
2	1.52	0.55	0.68	1.72	2.13	0.80	1.54	1.90	0.65
3	1.61	1.00	0.95	1.86	1.23	1.81	1.57	1.77	2.61
4	2.17	2.13	2.17	2.08	1.71	2.06	1.81	2.42	2.14
5	2.69	2.38	1.70	3.06	1.56	2.55	1.6	2.07	1.98

Moreover, d values were lower at Stations 1, 2, 3 and 5 in summer and at Station 4 only in the summer of 1984. This confirms the statement that this station is under the poorest impact from the land, as well as of the fact that defined structural changes, of either smaller or greater extent, have taken place at other stations. These changes are apparent at Station 5 as well, even though it was established to be under the poorest influence of the land.

The comparison of variations of diversity index for the 1970-1974 period to those for 1982-1985 shows a decrease in all months except in October which further proves the persistence of these changes for the last time.

Table XIII

Variations of species diversity index at Station 5 in the 1970-1974 and 1982-1985 periods.

Month Year	February	March	May	June	August	October	December
1970-74	2.87	2.95	2.65	2.26	2.77	2.68	3.28
1982-83	2.69		2.38	1.56	1.70	3.06	
1984		2.55			1.6		2.07
1985			1.98				

The observations of species diversity index d at Station 5, which is under the poorest influence of the land pollution, has proved to be the most sensitive method for registration of changes that have occurred over a rather long time interval.

5. CONCLUSIONS

All that has been described here indicates that the changes affected by the pollution from the land were most intensive at Stations 2 and 3 which are under the strong and direct influence of harmful chemical substances, heavy metals, food industry wastes, breaker's yard, shipyard, cement factory, etc.

These changes are not shown by the number of species but by percentage relations between them and the high percentage of _Acartia clausi_ (up to 93%).

The rhythm of seasonal variations in the number of species was changed; instead of having the usual spring maximum, the number of individuals increased from spring to summer.

The species diversity index was considerably lower than at other stations, its values being even lower during summer.

With respect to the intensity of changes Station 1, affected by agricultural wastes, follows Stations 2 and 3. All changes mentioned were recorded from this station, as well. Only _Acartia clausi_ was somewhat less represented than at those two stations (up to 50%).

Properties of Station 4 differ considerably from those of other stations probably due to its peculiar position at the entrance to the Kastela Bay. Species composition and number, dominant copepod species, usual rhythm of annual variations with low values in summer and high species diversity index d indicate that it is not being affected by the land.

Finally, Station 5 in the middle of the bay shows characteristics similar to those of Station 4. However, Station 5 is more affected by the land which is evident from the comparison of results of long-term studies (particularly species diversity index $\underline{d}$).

On the basis of all that has been brought out it is held that the areas exposed to land effects should be observed over a long period since no short-term observations could establish the changes that these effects produce.

6. REFERENCES

Buljan, M. and M. Zore-Armanda, Introduction to oceanography and marine meteorology,
1971   Split, Institut za Oceanografiju i Ribarstvo, 425 p. (in Croatian)

Margalef, R., Diversidad de especies. En las comunidades naturales. Publ.Inst.Biol.Apl.,
1951   Barc., 9:5-27

Moraitou-Apostolopoulou, M., The annual cycle of zooplankton in Elefsis Bay (Greece).
1981   Rapp.P.-V.Réun.CIESM, 27(7):105-6

Pucher-Petkovic et al., Developpement des population phytoplanctoniques caracteristiques
1980   pour un milieu eutrophise (Baie De Kastela, Adriatique Centrale). Acta Adriat., 21(2):79-93

Regner, D., The seasonal distribution of copepods in the central Adriatic in 1971.
1973   Ekologija, 8(1):139-46

_____, On the copepod diversity in the Central Adriatic in 1971. Rapp.P.-V.Réun.
1976   CIESM., 23(9):95-6

_____, The oscillations of copepod number in the Kastela Bay (Central Adriatic) in
1977   relation to some ecological factors. Rapp.P.-V.Réun.CIESM, 24(10):165-6

_____, The influence of eutrophication on the copepods in the coastal area of
1978   Split (Central Adriatic). Rapp.P.-V.Réun.CIESM, 25/26(8):101-2

_____, Seasonal and multiannual dynamics of copepods in the Middle Adriatic.
1979   Disertacija, Zagreb, 188 p.

_____, The changes in seasonal oscillations of copepods in the Central Adriatic.
1981   Rapp.P.-V.Réun.CIESM, 27(7):177-9

_____, The oscillations of copepod density in the Kastela Bay as influenced by
1982   some environmental factors. Acta Adriat., 23(1/2):137-52

_____, Seasonal and multiannual oscillations of copepod density in the Central
1984   Adriatic. Crustaceana, 7:351-359.

_____, Copepod community of the Kastela Bay as influenced by different level of
1985   pollution. Rapp.P.-V.Réun.CIESM, 29(9):245-7

Regner, D. and T. Vucetic, Seasonal and multiannual fluctuations of copepods in the
1980   Kastela Bay (1960-1969). Acta Adriat., 21(2):101-22

Vukadin, I. et al., Kastela Bay - Some oceanographic properties and pollution with heavy metals, III Konferencija o zastiti Jadrana. (in Croatian)(in press).

Zore-Armanda M. et al., Oceanographic studies in the area of Split. Stud.Elaborati Inst.Oceanogr.Ribar., Split, (5)

_____, Oceanografsko-bioloska svojstva mora i epidemiolosko-bakterioloska svojstva
1976   otpadnih voda sjevernog slivnog podrucja Splita. Stud.Elaborati Inst.Oceanogr.Ribar., Split,

EFFECTS OF POLLUTION ON FISH POPULATIONS IN EGYPTIAN WATERS

by

H.H. SALEH and A.F. EL KARASHILY
Alexandria Institute of Oceanography and Fisheries,
Alexandria, Egypt

1. INTRODUCTION

During the last few years, as a result of increasing activities mainly in Alexandria, the Egyptian Mediterranean coastal area has been receiving different wastes such as industrial wastes, agricultural drainage, petroleum hydrocarbons and sewage.

Saleh(1983,1985) ivestigated the lethality to various marine organisms of certain industrial waste discharged into the sea off the Egyptian coast. He also investigated (Saleh, 1985a) their sublethal effects when diluted. A considerable amount of agricultural drainage which contains pesticides is channelled into the sea (El Sebae and Abo ElAmayem, 1979).

El Sharkawi(1979) referred to the contamination of the Alexandria coast due to sewage discharge.

Wahby and El Deeb(1981), and Aboul Dahab and Halim(1981) have reported on the pollution of the Mediterranean coast of Egypt by petroleum hydrocarbons.

This study is an attempt to show the effect of pollution on the population and ecosystem of the Egyptian Mediterranean coastal area.

2. MATERIALS AND METHODS

The annual statistical reports of the catch population from the Mediterranean sea in front of Egypt are collected and computed. The catch population is divided into pelagic fishes and bottom fauna.

Annual variations in the ratio of bottom fauna to pelagic fishes are calculated for the whole catch population, and also for selected zones e.g. El Mex zone, Eastern Harbour zone, Abu Qir zone, and Port Said zone which have, to some extent, different contamination levels.

The individual percentage in the bottom fauna or pelagic fishes is calculated for each year to show the annual variations in the ecosystem of the Mediterranean sea in front of Egypt, taking into consideration the fact that industrial activities, massive use of pesticides in agriculture and oil extraction in Egypt started in about 1965.

3. RESULTS AND DISCUSSION

In spite of the use of advanced equipment in fishing operations in recent years, statistical reports show that there is a certain decline in the catch population of the Mediterranean sea in front of Egypt (Fig. 1). Such a decrease in the fisheries potential could probably be attributed to the spread of pollution in the sea environment (Sindermann, 1980). The sharp drop in the catch during the period 1967-1978 was mainly due to irregularities in fishing activities due to war conditions.

It could be noticed also that the annual decrease of the bottom fauna is more than that of the pelagic fishes (Fig. 1). This may be explained by the fact that the sediments of the Mediterranean sea in front of Egypt became more polluted (El Sokkary, 1985). In such a case, the sediments could be considered as a vast reservoir of pollutants, and animals living on the sea floor are constantly exposed to sedimented material found below the water-sediment interface, subjecting it to a highly polluted water column, i.e. the animals living in the vicinity of the polluted

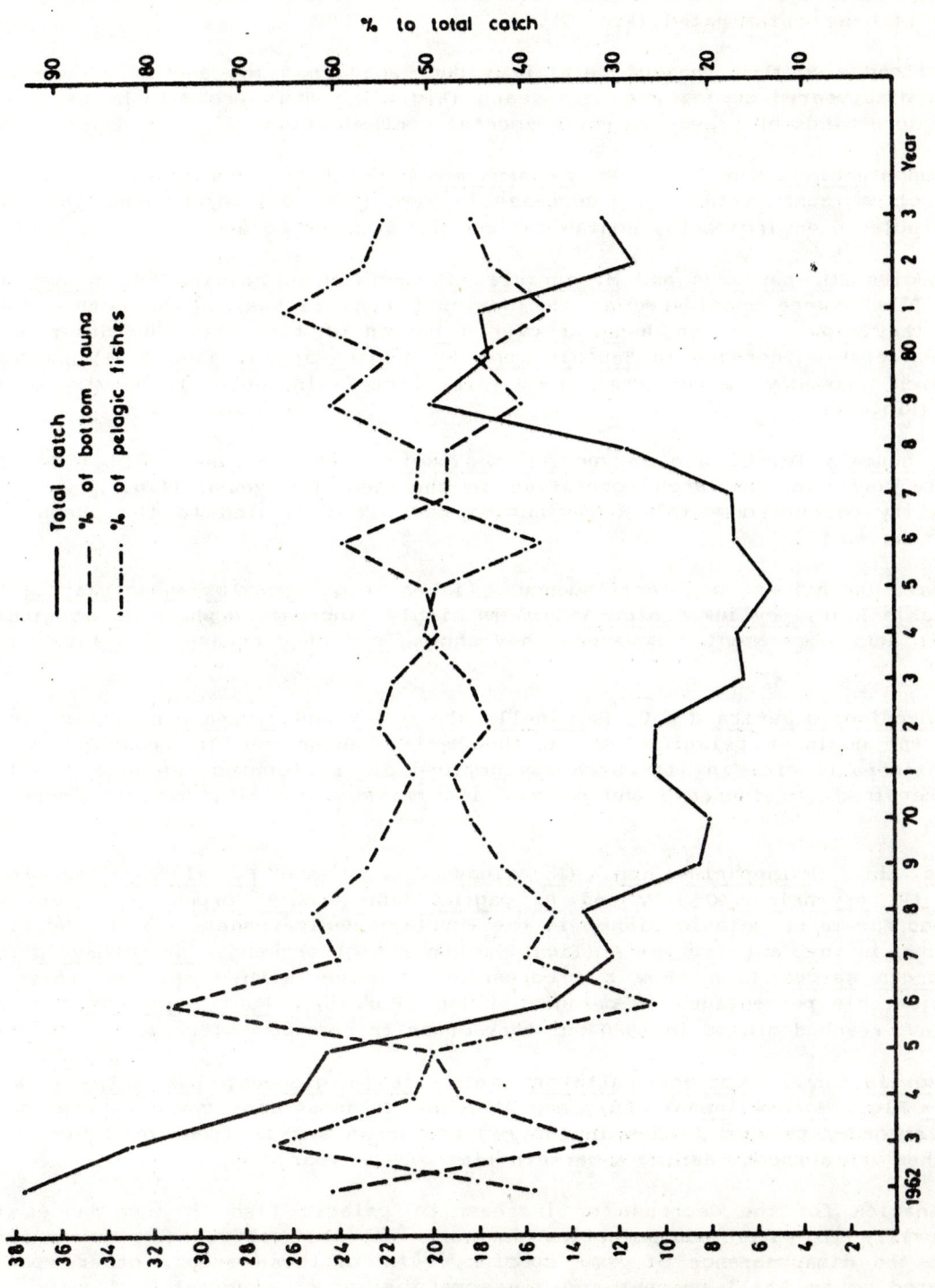

Figure 1. Annual variation of total catch, % of bottom fauna, and % of pelagic fish in the Mediterranean sea in front of Egypt.

sediments would be exposed to substances released by such polluted sediments (Bezard et al., 1985), which means high lethal and sublethal effects of environmental pollution on the bottom fauna (Saleh, 1983).

Annual investigation of individuals in the bottom population of the Mediterranean sea in front of Egypt showed that there is a significant increase in crabs Portunus pelagicus, and prawns caught from the deep sea, which means that Portunus pelagicus is only slightly vulnerable to environmental contamination and the migration of prawns to deep sea which is less contaminated (Fig. 2).

It was noticed also that Penaeus kerathurus decreased in time, and it became very scarce or even disappeared during the last years (Fig. 2). This probably means a high sensitivity of such kinds of prawns to environmental contamination.

Metapenaeus stebbingi Nobili and Metapenaeus monoceros Fabricius represent constant values in the bottom fauna, with slight decrease in time (Fig. 2), which means that they are less vulnerable to environmental contamination than Penaeus kerathurus.

Mullus species (M. barbatus and M. surmuletus) and Saurus species (S. myops Schi and S. tumbil Bloch) were considered as the dominant bony fishes in the bottom fauna (Fig. 3), but they show a certain decrease over a period of time. On the other hand, there is a considerable increase in Teuthis spp. (T. sigana and T. java Lin.) and Solea solea Lin. which probably means that they are less vulnerable to environmental contamination (Fig. 3).

Trichurus haumela Forsk. and Merluccius merluccius Lin. decrease with time, and they are rarely found in the catch population in the last few years (Fig. 3). Their high vulnerability to environmental contamination may be attributed to the absence of scales on their body.

The annual variations of cartilagenous fishes e.g. Squalus acanthias Lin., Ancanthia vulgaris Riss, Seylium canicula Cuv. is highly fluctuating which is attributed mainly to their active movement. However, they show a slight decrease with time (Fig. 4).

Sardine (Sardinella aurita C & V, Sardinella eba C & V and Clupea poncita) could be considered as the dominant pelagic fish in the Mediterranean sea in front of Egypt. However, a considerable drop in its catch was noticed after stopping the Nile flood in 1965, but it regained its increase and reached its maximum in 1979, then it decreased again (Fig. 5).

Box loops Lin., Epinephelus spp. (E. gigas C & V and E. alexandrinus Bod), Pagellus spp. (P. erythrinus C & V and P. pagrus Linn and P. orphus C & V) were considered among the major pelagic fishes in the Egyptian Mediterranean for a long time, but they declined in the last few years (Fig. 5). On the other hand, Chrysophrys aurata C & V and Sargus sargus Linn show a progressive increase with time, and they now represent considerable percentages in pelagic fishes (Fig. 5). Mugil cephalus Lin. and Mugil capito Cuv. reached minima in 1980 but they began to increase afterwards (Fig. 5).

Sciaena aquila Cuv., Temnodon saltator Lin., Lichia glaucus, Box salpa C & V, Morone punctata Blg., Morone labrax Blg., and Atherina hepsetus Lin. could be considered as minor and temporary pelagic fishes in the Mediterranean sea in front of Egypt (Fig. 6). Most of them are abundant during a certain time of the year.

The explanation for the decrease or increase of pelagic fish in accordance with their vulnerability to water contamination is hard and not valid. Sindermann (1980) explained that the disappearance of some species, even in these areas, other species might be expected to be still present and, in some instances, abundant. This is true particularly of a number of coastal estuarine dependent fish species, many of whom spent much of their life cycles in waters that are to some extent contaminated.

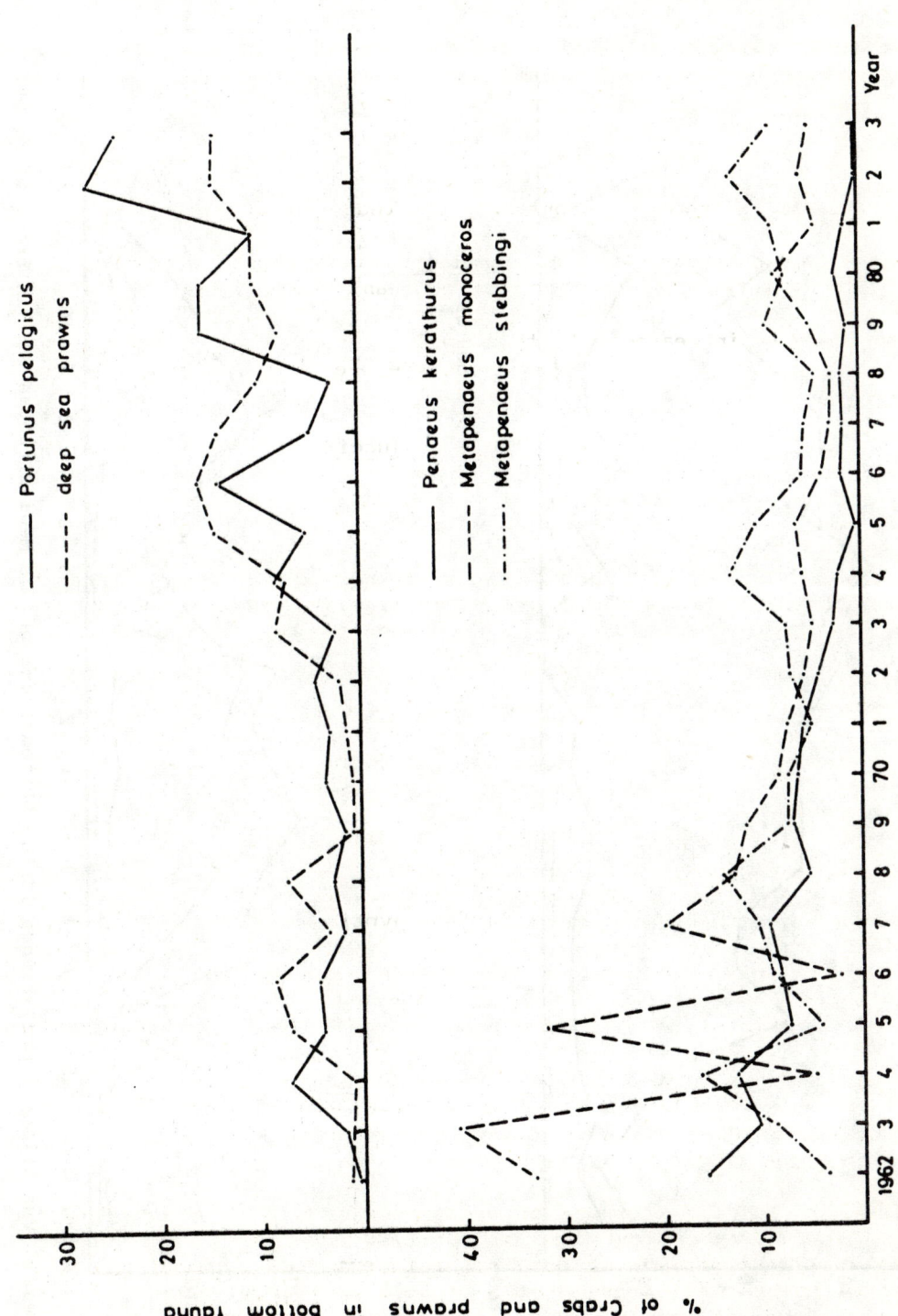

Figure 2. Annual variation in the percentages of prawns and crabs in the bottom fauna of the Mediterranean sea in front of Egypt.

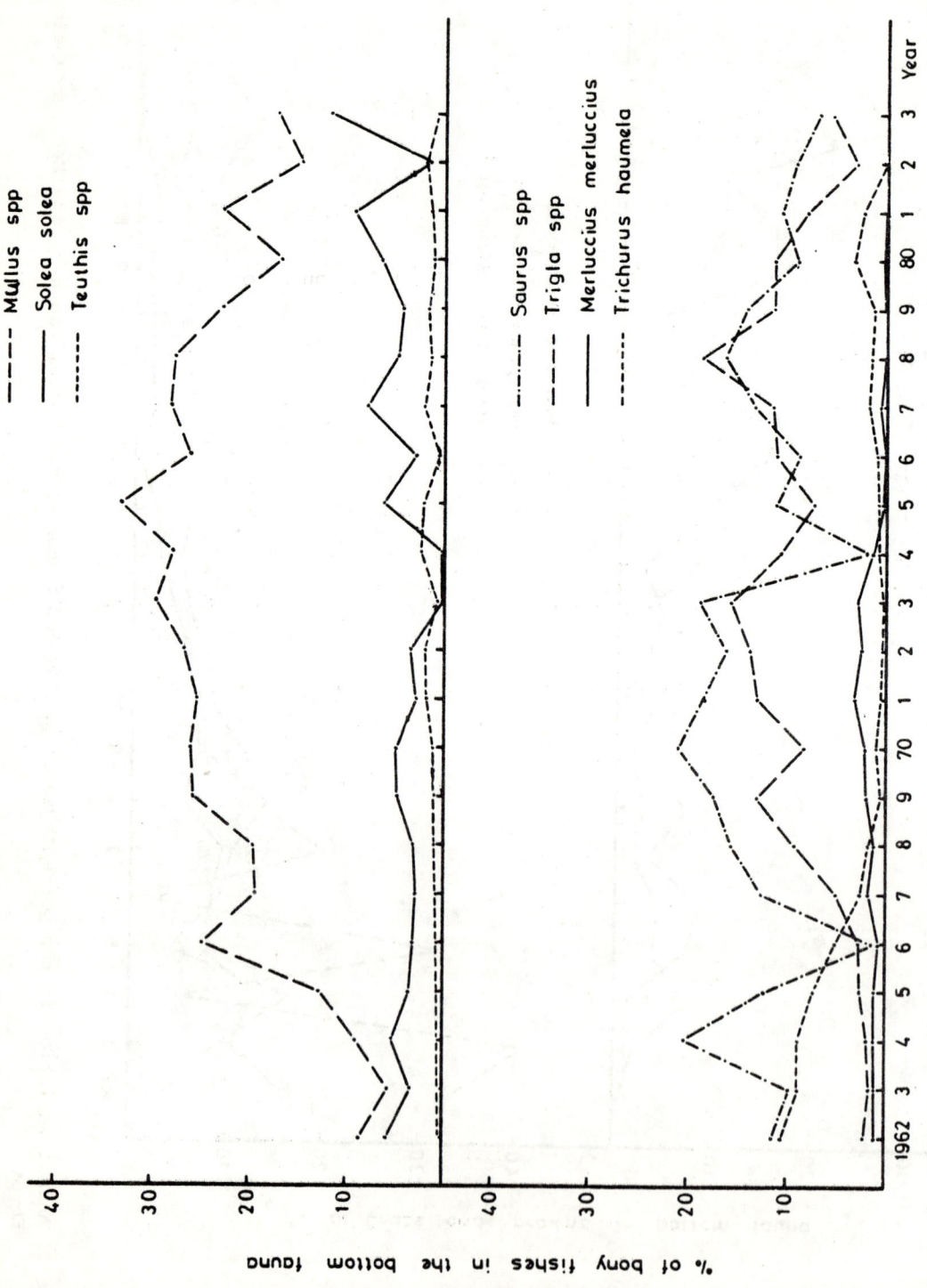

Figure 3. Annual variation of bony fishes in the bottom fauna of the Mediterranean sea in front of Egypt.

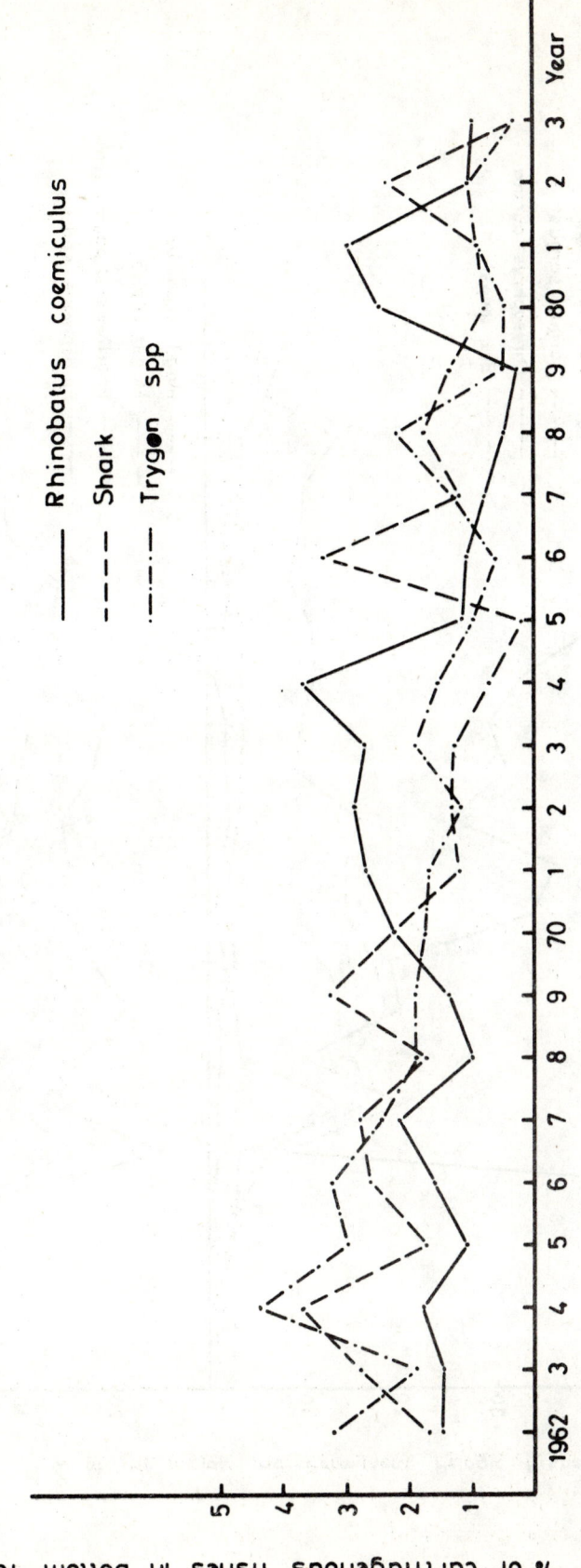

Figure 4. Annual variation of cartillagenous fishes in the bottom fauna in the

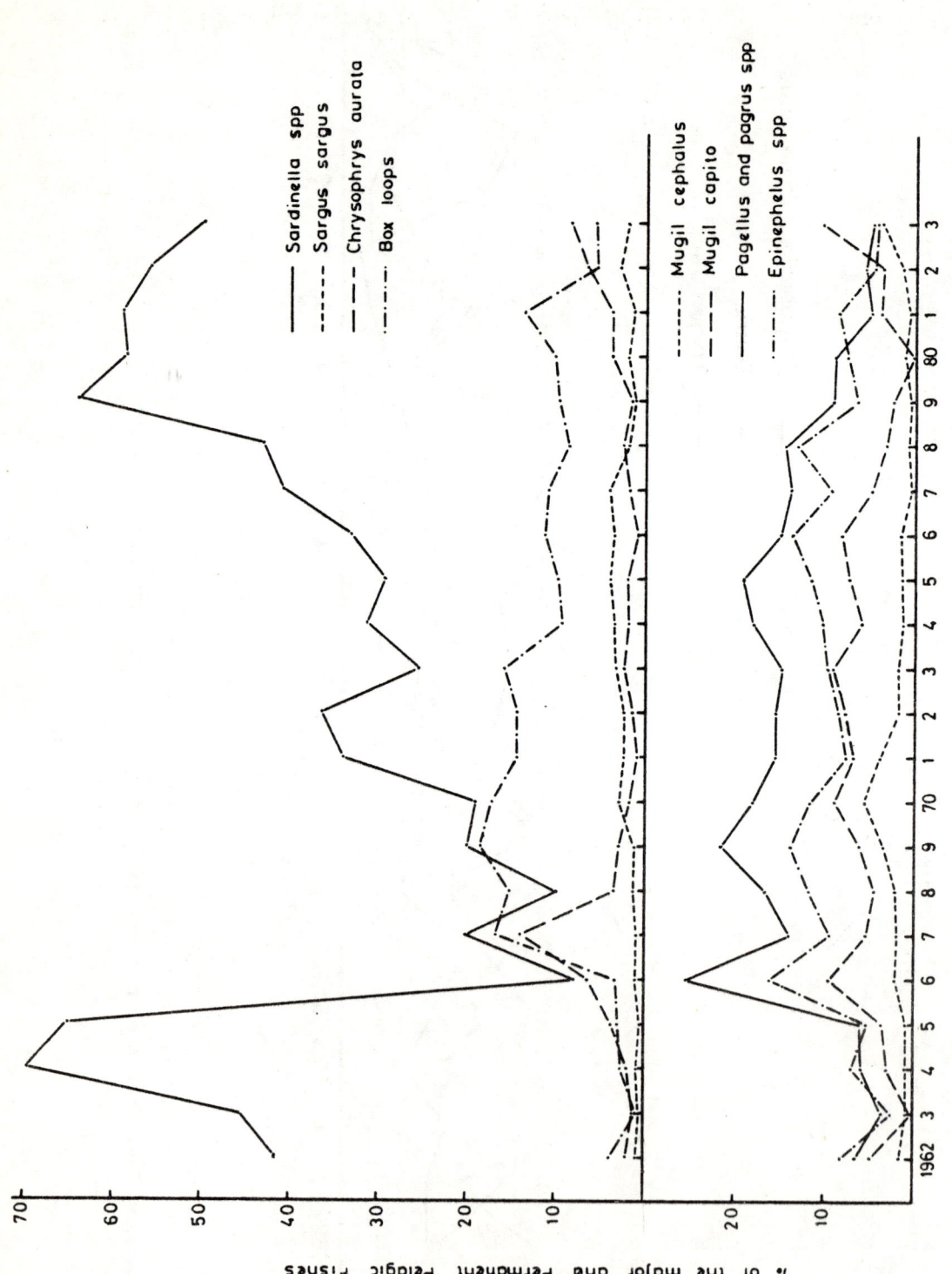

Figure 5. Annual variation of the major and permanent pelagic fishes in the Mediterranean sea in front of Egypt.

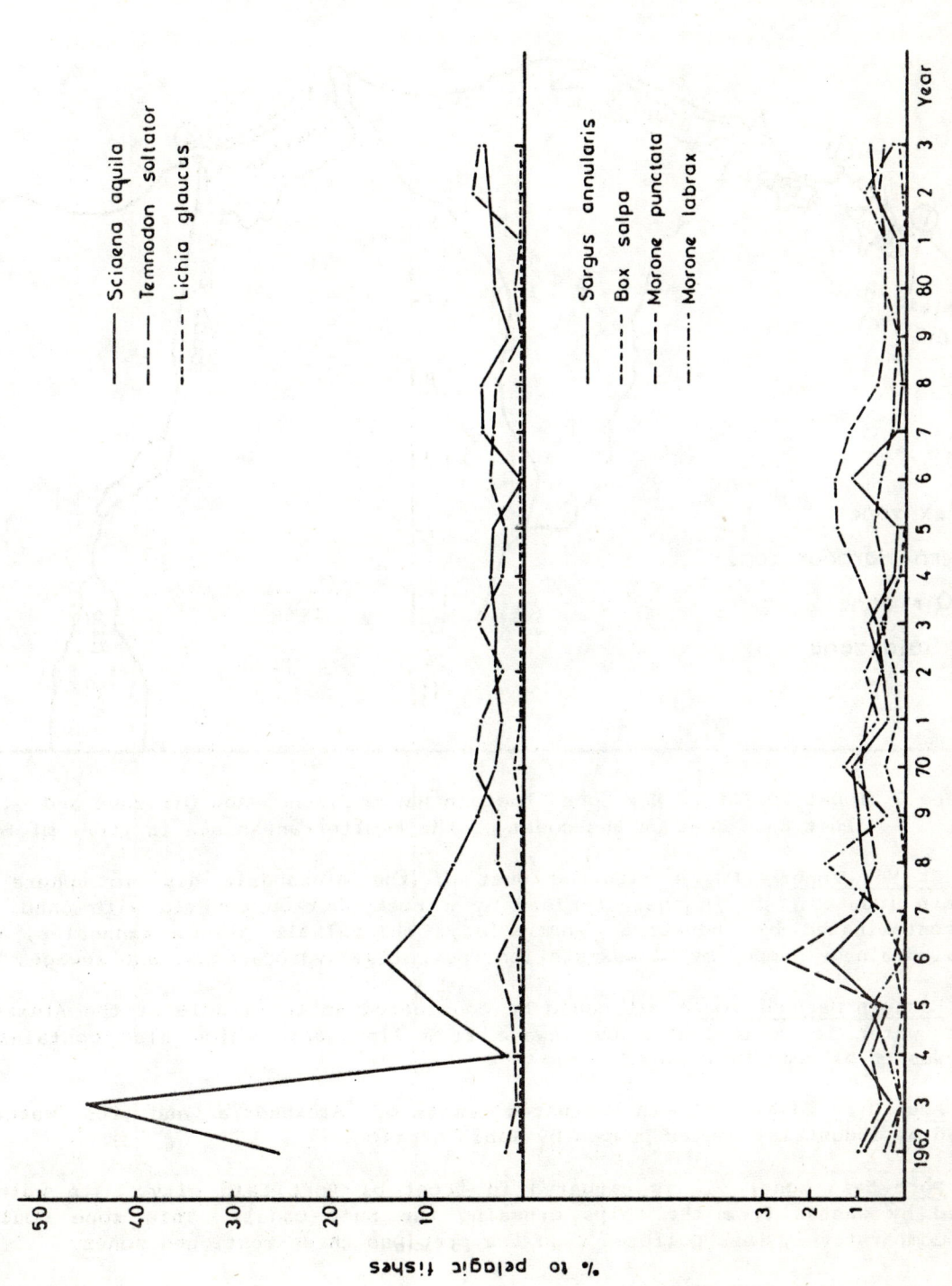

Figure 6. Annual variation of minor or temporary pelagic fishes in the Mediterranean sea in front of Egypt.

On the basis of the idea that bottom fauna are more affected by environmental contamination, the Mediterranean sea in front of Egypt is divided into four zones which are, to some extent, different in water contamination (Fig. 7).

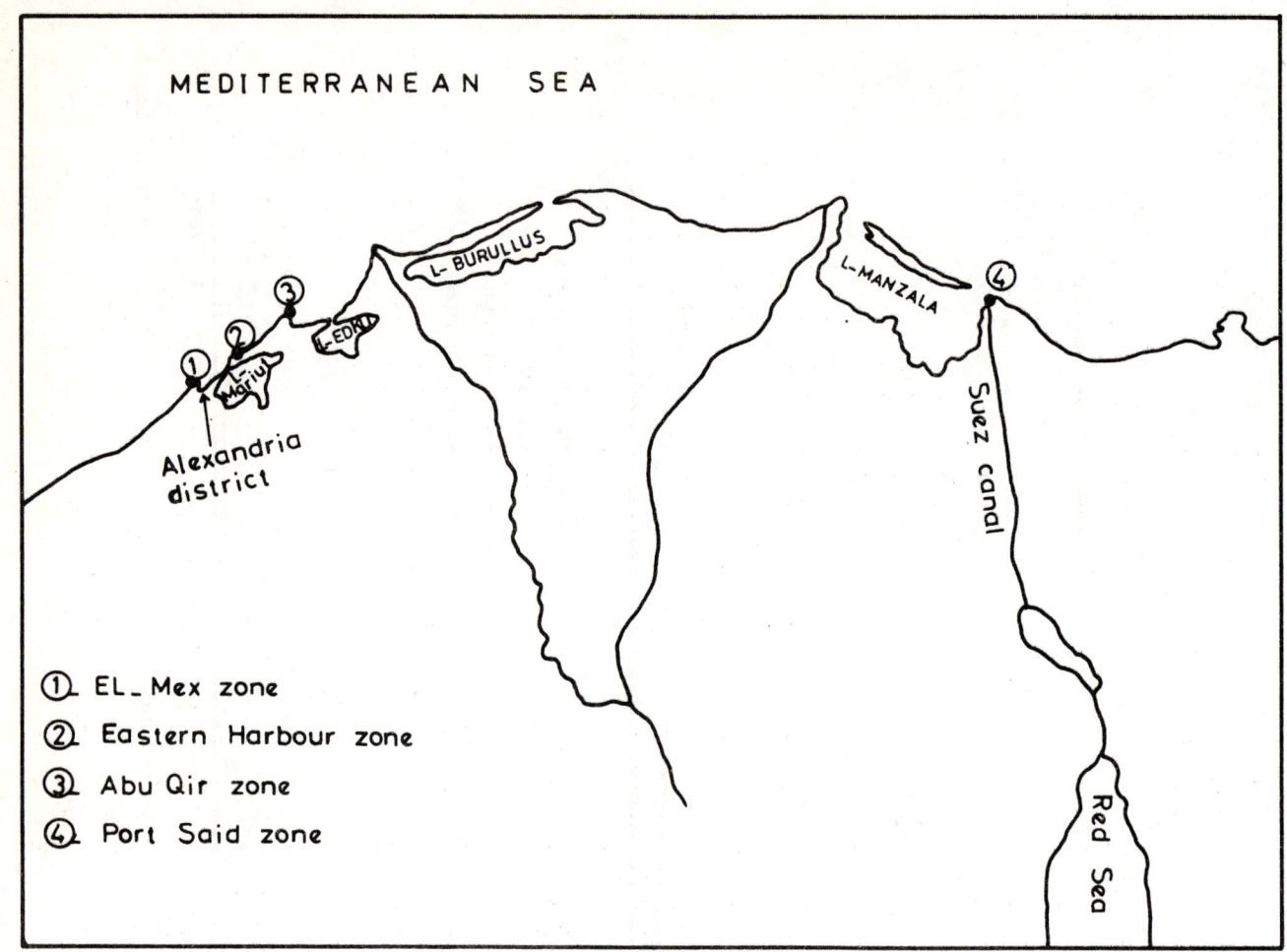

Figure 7. Locations of El Mex zone, Eastern Harbour zone, Abu Qir zone and Port Said zone on the coast of the Mediterranean sea in front of Egypt.

(a) <u>El Mex zone</u>: It is situated west of the Alexandria district where many factories are located. It is characterised by a rocky bottom covered with sand. Its water is contaminated by industrial wastes e.g. chloralkali plant, tanneries, etc., agricultural drainage pumped by El Mex station, petroleum hydrocarbons, and sewage.

(b) <u>Eastern Harbour zone</u>: It could be considered as the middle of the Alexandria coast. Its water is contaminated by sewage from Alexandria which also contains the industrial wastes of some factories.

(c) <u>Abu Qir zone</u>: It is situated east of Alexandria and its water is contaminated by industrial wastes pumped by Tabia station.

(d) <u>Port Said zone</u>: It is situated in front of Port Said city. Its water is contaminated by wastes from the ships crossing the Suez canal. This zone could be considered comparatively less polluted than the previous three mentioned zones.

Catch population in the El Mex zone which is a comparatively more contaminated zone is represented mainly by pelagic fishes, while bottom fauna are poor and decreasing (Fig. 8).

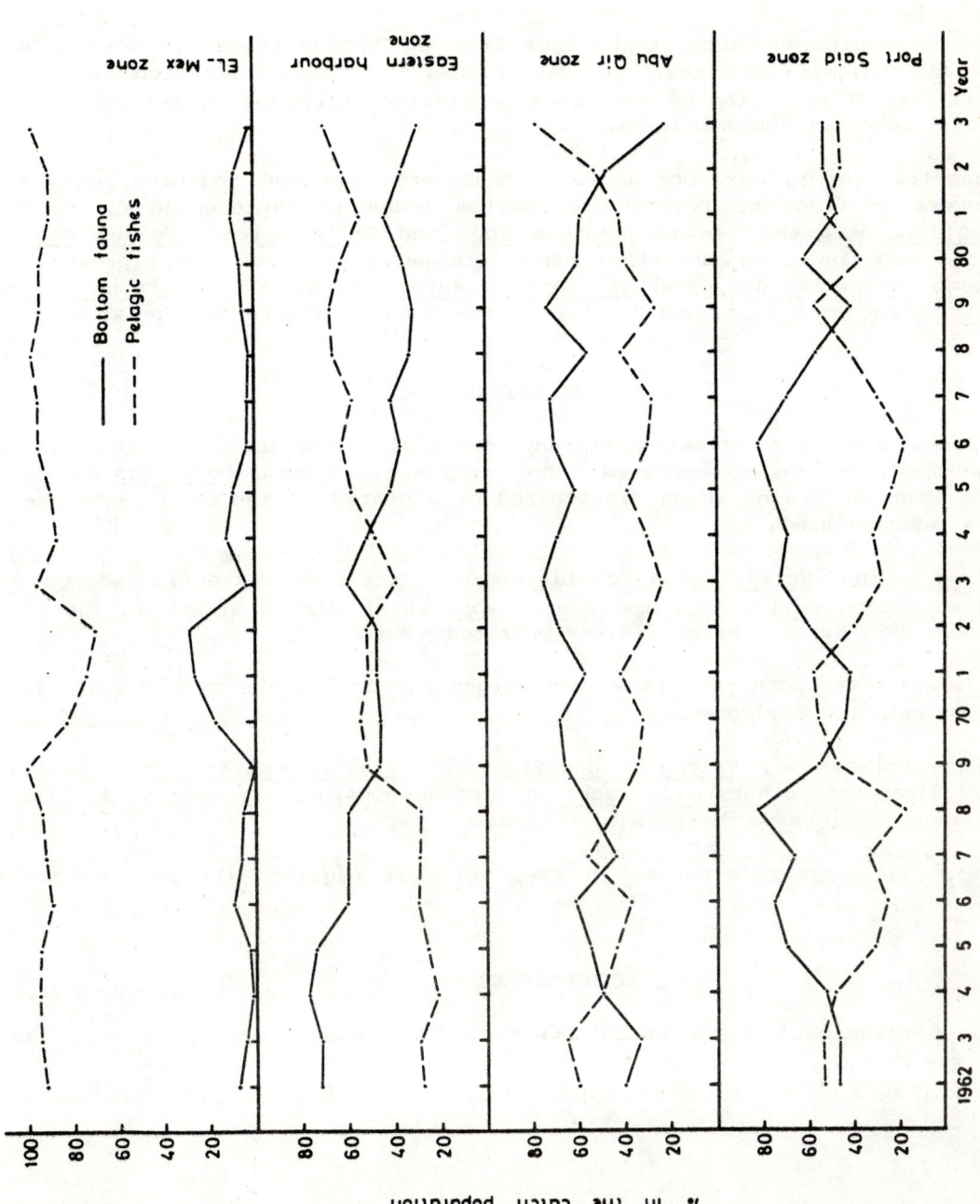

Figure 8. Annual variation in pelagic fishes and bottom fauna in El Mex zone, Eastern harbour zone, Abu Qir zone, and Port Said zone.

Bottom fauna in the catch population of Eastern harbour zone were found in bigger quantities than pelagic fishes. Recently, the reverse applies, and pelagic fishes become dominant in the catch population, which may be attributed to contamination of the bottom (Fig. 8).

Bottom fauna are dominant in the catch population of Abu Qir zone, unless it begins decreasing which may be an indicator of bottom contamination (Fig. 8).

Bottom fauna are dominant in the catch population of Port Said zone which is considered a less polluted area, unless the difference between bottom fauna and pelagic fishes becomes small in the last years (Fig. 8).

The previous results may support the fact that the bottom fauna are more affected by environmental contamination than pelagic fishes, i.e. contamination of aquatic environments firstly affected the bottom fauna population which may be considered as an indicator of environmental contamination.

This means that the El Mex zone could be considered the most polluted zone of the Mediterranean sea in front of Egypt. Its bottom fauna is represented by the crab Portunus pelagicus, deep sea prawns, Teuthis spp. and Solea solea. Mullus spp. and Saurus spp. are very few. On the other hand, its pelagic fishes are represented by Sardinella spp., Sargus sargus, and Chrysophrys aurata (Fig. 9). Epinephelus spp., Morone punctata, Temnedon saltar, and Mugil spp. are minor in the catch population.

4. CONCLUSIONS

Bottom fauna are more affected by environmental contamination. Most of its individuals decreased or even disappeared. The crabs e.g. Portunus pelagicus can resist pollution more than the prawns which disappeared or migrated to the deep sea where the environment is less polluted.

Teuthis spp. and Solea solea could resist, to some extent, environmental contamination more than Mullus spp. and Saurus spp. which were dominant in the bottom fauna and becomes very scarce, mainly in the polluted zones.

Sardinella spp., Chrysophrys aurata, and Sargus sargus are the main pelagic fishes found in the contaminated environment.

Some fish species, e.g. Trichurus haumela and Merluccius merluccius become very scarce or even disappear, others e.g. Pagellus species considerably decreased, although it was the dominant fish among the pelagic fishes.

Mugil spp. considerably decreased in 1980, then it regained its position in the catch population.

5. ACKNOWLEDGEMENT

The authors thank Mrs. E.M. Mesbah and Mrs. Z.A. Helmy, for their help during collection of the data.

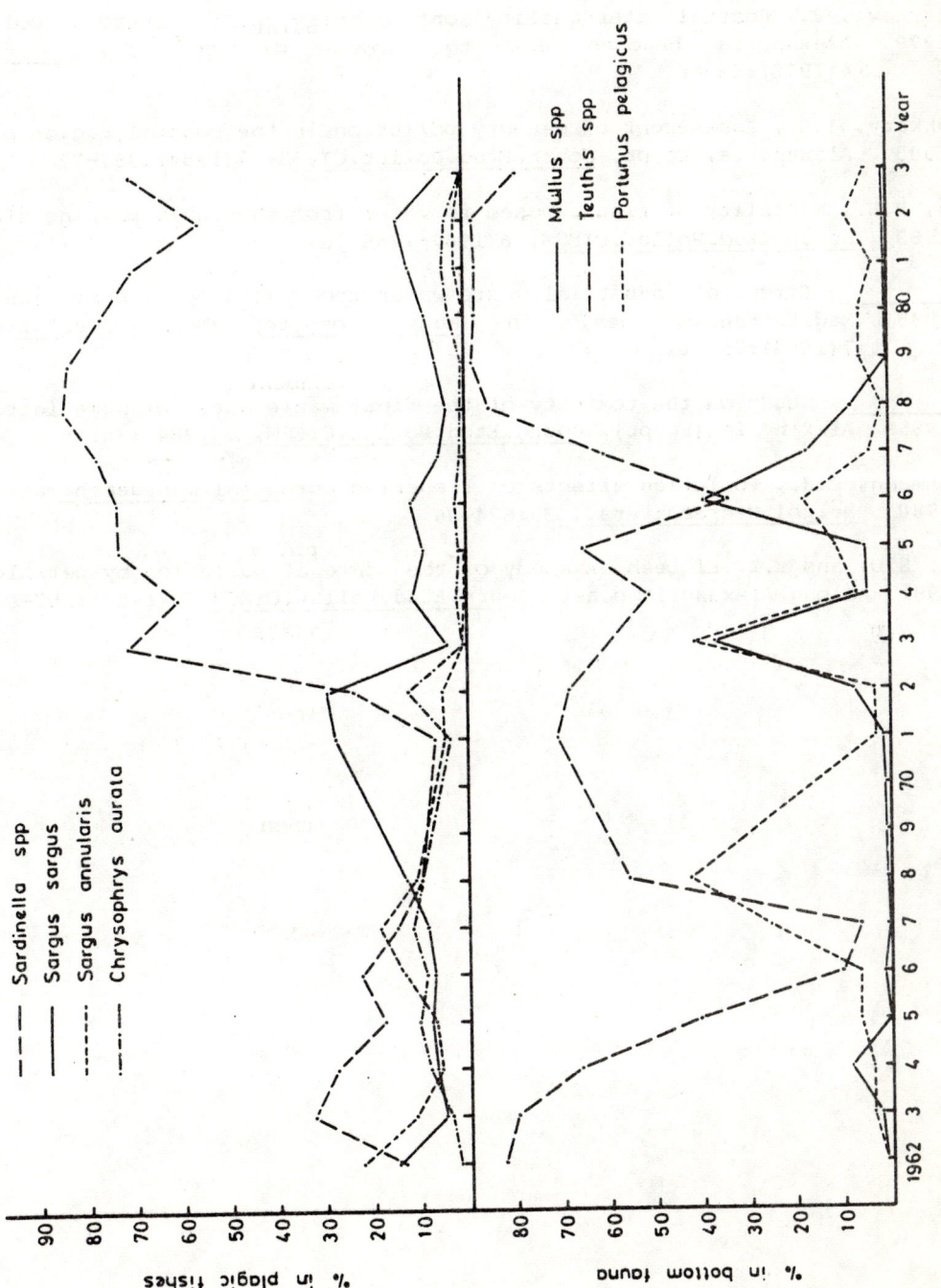

Figure 9. Annual variation of the dominant fishes and crustaceans in El-Mex zone, Alexandria.

6. REFERENCES

Aboul Dahab, O. and Y. Halim, Oil pollution of the marine environment in the area of
1981   Alexandria, Journ.Etud.Pollut.CIESM, 5(1980):201-8

Bezard, D., A. Veglia and R. Vaissiere, Mercury and cadmium concentrations in three
1985   marine benthic organisms. Variations according to the sampling areas in the
         canal de Corse. Journ.Etud.Pollut.CIESM, 7(1984):309-15

El Sebae, A.H. and M. Abo ElAmayem, A survey to determine potential pollution of the
1979   Mediterranean by pesticides from the Egyptian region. Journ.Etud.Pollut.
         CIESM, 4(1978):149-54

El Sharkawi, F., Coastal water quality control, Study of the state of pollution of
1979   Alexandria beaches due to sewage discharge. Journ.Etud.Pollut.CIESM,
         4(1978):523-6

El Sokkary, I.H., Assessment of mercury pollution in the coastal region of El Mex at
1985   Alexandria, Egypt. Journ.Etud.Pollut.CIESM, 7(1984):367-72

Saleh, H.H., Mortality of fish exposed to water from the Tabia pumping station, Egypt.
1983   Journ.Etud.Pollut.CIESM, 6(1982):765-70

_____, Effect of industrial waste water from Tabia pumping station in Abu Qir Bay,
1985   Mediterranean Sea, on Mugil capito Cuv. Journ.Etud.Pollut.CIESM,
         7(1984):757-61

_____, Study on the toxicity of the final waste water of some factories in
1985a   Alexandria (Egypt). Journ.Etud.Pollut.CIESM, 7(1984):753-6

Sindermann, C.J., Pollution effects on fisheries potential management activities.
1980   Helgol.Meeresunters., 33:674-86

Wahby, S.D. and K.Z. El Deeb, A study of the state of pollution by petroleum hydrocarbons
1981   along Alexandria coast. Journ.Etud.Pollut.CIESM, 5(1980):257-62

L'EXPERIMENTATION in situ EN TANT QUE CRITERE D'EVALUATION DES EFFETS DE LA
POLLUTION: ETUDE DES MECANISMES EVOLUTIFS ET DU REPEUPLEMENT DE SEDIMENTS POLLUES

par

G. STORA*, A. ARNOUX** et C. DIANA**

* Centre d'Océanologie de Marseille-Station marine d'Endoume, Rue de la batterie
  des Lions, 13007 Marseille, France.

** Laboratoire d'Hydrologie et de Molysmologie Aquatique, Faculté de Pharmacie,
   13385 Marseille cedex 4, France.

1. INTRODUCTION

Depuis de nombreuses années, les études visant à la compréhension des effets et des mécanismes d'action de différents facteurs altéragènes sur des écosystèmes benthiques marins et, plus particulièrement, en Méditerranée, utilisent deux voies de recherches complémentaires que sont les recherches dans le milieu naturel (Bellan, 1985) et les études expérimentales in vitro. Il apparait que pour ces dernières, se pose avec acuité le problème des relations entre les résultats obtenus en laboratoire et le milieu naturel. Si des réponses peuvent y être apportées (Stora, 1984), l'expérimentation in situ véritable compromis entre les études classiques dans le milieu naturel et les recherches en laboratoire offre l'avantage indéniable d'appréhender avec la plus grande fiabilité les processus évolutifs des écosystèmes benthiques marins soumis à perturbation (Hansen, 1982).

D'une manière générale, parmi les études expérimentales in situ dont une revue a été établie récemment par Dauvin(1984), rares sont celles réalisées dans un but écotoxicologique. On peut citer les études de colonisation de substrats en milieu pollué de Reish(1961), MacCall(1977), Hannan(1981) ou celles sur la dynamique de populations dans des sédiments sélectivement contaminés réalisées dans des bassins de simulation par Kuiper et al.(1984). Les travaux de Kalke et al. et de Tagatz et Deans(1983) sont intermédiaires entre les expérimentations in situ et in vitro, des modules colonisés dans le milieu naturel par des populations benthiques étant ramenés et contaminés selectivement en laboratoire par divers altéragènes.

Dans les expériences faisant l'objet du présent travail, nous avons choisi d'étudier les possibilités de repeuplements de milieux globalement et "naturellement" pollués caractéristiques d'aires côtières dégradées où la somme des facteurs d'altération supprimait toute possibilité de colonisation. Le but des expériences entreprises visait à étudier les potentialités de réorganisation de peuplements macrobenthiques en fonction de la cinétique des polluants à la suite d'une réduction notoire des sources d'altéragènes. Cette simulation expérimentale de mise en condition plus favorable a pu être obtenue en implantant dans une aire moins perturbée des modules expérimentaux remplis de sédiments provenant de biotopes dégradés.

Trois séries d'expériences ont été réalisées. Dans un but de commodité, nous les avons respectivement désignées par deux noms correspondant aux lieux de prélèvements des sédiments et d'implantation des modules:

- Fos-Fos (Stora 1982, 1983)
- Vaine-Fos (L.H.M.A., 1983)
- Berre-Tamaris (L.H.M.A., 1983, Diaz-Castaneda 1984).

Dans ce travail sont étudiées plus particulièrement la colonisation et l'évolution de la faune Malacologique en fonction de la cinétique des polluants.

2. METHODOLOGIE

Matériel sédimentaire expérimental

La première expérience était destinée à mettre en évidence l'influence propre des conditions expérimentales sur les processus et la dynamique de la recolonisation par la macrofaune benthique d'un sédiment préalablement dépeuplé par chauffage à l'étuve à 50°C pendant 4 jours. Elle a été réalisée avec un matériel expérimental prélevé dans le site d'étude, le golfe de Fos. La deuxième et troisième expérience avaient pour but d'étudier le repeuplement de sédiments provenant de l'étang de Vaine et de Berre (Fig. 1), fortement perturbés par la pollution chimique (Arnoux et al., 1981) (Tab. I) et par une dessalure artificielle due aux rejets d'une usine hydroélectrique dans le nord de l'étang de Berre. Dans le sédiment de Vaine seul Cardium glaucum à l'état juvénile était faiblement représenté, le sédiment de Berre était azoïque du point de vue macrobenthos.

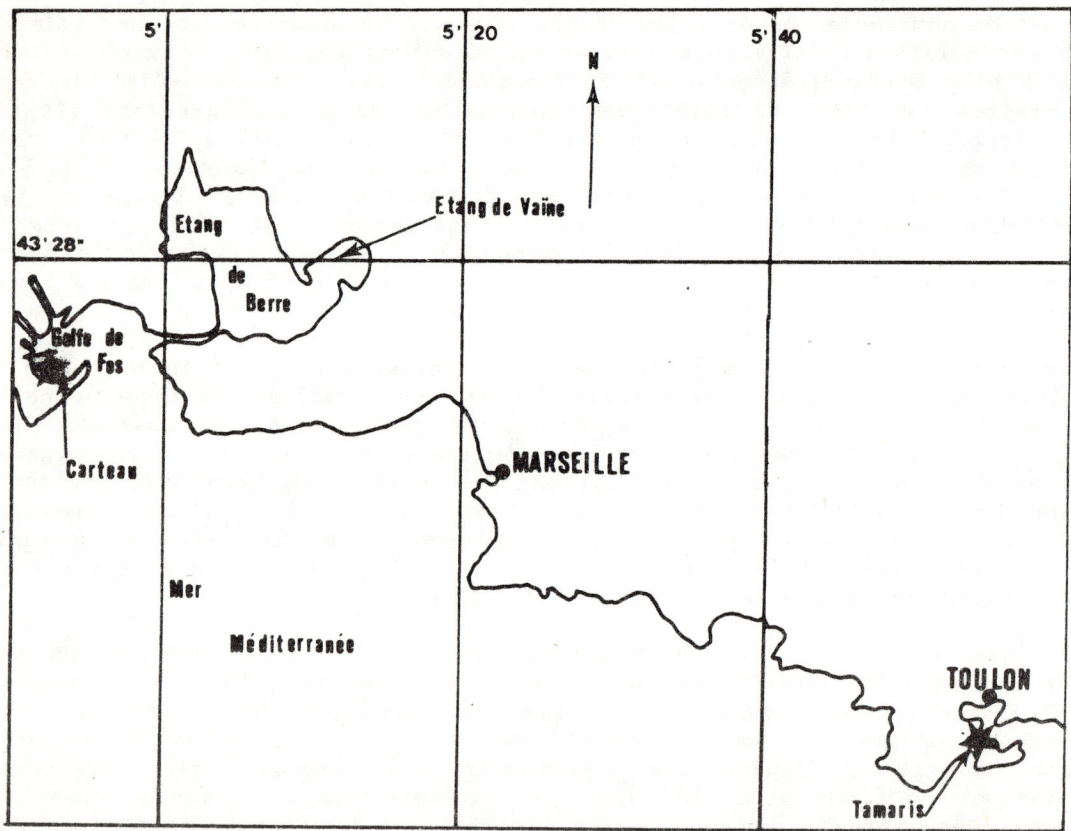

Figure 1. Lieux de prélèvements et d'implantations des sédiments.

Les sites d'études

Les sites d'études (Fig. 1) se situent dans les petits fonds de l'anse de Carteau à l'ouest du golfe de Fos et à Tamaris dans la baie du Lazaret au sud ouest de la rade de Toulon. Leurs choix reposent tout d'abord sur des caractéristiques communes aux deux sites:

- Le faible hydrodynamisme des biotopes évitant toute perturbation des modules expérimentaux tout en assurant un bon renouvellement des eaux surnageantes.

- La nature identique des peuplements en place correspondant à un assemblage de sables vaseux de mode calme (SVMC) (Peres et Picard, 1964). Leur richesse qualitative et quantitative et leur homogénéité permettent une étude représentative sur de faibles quantités de sédiments (Stora, 1982). De plus, ces peuplements sont semblables à ceux qui se developpaient dans les sables vaseux de l'étang de Vaine et de Berre (Febvre, 1968) avant que n'intervienne leur dégradation totale. Bien que présentant des distributions granulométriques quelques peu différentes (Tab. I), la nature physique des sédiments expérimentaux, contrairement à leur composition chimique ne constituait pas de ce fait, un facteur de perturbation majeur dans les processus de colonisation.

Les différences entre les sites d'études tiennent à:

- La moindre pollution des sédiments de Carteau par rapport à ceux de l'étang de Vaine (Tab. I) soumis neammoins à une influence rhodannienne génératrice de dessalure du milieu (Diana, 1983).

- La pollution plus importante des sédiments de Tamaris par rapport à ceux de l'étang de Berre (Tab. I) mais une salinité plus élevée et plus stable du milieu par rapport au site de prélèvement.

Ces différences entre les deux biotopes permettaient ainsi d'appréhender l'incidence respective de la pollution et de la stabilité du milieu environnant sur les processus de colonisation et la dynamique des peuplements des modules.

Tableau I

Caractéristiques granulométriques et chimiques (Phase solide) des sédiments de l'étang de Berre et de Vaine, du golfe de Fos et de la baie de Tamaris.

	Fractions granulo.			C.organique %	Hydrocarbures totaux mg/100	Zn	Pb	Cu	Cd	Hg
	$>$2mm	2$<$f$<$63	$<$63µ			---------µg g^{-1}------------				
Vaine	.45	21.1	78.4	7.17	652	400	273	104	5.4	1.12
Fos	1.4	41.6	57	0.90	14.1	207	33	20.7	1.7	0.26
Berre	3.1	21.5	75.4	2.14	21.8	394	138	29.5	3.2	0.48
Tamaris	27.8	36.5	35.7	4.26	79.5	661	185	170	.48	1.36

Techniques expérimentales

Les modules utilisés pour l'étude de la colonisation des sédiments pollués sont constitués par des récipients en polyéthylène de section carrée de 25,5 cm de côté et de 11 cm de hauteur (Stora, 1982). Le choix de ces modules se justifie par leur maniabilité en plongée et par la surface et le volume de sédiments qu'ils permettent d'utiliser (0.065m^2 et 6 dm^3). Cette surface et ce volume sont proches de ceux obtenus par la benne Orange peel (0.085 m^2-4.5 dm^3) qui correspondent à une aire et un volume minimal susceptibles de permettre la récolte de la majorité des espèces dans les biotopes étudiés. L'expérience Fos-Fos a débuté le 31 juillet 1976, celle de Vaine-Fos le 16 juillet 1981 et celle de Berre-Tamaris le 3 Mars 1982. Le point de départ de chaque expérience a consisté à la mise en place de 24 modules contenant les sédiments expérimentaux préalablement homogénéisés, sur des fonds de 3 à 5 m (Fos-Fos 3m, Vaine-Fos 4m,, Berre-Tamaris 5m). Dans les deux derniers cas, l'intervalle de temps entre le prélèvement des sédiments et la mise en place des modules n'a pas excédé 24 heures.

Le programme d'étude établi dans chaque cas pour couvrir un cycle annuel (Tab. IV) prévoyait le relèvement mensuel de deux modules destinés uniquement à l'analyse biologique dans la première expérience (Fos-Fos) et, à l'analyse chimique et aux déterminations biologiques dans les deux autres. Les relèvements de modules ont été chaque fois accompagnés du prélèvement de sédiment témoin (4.5 dm^3-0.085 m^2) collecté à quelques mètres de chaque site expérimental afin de suivre l'évolution du peuplement en place pris comme référence. Ce sédiment témoin a fait l'objet d'analyses chimiques, non systématiques dans l'expérience Vaine-Fos et régulières à Tamaris. La destruction des modules en Janvier 1982, vraisemblablement provoquée par un chalutage sur le site d'expérience nous a contraint d'interrompre l'étude Vaine-Fos après 6 mois d'expérimentation.

Analyse chimique

Les différents paramètres physiques et chimiques analysés sont donnés dans les tableaux I, II et III. L'eau interstitielle a été obtenue après centrifugation des sédiments à 5000 g et filtration sur filtres WHATMAN GF/C préalablement lavés à l'acide nitrique et rincés à l'eau distillée haute pureté. Les techniques analytiques employées sont décrites par Arnoux et al.(1981a) pour la phase solide des sédiments et par Jerome(1982) pour l'eau interstitielle.

Analyse biologique

Les prélèvements ont été traités selon les méthodes standards (Stirn, 1982), le tamisage des sédiments naturels et expérimentaux étant effectué sur un tamis de maille diagonale de 1 mm. Dans l'expérience Fos-Fos, le tri de deux modules n'a été réalisé que sur les prélèvements des cinq premiers mois. Différentes méthodes analytiques et synthétiques ont été employées pour étudier et comparer la faune malacologique des modules et du milieu naturel de référence:

(a) Coefficients d'affinité

  (i) Coefficient qualitatif de Sorensen (Dajoz, 1972)

$$C = (2c/a+b) \ 100$$

  c = Nombre d'espèces communes aux prélèvements a et b.
  a = Nombre d'espèces du prélèvement a
  b = Nombre d'espèces du prélèvement b

  (ii) Coefficient quantitatif de Sanders (1960)

Chaque espèce commune à deux prélèvements est affectée d'une valeur égale à la dominance la plus faible obtenue dans l'un ou l'autre des deux prélèvements. Le degré d'affinité est obtenu en additionnant les valeurs des dominances minimales de toutes les espèces communes, il est exprimé en pourcentage (Guille, 1970).

(b) Diversité

Nous avons utilisé l'indice de Shannon-Weaver(1963), le plus couramment employé. Il est exprimé en bits/individus:

$$H' = - \sum_{i=1}^{i=s} p_i \log p_i$$

où S est le nombre d'espèces et p_i la fréquence relative de chaque espèce.

(c) Analyse factorielle des correspondances (AFC)

Ce type d'analyse tend à se developper pour l'étude du domaine benthique, Guille et Ponge(1975), Reys(1976), Desrosiers et Brethes(1984), Hily(1984), Dauvin(1984a). L'analyse factorielle des correspondances a été effectuée à partir des logiciels du service informatique du centre d'océanographie de Marseille sur miniordinateur HP 1000. Nous n'entrerons pas dans les détails de calcul developpés par différents auteurs (Benzecri, 1973; Legendre et Legendre 1979; Lebart et al., 1982), mais rappellerons simplement quelques principes.

La base de cette méthode est de fournir la représentation simultanée dans un seul plan de projection de l'ensemble des lignes et des colonnes d'un tableau de contingence. La projection des nuages des points observations et des points espèces sur le plan des deux premiers axes factoriels donne la meilleure représentation plane possible (Daget, 1976). L'analyse des résultats est fondée sur l'examen des contributions absolues et relatives et sur la représentation graphique des nuages de points.

3. RESULTATS

Les polluants

L'analyse des concentrations de polluants minéraux de la phase solide ne permet pas de mettre en évidence une évolution marquée des sédiments transplantés. Cela était prévisible au moins pour les métaux dont les teneurs sont très importantes par rapport à celles éventuellement libérées dans les eaux interstitielles (L.H.M.A., 1983). Comme le montrent les tableaux et les figures 2 et 3, l'eau interstitielle des sédiments présente des variations relativement plus importantes.

(a) Expérience Vaine-Fos

Dans l'expérience Vaine-Fos, on peut distinguer une évolution de l'eau interstitielle des modules en deux périodes. La première phase, au cours des premiers mois se singularise par une diminution rapide de l'ammonium et une augmentation très progressive de la salinité, démontrant la lenteur d'imprégnation des sédiments des modules par l'eau surnageante et, par conséquent, le remplacement progressif de l'eau interstitielle d'origine. Le deuxième mois est marqué par un relargage important du zinc, du cuivre et du plomb dont les concentrations atteignent des valeurs extrêmement élevées. Durant cette période, seul le cadmium participe peu à l'enrichissement en métaux de l'eau interstitielle.

Nous attribuons ce phénomène général apparaissant durant cette phase initiale à un processus de mise en équilibre du sédiment expérimental avec le milieu ambiant.

La deuxième phase plus tardive est marquée en décembre par un nouveau relargage de métaux lourds se caractérisant par un pic important des concentrations de cuivre et de cadmium. Le plomb ne participe que modérement a ce relargage. Durant cette période les taux d'ammonium, de nitrates et de phosphates hydrolysables augmentent sensiblement (Tab. II). Il semblerait que l'on assiste ici à un mécanisme différent de la première phase avec intervention d'une oxydation du sédiment favorisant un nouveau relargage plus selectif que le premier.

Les mesures ponctuelles réalisées dans le milieu naturel montrent de juillet à octobre, une évolution concomitante de l'ensemble des métaux lourds dans l'eau interstitielle des sédiments de référence et des modules. D'octobre à janvier, ceci s'observe essentiellement pour le cuivre et le plomb (Arnoux et al., 1985).

(b) Expérience Berre-Tamaris

Dans l'expérience Berre-Tamaris une évolution chimique très variable dans le temps apparait aussi bien dans l'eau interstitielle des modules que dans celle du sédiment de Tamaris (Tab. III et Fig. 3).

Tout comme dans l'expérience précédente on distingue une première phase de 4 mois correspondant à la mise en équilibre des sédiments des modules avec le milieu ambiant avec l'augmentation de la salinité, la diminution de l'ammonium et le relargage marqué de zinc, du plomb et du cuivre. Par la suite l'évolution des métaux lourds est beaucoup plus complexe car des phénomènes de relargage se produisent, non seulement dans les modules, mais également dans le sédiment de Tamaris dont la charge en métaux lourds est sans aucun doute en relation avec la forte pollution subie par la région littorale toulonnaise.

A l'exception des 4 premiers mois on peut constater que si le zinc évolue de la même manière dans le milieu naturel et expérimental, les autres métaux indépendamment des concentrations analysées évoluent quelquefois avec un certain décalage dans le temps. Les pics de plomb observés dans les modules en juillet, septembre et janvier, précédent respectivement ceux d'août, novembre et mars dans le milieu naturel. Le phénomène inverse s'observe pour le cuivre, les fortes teneurs analysées en juillet et septembre précédent le pic d'octobre du milieu expérimental. De même si le cadmium évolue de manière similaire du début de l'expérience jusqu'en octobre, le pic de novembre dans le milieu naturel précède celui de janvier dans les modules. Nous reviendrons sur ces décalages au cours de la discussion.

Tableau II

Expérience Vaine-Fos. Composition de l'eau interstitielle des sédiments des modules et du milieu naturel (M= Modules, MN= Milieu naturel).

FOS-VAINE		JUILLET	AOUT	SEPTEMBRE	OCTOBRE	NOVEMBRE	DECEMBRE
pH	MN	7.36				7.06	7.34
	M	7.34				6.9	6.98
rH	MN	15.75				14.05	20.3
	M	9				19.94	23.1
Salinité	MN	34.6			35.2		
g NaCl l^{-1}	M	12.9	27.9		36.3	32	32.9
NH_4^+	MN	167			160		125
µg-at N l^{-1}	M	501	375	210	255	365	205
NO_2^-	MN	2.52			.15		.38
µg-at N l^{-1}	M	.34		.34	.35	.45	1
$O.PO_4^{3-}$	MN	2.98			6.4		7.6
µg-at P l^{-1}	M	22.9		23.7	20.8	22.3	21.6
PO_4^{3-} Hydrol.	MN	2.9			2.1		3.9
µg-at P l^{-1}	M	26.6		9.5	4	13.7	12.8
Zinc µg l^{-1}	MN	246			500		632
Plomb µg l^{-1}	MN	18			18.5		7.8
Cuivre µg l^{-1}	MN	50			4.8		20
Cadmium µg l^{-1}	MN	.83			2.2		1.1

La faune malacologique

(a) Evolution générale

La faune malacologique présentée dans les sites expérimentaux aussi bien dans les prélèvements de référence que dans les modules, correspond à celle d'un assemblage de sables vaseux de mode calme, marquée plus particulièrement par la présence de deux espèces caractéristiques exclusives de cet assemblage, Venerupis aurea et Loripes lacteus (Peres et Picard, 1964).

Dans le milieu naturel, on retrouve le plus souvent en position dominante ou subdominante (Sanders et al., 1982) (Tab. IV), classés en fonction de leur constance au sein des prélèvements, les mollusques Venerupis aurea, Abra alba, Corbula gibba, Loripes lacteus, Abra ovata, Parvicardium exiguum, Mysella bidentata, Nassa corniculum, Spisula subtruncata, Venus verrucosa et Gouldia minima.

Dans les modules (Tab. IV) les espèces les mieux representées, classées comme précedemment sont Venerupis aurea, Parvicardium exiguum, Abra alba, Corbula gibba, Abra ovata, Loripes lacteus, Nassa corniculum et Tricolia pulla.

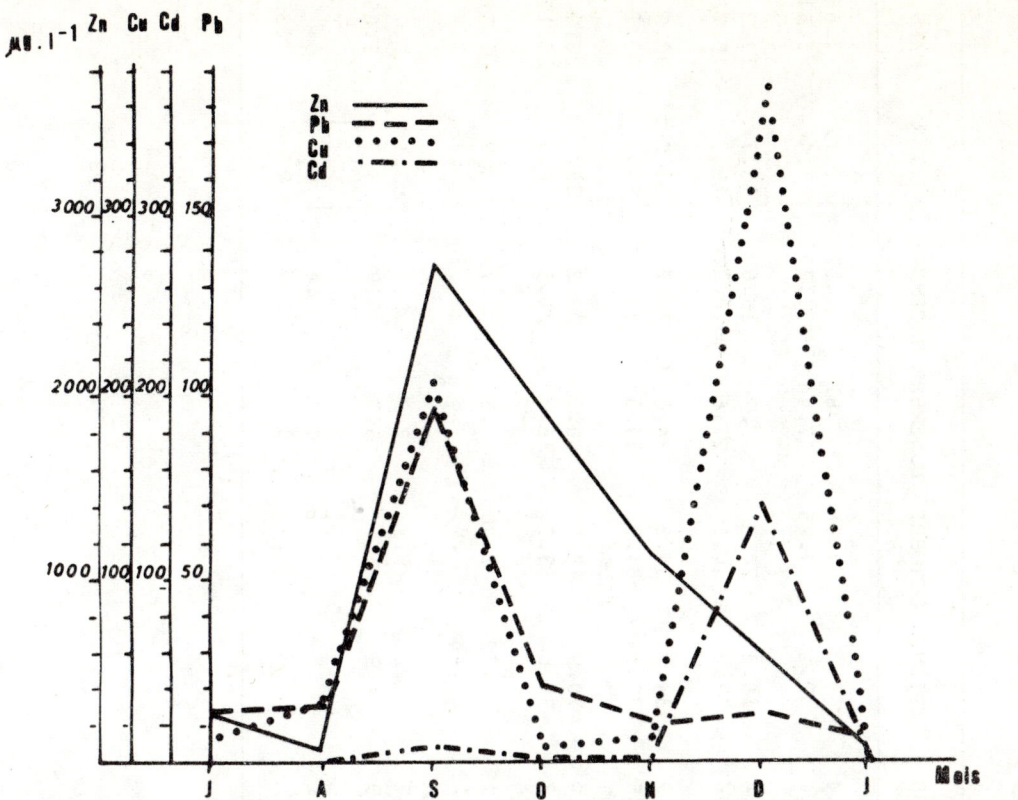

Figure 2. Expérience Vaine-Fos. Variations des métaux lourds dans l'eau interstitielle des Modules

D'un point de vue dynamique, indépendamment des sites d'études, la richesse spécifique et la densité des peuplements dans le milieu naturel sont extrèmement variables, marquées par une succession de phases d'enrichissement et d'appauvrissement de la faune (Fig. 4).

Dans les modules, pour la première expérience Fos-Fos, la colonisation du sédiment est immédiate et très rapidement la richesse spécifique et l'effectif total du peuplement des modules évoluent comme ceux du milieu naturel. (Correlation positive significative $r_S=0.75$, $r_S 0.05=0.63$ par le coefficient de rang de Spearman entre la richesse spécifique du milieu naturel et les modules).

Contrairement à la première expérience, le processus de colonisation des sédiments de Vaine placés à Fos ou ceux de Berre à Tamaris est extrèmement lent. Les premiers mois sont marqués par rapport au milieu naturel par un très faible nombre d'espèces à Fos et une absence totale de Mollusques à Tamaris. Dans l'expérience Vaine-Fos, le nombre d'individus, peu élevé de juillet à octobre, augmente régulièrement pour arriver à une valeur maximale de 146 individus pour 0.1 m^2 en novembre. De novembre à décembre, comme dans le milieu naturel l'effectif total tend à baisser.

A Tamaris la richesse spécifique du peuplement des modules reste faible jusqu'en novembre, puis dépasse celle du milieu naturel en décembre. La densité de la faune malacologique après une faible croissance de juillet à août présente une évolution parallèle à celle du milieu naturel. A partir du mois de novembre l'effectif total des mollusques des modules présente une brusque croissance pour atteindre 143 individus pour 0.1 m^2 en janvier. Au cours de cette période celui du milieu naturel reste appauvri.

Du fait de la variabilité de la richesse spécifique et de l'effectif des populations, la structure des peuplements des modules et du milieu naturel marquée par les indices de diversité évolue en dent de scie (Fig. 4). On peut constater cependant que dans l'expérience Fos-Fos, les indices de diversité dans le milieu naturel et les

Tableau III

Expérience Berre-Tamaris. Composition de l'eau interstitielle des sédiments des modules et du milieu naturel (M= Modules, MN= Milieu naturel).

BERRE-TAMARIS		MARS	AVRIL	MAI	JUIN	JUILLET	AOUT	SEPTEMB.	OCTOBRE	NOVEMBRE	DECEMBRE	JANVIER	MARS
pH	MN	7.48			7.46	7.3	7.4	7.3	7.26	7.62	7.12	7.12	7.66
	M	7.47	7.4	7.13	7.16	7.16	7.2	7.2	7.25	7.44	7.35	7.46	7.51
rH	MN	17.3			16.5	14.5	16.4	18.3	20.3	22.7	18.6	18.7	17.3
	M	11.5	17.53	17.93	16.9	16.95	19.03	17.83	16.5	22.6	20.13	19.33	15.28
Salinité	MN	23.2		33.5	33.7	34.6	34.9	34.9	34.3	34.6	35.2	33.8	33.9
g NaCl l^{-1}	M	36			34.3	34.9	35.5	35.5	34.3	34.3	34.6	34.3	33.2
NH$_4^+$	MN	54.8		202	55.7	82.5	61.4	70.9	62.1	46.2	28.2	28.7	26.9
µg-at N l^{-1}	M	410	174		158	85	36.8	30.6	18.2	38.3	21.5	28.7	20.15
NO$_2^-$	MN	.52			.66	.42	.19	.72	1.00	.85	.75	.98	.7
µg-at N l^{-1}	M	.24	.48	.64	.26	.24	.19	.43	.83	.8	1.08	.98	1.47
o.PO$_4^{3-}$	MN	1.3			1.66	2.25	3.7	2.1	2.9	2.8	2.5	3.1	1.7
µg-at P l^{-1}	M	25.8	12.1	15.7	9.75	11.1	8.9	4.9	6.6	5.7	2.9	3.1	2.05
PO$_4^{3-}$ Hyd	MN	1.74			1.5	.88	.1	.5	0	0	.4	0	0
µg-at P l^{-1}	M	7.2	1.8	1.9	5.5	0	0	.3	.4	0	0	0	.35

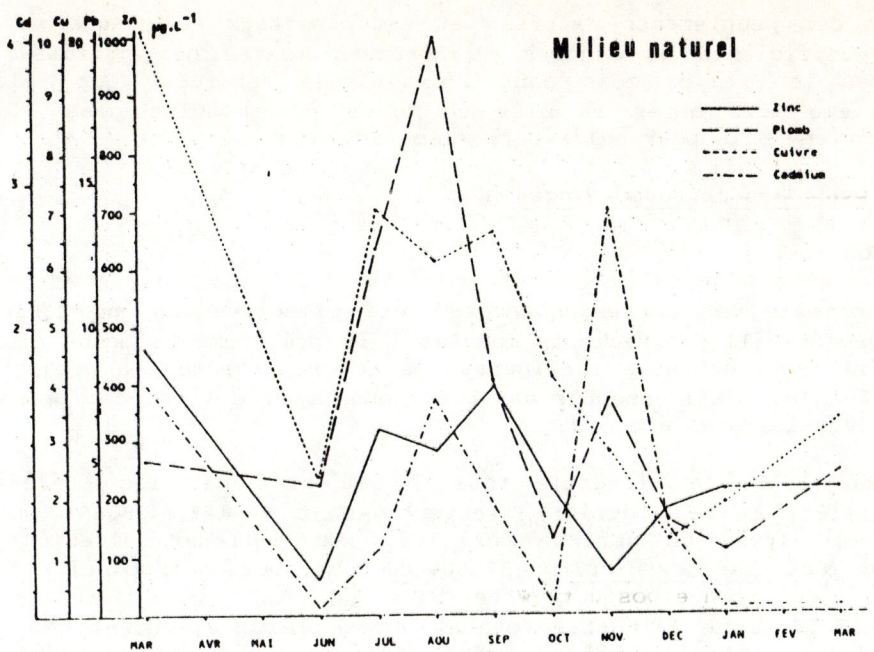

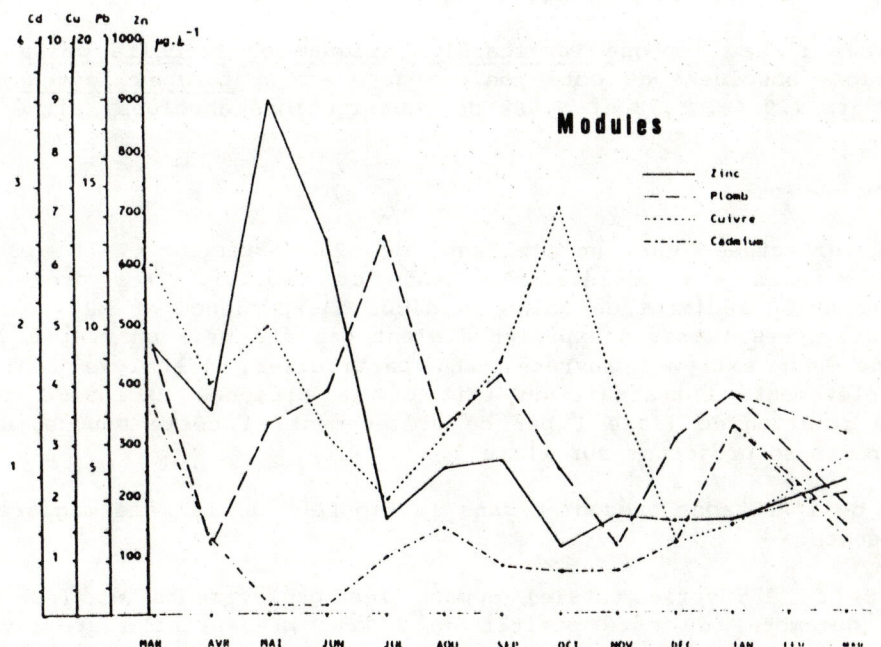

Figure 3. Expérience Berre-Tamaris. Variations des métaux lourds dans l'eau interstitielle des Modules et du sédiment de référence.

modules présentent des variations concomitantes (Coefficient de rang de Spearman, correlation positive près du seuil de signification $r_s=0.65$ $r_s0.05=0.66$), démontrant des phases de destructuration et restructuration similaires.

Dans l'expérience Vaine-Fos les indices de diversité du peuplement des modules tendent à croitre progressivement et bien que toujours inférieurs, ils évoluent de la même manière que ceux du peuplement du milieu naturel à partir du mois d'octobre.

A Tamaris, les indices de diversité du peuplement des modules présentent des variations extrèmement importantes par rapport à ceux du peuplement du milieu naturel, traduisant des phases de perturbation et de restructuration plus marquées.

La comparaison des peuplements naturels et expérimentaux de chaque site d'étude (Fig. 4) par les coefficients de Sanders et Sorensen montre des affinités marquées prochent de 50% dès le premier mois dans l'expérience Fos-Fos. Ces valeurs sont atteintes le quatrième mois après la mise en place des modules dans l'expérience Vaine-Fos et le cinquième mois pour celle Berre-Tamaris.

(b) Analyse factorielle des correspondances

- Expérience Fos-Fos:

L'analyse factorielle des correspondances a été effectuée sur un tableau de 23 espèces x 21 prélèvements (11 prélèvements modules + 10 prélèvements benne cf Tab. IV). Les résultats de l'AFC sont donnés à la figure 5 où est représentée la projection plane dans le plan factoriel I-II de l'ensemble des prélèvements et des espèces présentant les contributions relatives les plus élevées.

Le premier axe (28.6% de l'inertie totale) est créé par les prélèvements B8 (juillet), M10 (juillet) et M6 (février) (respectivement 46.15%, 28.65% et 5.63% de contribution absolue) situés en abcisse positive. Du même côté l'espèce la plus influente sur l'axe 1 est Nassa corniculum (87.50% de contribution absolue).

Le deuxième axe (20.3% de l'inertie totale) oppose les prélèvements M2 (octobre), M4 (décembre) et M1 (septembre) situés du côté positif présentant respectivement une contribution absolue de 42.28, 7.01 et 6.32% aux prélèvements B10 (octobre 2), B7 (juin), B9 (août), B4 (février) et B5 (mars) du côté négatif de l'axe 2 dont les contributions absolues sont de 11.19, 8.20, 6.25, 6.05 et 5.17%.

Pour les espèces l'axe 2 oppose Parvicardium exiguum et Tricolia pulla (40.17 et 4.07% de contributions absolues) du côté positif de l'axe à Loripes lacteus, Gastrana fragilis et Abra ovata (29.49, 5.73 et 5.56% de contributions absolues) situées du côté négatif de l'axe 2.

- Expérience Vaine-Fos:

L'AFC a été effectuée sur un tableau de 28 espèces x 10 prélèvements (4 prélèvements de modules + 6 prélèvements benne cf Tab.IV). Les prélèvements M1 (juillet) correspondant au sédiment de vaine en début d'expérience et M2 (août) premier prélèvement de module après 1 mois d'expérience n'ont pas été pris en compte dans cette analyse du fait de leur extrême pauvreté. En particulier, l'analyse réalisée avec l'ensemble des prélèvements entrainait du fait de la présence exclusive de Cardium glaucum dans M1 la création de l'axe 1 par ce prélèvement et cette espèce, les autres prélèvements et espèces se projetant sur l'axe 2.

Les résultats de l'AFC sont présentés dans la figure 5 de la même manière que dans l'expérience précédente.

L'axe 1 (32.3% de l'inertie totale) oppose les prélèvements M5 (novembre), M6 (décembre) et B6 (décembre) du côté positif de l'axe, présentant respectivement une contribution absolue de 38.49, 10.29 et 8.61% aux prélèvements B1 (juillet) et B2 (août) aux contributions absolues de 18.02 et 12.84%.

Pour les expèces, Corbula gibba, Nassa reticulata, Acanthocardia paucicostata et Venerupis aurea (18.34, 17.46, 7.50 et 4.61% de C.A.) en abcisse positive sont opposées à Spisula subtruncata, Abra alba, Acanthocardia aculeata et Parvicardium papillosum (16.12, 13.95, 2.93 et 1.76% de C.A.).

L'axe 2 (15.9% de l'inertie totale) est essentiellement créé par le prélèvement B4 (octobre) situé dans la partie positive de l'axe présentant une contribution absolue de 81.32%. Les espèces les plus influentes sur l'axe 2 sont Saxicava artica, Myrthea spinifera, Tellina sp., Murex trunculus et Lutraria sp. présentant respectivement une contribution absolue de 28.35, 25.80, 13.24, 4.90 et 2.71%, toutes situées du côté positif de l'axe.

Tableau IV

Espèces dominantes (Sp.Dom), espèces subdominantes (Sp.Subd.) et Dominance (D%) des Mollusques présents dans les modules (M1...M11) et du milieu naturel (B1...B12) au cours du cycle d'étude de chaque expérience réalisée.

		SEPTEMBRE 02/09/76	D %		OCTOBRE 30/09/76	D %		NOVEMBRE 03/11/76	D %
FOS-FOS		B1			B2				
Sp. Dom.		V. aurea	71.58		V. aurea	62.50			
Sp. Subd.		P. exiguum	10.53		A. ovata	15.28			
		A. ovata	6.32		L. lacteus	9.72			
		L. lacteus	5.26		M. bidentata	6.94			
		C. neritea	4.21						
		M1			M2			M3	
Sp. Dom.		V. aurea	52.78		V. aurea	42.54		V. aurea	51.28
Sp. Subd.		P. exiguum	30.56		P. exiguum	41.44		A. ovata	17.95
		T. pulla	5.56					G. minima	10.26
								P. exiguum	7.69
								T. pulla	5.13

		DECEMBRE 08/12/76	D %		JANVIER 11/01/77	D %		FEVRIER 01/02/77	D %
		B3						B4	
Sp. Dom.		V. aurea	59.56					V. aurea	47.18
Sp. Subd.		A. ovata	33.93					G. fragilis	15.90
		L. lacteus	33.93					A. ovata	15.38
		N. cornic.	6.62					L. lacteus	12.82
		M4			M5			M6	
Sp. Dom.		V. aurea	41.30		V. aurea	70.43		V. aurea	24.55
								N. cornic.	24.55
Sp. Subd.		P. exiguum	17.39		A. ovata	19.21		A. ovata	20.00
		T. pulla	15.38		P. exiguum	3.35		C. neritea	11.82
		A. ovata	6.52						

Tableau IV   (suite 1)

	MARS 14/03/77	D %	AVRIL 25/04/77	D %	JUIN 02/06/77	D %
FOS-FOS	B5		B6		B7	
Sp. Dom.	L. lacteus	88.24	V. aurea	33.33	L. lacteus	30.48
Sp. Subd.	V. aurea	11.76	A. ovata	22.22	V. aurea	20.00
			C. neritea	22.22	A. ovata	15.24
			L. lacteus	11.11	N. cornic.	13.33
			N. cornic.	11.11	M. gallopr.	9.52
	M7		M8		M9	
Sp. Dom.	V. aurea	81.82	A. ovata	43.55	V. aurea	47.83
Sp. Subd.	L. lacteus	9.09	V. aurea	40.32	M. gallopr.	23.48
	C. vulgatum	9.09	C. neritea	6.45	A. ovata	15.65
					N. cornic.	10.43

	JUILLET 08/07/77	D %	AOUT 29/08/77	D %	OCTOBRE 2 13/10/77	D %
	B8		B9		B10	
Sp. Dom.	N. cornic.	80.56	A. ovata	37.02	V. aurea	35.00
Sp. Subd.	V. aurea	16.67	L. lacteus	27.32	L. lacteus	31.50
			V. aurea	22.12	A. ovata	15.50
			P. exiguum	7.69	G. fragilis	7.50
	M10		M11			
Sp. Dom.	N. cornic.	66.67	V. aurea	38.00		
Sp. Subd.	V. aurea	26.09	A. ovata	20.00		
			N. cornic.	18.00		
			P. exiguum	12.00		
			L. lacteus	10.00		

	JUILLET 16/07/81	D %	AOUT 18/08/81	D %	SEPTEMBRE 23/09/81	D %
VAINE-FOS	B1		B2		B3	
Sp. Dom.	V. aurea	23.47	V. aurea	29.13	V. aurea	36.00
Sp. Subd.	S. subtrun.	22.45	S. subtrun.	22.33	S. subtrun	20.69
	A. alba	22.45	A. alba	21.36	G. minima	10.34
	C. gibba	10.20	S. artica	7.77	A. alba	6.90
					Haminea sp.	6.90
	M1		M2		M3	
Sp. Dom.	C. glaucum	100	N. retic.	62.50	V. aurea	90.00
Sp. Subd.			T. communis	12.50	A. ephippium	10.00
			A. ephippium	12.50		
			L. lacteus	12.10		

Tableau IV   (suite 2)

	OCTOBRE 20/10/81	D %	NOVEMBRE 18/11/81	D %	DECEMBRE 23/12/81	D %
	B4		B5		B6	
Sp. Dom.	V. aurea	30.00	V. aurea	33.33	C. gibba	38.34
Sp. Subd.	S. artica	17.50	C. gibba	16.00	V. aurea	27.28
	A. alba	7.50	S. subtrun	16.00	C. paucic.	13.89
			P. exiguum	9.33		
			G. minima	6.67		
	M4		M5		M6	
Sp. Dom.	V. aurea	50.00	V. aurea	47.67	C. gibba	43.24
Sp. Subd.	C. gibba	18.18	C. gibba	22.34	V. aurea	39.19
	G. minima	9.09	N. retic.	20.21	A. alba	9.46

	MARS 02/03/82	D %	AVRIL 01/04/82	D %	MAI 03/05/82	D %
BERRE-TAMARIS	B1		B2		B3	
Sp. Dom.	A. alba	76.6	A. alba	65.31	A. alba	51.82
Sp. Subd.	V. aurea	4.26	C. gibba	18.37	C. gibba	21.90
	C. gibba	4.26	P. exiguum	6.12	V. aurea	10.95
	V. verrucosa	6.38	V. aurea	4.08	M. bidentata	6.57
					P. exiguum	5.84
	M1		M2		M3	
Sp. Dom.	-	-	-	-	-	-
Sp. Subd.	-	-	-	-	-	-

	JUIN 01/06/82	D %	JUILLET 28/06/82	D %	AOUT 30/07/82	D %
	B4		B5		B6	
Sp. Dom.	A. alba	69.81	V. aurea	45.12	A. alba	43.48
Sp. Subd.	C. gibba	16.98	A. alba	28.05	C. gibba	30.43
	V. aurea	5.66	C. gibba	9.76	V. aurea	8.70
	M. bidentata	5.66			V. verrucosa	8.70
	M4		M5		M6	
Sp. Dom.			A. alba	75.00	V. aurea	33.33
Sp. Subd.			C. gibba	25.00	C. gibba	25.00
					A. alba	16.67
					P. exiguum	16.67
					Chiton	-

Tableau IV   (suite 3)

	SEPTEMBRE 08/09/82	D %	OCTOBRE 12/10/82	D %	NOVEMBRE 15/11/82	D %
	B7		B8		B9	
Sp. Dom.	A. alba	60.00	A. alba	50.00	C. gibba	55.00
Sp. Subd.	M. bidentata	20.00	L. Lacteus	14.29	A. alba	20.00
	P. exiguum	13.33	P. exiguum	14.29	V. aurea	15.00
	V. verrucosa	6.67	C. gibba	14.29	G. minima	5.00
					Chiton	5.00
	M7		M8		M9	
Sp. Dom.	A. alba	100	A. alba	54.55	A. alba	100
Sp. Subd.			C. gibba	27.27		
			V. aurea	18.18		

	DECEMBRE 22/12/82	D %	JANVIER 24/01/83	D %	MARS 2 11/03/83	D %
BERRE-TAMARIS	B10		B11		B12	
Sp. Dom.	A. alba	50.00	A. alba	37.50	A. alba	54.55
	C. gibba	50.00				
Sp. Subd.			C. gibba	25.00	C. gibba	18.18
			V. verrucosa	12.50	Haliotidae	9.09
			Haliotidae	12.50	V. aurea	9.09
			P. exiguum	12.50	M. bidentata	9.09
	M10		M11		M12	
Sp. Dom.	A. alba	43.14	A. alba	43.01	A. alba	41.52
Sp. Subd.	V. aurea	21.57	C. gibba	32.26	C. gibba	38.60
	P. exigum	15.69	V. aurea	13.96	V. aurea	13.45
	C. gibba	13.73	P. exiguum	9.68		

- Expérience Berre-Tamaris:

Aucune espèce de Mollusques n'étant présente dans les prélèvements M1 (mars), M2 (avril), M3 (mai), M4 (juin), l'AFC a été effectuée sur un tableau de 16 espèces x 20 prélèvements (8 prélèvements modules + 12 prélèvements benne). Les résultats de l'AFC sont présentés dans la figure 5.

L'axe 1 (26.5% de l'inertie totale) oppose les prélèvements B5 (juillet) et M6 (août), 52.71 et 4.65% de C.A. situés du côté positif de l'axe aux prélèvements B7 (septembre), M9 (novembre), B4 (juin), B2 (avril) et M7 (septembre) présentant une contribution absolue respective de 4.26, 4.34, 4.07, 3.65 et 2.31%. Les espèces qui créent l'axe 1 sont Venerupis aurea, Chiton sp., Gouldia minima et Cerithium vulgatum (51.07, 7.71, 4.37 et 4.41% de C.A.) situées du côté positif de l'axe et opposées à Abra alba (C.A. 20.41%).

L'axe 2 (15% de l'inertie totale) oppose les prélèvements M12 (mars 2), M11 (janvier) et B9 (novembre) présentant une contribution absolue respective de 20.61, 14.39 et 11.83% du côté négatif de l'axe au prélèvent B1 (mars) (C.A. 9.18%). Les espèces contribuant à la formation de cet axe sont <u>Corbula gibba</u> et <u>Acanthocardia paucicostata</u> (C.A. 52.50 et 1.65%) situées du côté négatif et opposées à <u>Mysella bidentata</u> et <u>Modiolus sp.</u> (C.a. 8 et 5.51%).

4. DISCUSSION

La méthodologie

Toute étude écotoxicologique qu'elle soit réalisée in situ ou in vitro doit être basée sur une méthodologie rigoureuse garantissant des résultats dépendant bien des facteurs étudiés et non des techniques expérimentales employées. Les résultats de l'expérience Fos-Fos montrent que la méthodologie employée n'altère pas la dynamique et la structure du peuplement expérimental qui évolue comme le peuplement de référence.

Dans le cadre des sites choisis et des sédiments utilisés. L'isolement de fractions de biotopes, du fait du caractère semi clos des modules n'entraine aucune autopollution particulière des sédiments expérimentaux par rapport aux sédiments de référence. Les valeurs similaires de Rh obtenues respectivement dans les modules et le milieu naturel en fin d'étude pour les expériences Vaine-Fos ou Berre-Tamaris (Tab. II et III) en témoignent.

En ce qui concerne la representativité de l'échantillonage, il est évident que dans toute étude classique sur le terrain il est préconisé une multiplication des prélèvements afin de tenir compte de la variabilité spatiale des peuplements étudiés. Cependant dans les expérimentations in situ une multiplication extrême des prélèvements et en particulier des modules peut entrainer des contraintes expérimentales ne permettant pas de mener à bien une expérience. Ces contraintes peuvent être liées aux risques présentés par la mise en place d'un champ expérimental trop étendu plus facilement repérable et endommageable de manière volontaire ou involontaire par l'homme dans des milieux non protégés, aux difficultés de mise en place en plongées et de repérage des modules ou du coût de l'expérimentation lui même (Hansen, 1982).

Pour les biotopes et les peuplements étudiés le choix des volumes et des surfaces de sédiments utilisés entrainant la prise d'un seul module est parfaitement justifié. D'une part dans l'expérience Fos-Fos, la moyenne des effectifs calculée à partir du tri de deux modules présentent des écarts faibles (Fig. 4), d'autre part l'évolution temporelle similaire des Mollusques, obtenue à partir de deux techniques de prélèvements différentes (module-benne) tend bien à montrer que cette évolution tient à des facteurs indépendants des techniques de prélèvements et du nombre de prélèvements réalisés.

Etude des processus de colonisation et de la cinétique des polluants

Si la colonisation par les Mollusques est immédiate dans les modules contenant des sédiments de même nature physique et chimique que ceux du milieu naturel, il n'en est pas de même pour les modules contenant des sédiments pollués transplantés. Le processus de colonisation ne commence véritablement qu'entre le deuxième et troisième mois pour les sédiments de Vaine à Fos et le quatrième et cinquième mois pour ceux de Berre à Tamaris. Plusieurs facteurs peuvent induire ces retards de colonisation: le caractère saisonnier de l'installation des populations SVMC, les conditions générales défavorables du milieu ambiant ou la charge polluante des sédiments expérimentaux.

L'influence saisonnière ne peut être retenue. L'évolution des peuplements SVMC en milieu perturbé est indépendante des saisons (Stora, 1982), de plus l'expérience Vaine-Fos débute en juillet comme l'expérience précédente Fos-Fos. Si les conditions ambiantes générales défavorables, illustrées par la chute des espèces et de l'effectif total observée pour le peuplement de référence dans l'expérience Vaine-Fos, ont pu jouer un role sur le retard de colonisation, celui ci s'observe de même dans les modules de Tamaris alors que les conditions du milieu ne limitent pas particulièrement le

- 244 -

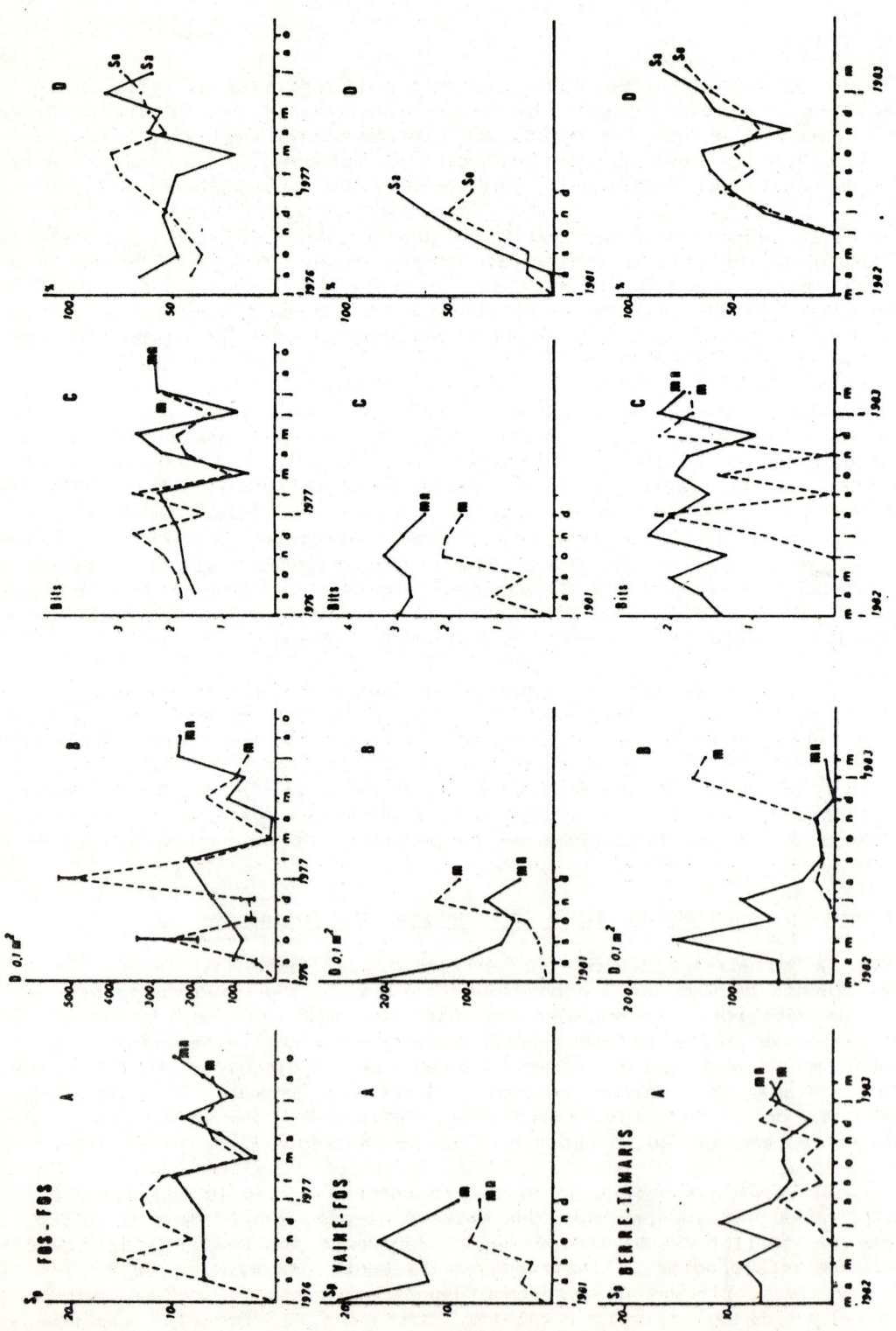

Figure 4. Evolution temporelle de la richesse spécifique (A), de la densité (B) et de la diversité (C) de la faune Malacologique présente dans les modules (m) et les sédiments de référence (Mn) dans les trois expériences réalisées. (D) : Coefficients d'affinité (Sanders Sa, Sorensen So) entre les peuplements naturels et expérimentaux.

développement de la faune malacologique du peuplement témoin. Ce dernier facteur ne peut avoir une influence déterminante sur les processus de colonisations retardés observés.

Comme le montrent les tableaux et les figures 2 et 3, les modules des deux sites d'études se caractérisent en début d'expérience par de fortes teneurs en ammonium ainsi qu'un relargage marqué de l'ensemble des métaux lourds dans l'eau interstitielle. Ces taux d'altéragènes élevés contribuent très probablement aux retards de colonisation observés.

Dans l'expérience Vaine-Fos, contrairement à celle de Berre-Tamaris, un deuxième relargage important et simultané peut être observé au cours du mois de décembre pour les métaux lourds étudiés, associé à une chute de l'effectif total des mollusques du sédiment expérimental. Comme nous l'avons vu précédemment, les taux de polluants dans le milieu naturel évoluent de la même manière que ceux des modules et en Décembre l'évolution de la faune malacologique du milieu naturel est semblable à celle des modules. On peut de ce fait attribuer à un ou des facteurs communs s'exerçant dans le milieu, ce relargage de polluants et son incidence sur la chute des effectifs dans les biotopes naturels et expérimentaux. Un de ces facteurs doit avoir son origine dans les variations de salinité du milieu dues aux très importantes fluctuations du débit du Rhone dont les eaux transitent par le canal de St. Louis ou par le they de la Gracieuse. En effet, on constate que chaque pic de relargage des métaux lourds suit une période ou le Rhône avec un décalage nécessaire et suffisant passent de forts débits à d'autres beaucoup plus faibles (Arnoux et al., 1985).

L'incidence d'une diminution de débits d'une telle ampleur sur la mortalité de peuplements lagunaires ou similaires à ceux étudiés dans ces expériences ont pu être mis en évidence in situ (Stora et Arnoux, 1983; Stora, 1984). Cette action marquée du facteur salinité sur le relargage de polluants dans l'eau interstitielle permet d'expliquer les relargages similaires observés en début d'expérience dans les modules de Fos et Tamaris. L'imprégnation des sédiments provenant d'un milieu déssalé par des eaux à salinité plus élevée engendre expérimentalement le même type d'effets que ceux provoqués dans le milieu naturel par les variations de débits du Rhône.

Par la suite, l'observation de relargage dans les sédiments de Tamaris non soumis à des variations de salinité aussi importantes que ceux de Carteau ne peut avoir son origine que dans d'autres facteurs déclanchant tels que l'hydrodynamisme et l'oxydation des sédiments qui en résulte (Forstner et Wittmann, 1979; L.M.H.A., 1983).

Comme le montre Jerome (1984), la vitesse de relargage des métaux lourds dans les sédiments dépend plus de la texture du sédiment que de sa contamination propre. Pour un facteur déclanchant de même origine mais de plus faible intensité que les variations de salinité, la nature granulométrique différente des sédiments de Berre et de Tamaris serait en mesure d'expliquer les décalages selectifs des différents métaux observés dans les modules et le sédiment de référence.

Les différentes expériences réalisées permettent de constater que si la nature chimique des sédiments intervient essentiellement dans le laps de temps mis pour que la colonisation commence, le processus de colonisation de la faune malacologique en lui même reste identique pour des substrats "non perturbés" ou perturbés. Tout comme pour les Polychètes (Stora, 1984), cette colonisation de substrats par les Mollusques SVMC ne commence pas par l'arrivée d'espèces pionnières particulières mais s'effectue globalement à partir des espèces les mieux représentées dans le milieu naturel au moment de la mise en place des modules. Cette particularité aboutit très rapidement à la présence d'espèces en position dominante ou subdominante communes aux milieux naturels et expérimentaux (Tab. IV) expliquant les pourcentages élevés atteint rapidement pour les coefficients de Sanders et Sorensen. Cette colonisation globale à partir des espèces les plus abondantes présentes dans le milieu naturel correspond aux résultats obtenus par MacCall (1977) et Zajac et Whitlatch (1982).

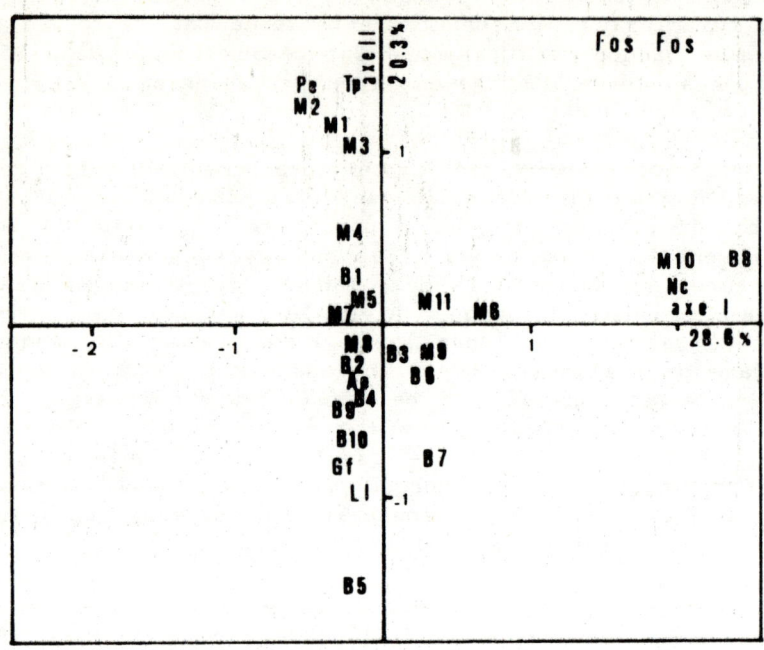

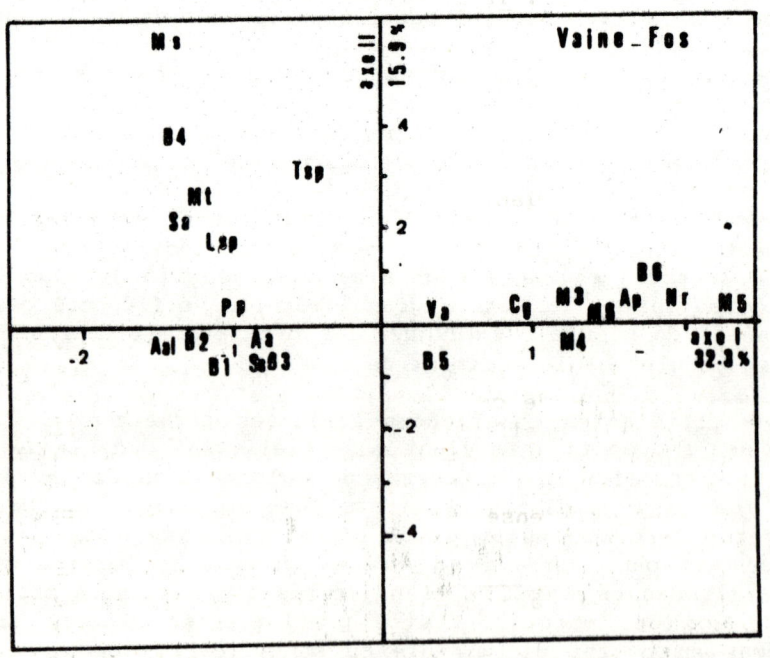

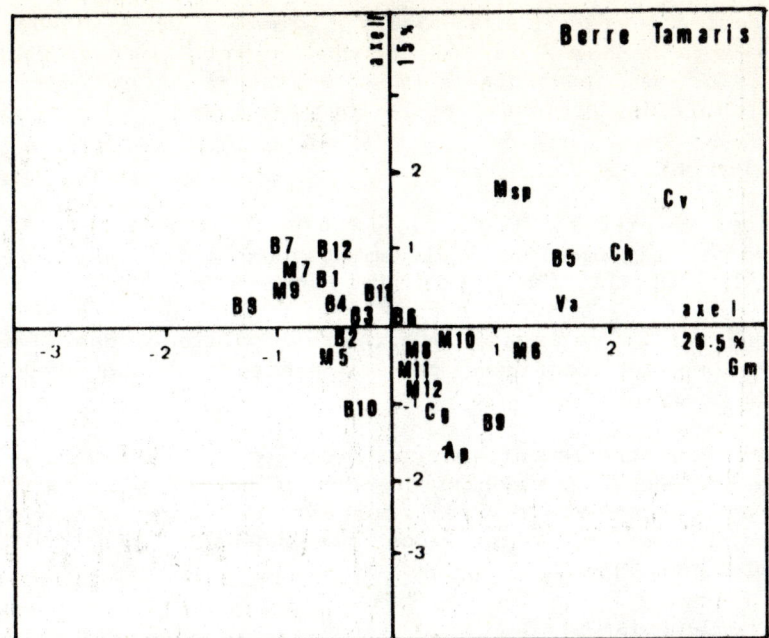

Figure 5: Résultats de l'analyse des Correspondances pour les trois expériences ; projections des espèces et des prélèvements dans les plans factoriels I et II. M 1...Mx : prélèvements modules, B1....Bx Prélèvements Benne (cf Tableau 4). A.a: Acanthocardia aculeata, A.p Acanthocardia paucicostata, A.al: Abra alba, A.o: Abra ovata, C.g: Corbulla gibba, Ch: Chiton indt., C.v Ceithium vulgatum, G.m: Gouldia minima, G.g Gastrana fragilis, Lsp: Luttaria sp, M.b: Mysella bidentata, M.s, Myrthea spinifera, M sp Modiolus sp, M.t: Murex trunculus, N.r Nassa reticulata, N.c: Nassa corniculum, P.e Parvicardium exiguum, P.p: Parvicardium papillosum, S.a Saxicava artica, S.s Spisula subtruncata, T sp: Tellina sp, T.p: Tricolia pulla, V.a: Venerupis aurea.

Dynamique des peuplements naturels et expérimentaux

L'examen simultané de la répartition des prélèvements et des espèces dans l'espace engendré par les premiers axes factoriels permet d'inférer la signification écologique des axes reconnus (Plante et al., 1983).

- Expérience Fos-Fos:

Dans l'expérience Fos-Fos (Fig. 5), la structure obtenue en AFC présente dans les plans des deux premiers axes une forme en parabole traduisant un effet Gutman, relation quadratique entre les facteurs qui engendrent ces axes (Plante et al., 1983). Le facteur commun responsable de ces évolutions est certainement le Rhône. En effet une étude de corrélation par le coefficient de rang de Spearman effectuée entre les débits moyens mensuels précédant le prélèvement et l'ordination des prélèvements le long de l'axe 1 montre une corrélation positive et significative ($r_s=0.49$; $r_s0.05=0.43$). Or l'axe 1 sépare pratiquement l'ensemble des prélèvements par rapport aux prélèvements de juillet des sédiments expérimentaux et naturels. De même l'axe 1 oppose l'ensemble des espèces structurantes à Nassa corniculum espèce indicatrice de perturbation maximale (Stora, 1982). Ce premier axe peut être reconnu comme marquant un gradient de perturbation engendré par les variations de débit. En effet au mois de juillet, les débits moyens du Rhône passent de 3289 m^3/s à 2348 m^3/s. Il est d'ailleurs fort probable que de ces variations aient induit un relargage de polluants entrainant la mortalité importante du peuplement en place et l'arrivée des Nasses nécrophages.

Le deuxième axe sépare des prélèvements réalisés en périodes de faibles et forts débits du Rhône. Du fait de la présence du côté négatif d'espèces de dessalure telles que Abra ovata et Gastrana fragilis l'axe 2 peut être assimilé à un gradient de salinité.

Bien que les premiers prélèvements modules soient plus éloignés sur l'axe 2 par rapport aux prélèvements bennes à la même époque, ce qui peut traduire l'influence du substrat dépeuplé, il est remarquable de constater l'évolution similaire de la trajectoire de l'ensemble des prélèvements le long des axes.

- Expérience Vaine-Fos:

Dans l'expérience Vaine-Fos (Fig. 5), l'axe 1 correspond à un gradient de perturbation. Il sépare les premiers prélèvements bennes de l'ensemble des prélèvements modules en cours de colonisation présentant une charge polluante importante. La possibilité d'un facteur perturbant commun s'exercant sur les milieux naturels et expérimentaux, marqué par un relargage important de polluants dans l'eau interstitielle des modules en décembre est confirmée par la présence du côté positif de l'axe du prélèvement benne de décembre.

L'axe 2 est créé essentiellement par le prélèvement d'octobre benne. Ce dernier correspondant à une période où les débits du Rhône sont les plus importants par rapport aux autres prélèvements, cet axe pourrait correspondre à un gradient de salinité, la première analyse ayant montré l'influence prépondérante des débits du Rhône sur l'évolution du peuplement SVMC.

- Expérience Berre-Tamaris:

Dans l'AFC réalisée à partir des résultats de l'expérience Berre-Tamaris (Fig. 5), la répartition des prélèvements et des espèces le long des axes semble plus difficilement interprétable. Contrairement à ce qui est obtenu dans l'expérience précédente, il n'y pas de séparation marquée entre les prélèvements modules et bennes dont les trajectoires évoluent de manière erratique le long des axes. En fait, à Tamaris aucun facteur "variation de débits" n'imprime sa marque sur la dynamique des peuplements. D'autre part, la charge polluante dans le site expérimental est plus importante que celle des modules. Les évolutions observées ne peuvent que traduire la perturbation des peuplements des sédiments naturels et expérimentaux. Les recherches de corrélation effectuées par le coefficient de rang de Spearman entre les différents paramètres analysés et l'ordination des prélèvements le long de l'axe 1 montre une corrélation positive et significative avec les taux de cadmium analysés ($r_S=0.46$; $r_S 0.05=0.46$) et pour l'axe 2 une corrélation positive et significative avec les taux de cuivre ($r_S=0.49$; $r_S 0.05=0.46$)), proche du seuil de signification pour le zinc ($r_S=0.43$; $r_S 0.05=0.46$)) du fait de l'évolution similaire de ces deux métaux ($r_S=0.56$; $r_S 0.05=0.46$)).

Il est bien évident qu'en l'état actuel de nos connaissances malgré les corrélations obtenues, on ne puisse attribuer directement aux facteurs Cd, Zn et Cu les répartitions des prélèvements et des espèces observées. Ces axes ne traduisent en fait que des gradients de perturbation liés à la synergie entre la charge polluante du milieu et des facteurs ayant été susceptibles de déclancher ces relargages sélectifs.

5. CONCLUSIONS

Les expérimentations in situ réalisées à partir d'une méthodologie éprouvée, permettant d'apprécier les processus de colonisation et la dynamique de peuplements macrobenthiques en fonction de la cinétique de polluants de sédiments extrêmement contaminés oligotypiques ou complètement azoiques du point de vue macrobenthos, sont riches d'enseignements.

Sur un plan fondamental ces expériences montrent que les concentrations de polluants de la phase solide des sédiments ne constituent pas nécessairement pour les peuplements étudiés, caractéristiques de biotopes côtiers, un facteur limitant majeur. Un sédiment extrêmement contaminé peut très rapidement être colonisé par un peuplement structuré lorsque les conditions du milieu s'améliorent.

Indépendamment de la charge solide, les processus de colonisation commençant plus ou moins rapidement sont identiques pour les différents biotopes étudiés. Cette colonisation s'opère de manière globale à partir des espèces présentes dans le milieu au moment où débutent les expériences. Comme le démontrent les AFC réalisées, la dynamique des peuplements est directement sous la dépendance de relargages de polluants dans l'eau interstitielle induits par l'oxydation des sédiments et les variations de salinité du milieu environnant. Pour ce dernier facteur, l'utilisation de sédiments provenant de biotopes soumis à dessalure et replacés dans un environnement marin permet la vérification expérimentale des processus de relargage retardant les possibilités de colonisation.

Pour les peuplements côtiers ces relargages représentent le facteur de dégradation majeur. Il apparait que même si les concentrations de polluants libérés ne sont pas léthales, les variations des conditions du milieu induisant ces relargages ne peuvent qu'accentuées par synergie les possibilités de dégradation des peuplements.

Si dans le cadre de ce travail, les expériences ont été réalisées à partir de sédiments globalement contaminés, il est évident que les expérimentations dans le milieu naturel offrent de grandes possibilités de champs d'investigations aussi bien à caractère fondamental qu'appliqué. L'expérimentation in situ peut faire intervenir une multiplicité de facteurs écologiques dont l'évolution naturelle lente ou rapide et les plus souvent difficile à prévoir, ne permet pas toujours d'en apprécier l'influence directe sur la faune, particulièrement en milieu perturbé où s'ajoutent ou se compensent les effets d'altéragènes tant physiques que chimiques. Ainsi, malgré les difficultés inhérentes à cette complexité de la situation générale, il est possible de provoquer certaines conditions particulières telles que des contaminations sélectives ou combinées de polluants, dont on pourra en vraie grandeur apprécier l'incidence directe ou indirecte sur la structure et le développement des peuplements benthiques.

6. BIBLIOGRAPHIE

Arnoux, A., L.P. Nienchewski et J. Tatossian, Comparaison de quelques méthodes d'attaque
1981   des sédiments marins pour l'analyse des métaux lourds. J.Fr.Hydrol., 12(34):29-48

Arnoux, A., G. Stora et C. Diana, In situ experimental study of the evolution and
1985   recolonization of polluted sediments. Mar.Pollut.Bull., 16(8):313-8

Arnoux, A. et al., Evolution et bilan de la pollution des sédiments de l'étang de Berre.
1981   Journ.Etud.Pollut.CIESM, 5(1980):433-66

Bellan, G., Effects of pollution and man-made modifications on marine benthic communities
1985   in the Mediterranean: a review. In Mediterranean marine ecosystems. edited by M. Moraitou-Apostolopoulou and V. Kiortsis, New York, Plenum Publishing Corporation: pp.163-94

Benzecri, J.P., L'analyse des données. Tome 2. L'analyse des correspondances. Paris,
1973   Dunod, 619 p.

Daget, J., Les modèles mathématiques en écologie. Paris, Masson et Cie,
1976   Collection d'écologie no. 8:172 p.

Dajoz, R., Précis d'Ecologie. Paris, Dunod, 434 p.
1972

Dauvin, J.C., Revue des principales techniques utilisées pour l'étude expérimentale de
1984   peuplements macrobenthiques subtidaux de sédiment meuble. Premiers résultats des expérimentations réalisées en baie de Morlaix. Océanis, 10(3):207-317

_____, Dynamique d'écosystèmes macrobenthiques des fonds sédimentaires de la baie
1984a   de Morlaix et leur perturbation par les hydrocarbures de l'Amoco Cadix.
Thèse d'Etat. Université Pierre et Marie Curie, Paris VI, 468 p.

Desrosiers, G. et J-C.F. Brêthes, Etude bionomique de la communauté à <u>Macoma balthica</u> de
1984   la batture de Rimouski. <u>Sciences et techniques de l'eau</u>, 17(1):25-31

Diana, C., Etude de la qualité des eaux du golfe de Fos. Thèse d'Université Pharmacie
1983   Aix-Marseille II. 132 p. + annexe.

Diaz-Castaneda, V., Etude de la colonisation benthique de sédiment mis en enceintes
1984   expérimentales dans la baie du Lazaret (Toulon, France). Thèse 3ème cycle.
Université de Aix-Marseille II. 230 p.

Febvre, J., Etude bionomique des substrats meubles de l'étang de Berre. <u>Rec.Trav.Stn.</u>
1968   <u>Mar.Endoume</u>, 44:297-355

Förstner, U. et T.W. Wittmann, Metal pollution in the aquatic environment. Berlin,
1979   Springer-Verlag, 486 p.

Guille, A., Bionomie benthique du plateau continental de la côte catalane française.
1970   2.- Les communautés benthiques de la macrofaune. <u>Vie Milieu(B Biol.Mar.).</u>,
21(1):149-280

Guille, A. et F. Ponge, Application de l'analyse des correspondances à l'étude des
1975   peuplements benthiques de la côte catalane française.
<u>Ann.Inst.Oceanogr.Monaco,</u> 51:223-35

Hannan, C.A., Polychaete larval settlement: correspondance of patterns in suspended jar
1981   collectors and in the adjacent natural habitat in Monterey Bay, California.
<u>Limnol.Oceanogr.</u>, 26(1):159-71

Hansen, D.J., Utility of toxicity tests to measure effects of substances on marine
1982   organisms. Gulf Breeze, Florida, Environmental Protection Agency, EPA/ERL
(EPA 600/D-82-359)

Hily, C., Variabilité de la macrofaune benthique dans les milieux hypertrophiques de la
1984   rade de Brest. Thèse d'Etat, Université de Bretagne Occidentale, 359 p.

Jerome, C., Contribution à l'étude de l'évolution de l'eau interstitielle des sédiments
1982   en milieu naturel et en milieu expérimental. Rapport DEA. Université de
Aix-Marseille II, 41 p.

_____, Evolution récente des sédiments de l'étang de Berre. Thèse. 3ème cycle.
1984   Université de Aix-Marseille II, 141 p.

Kalke, R.D. et T.A. Duke, Weathered IXTOC I oil effect on estuarine benthos. <u>Estuar.</u>
1982   <u>Coast.Shelf.Sci.</u>, 15(1):75-84

Kuiper, J., P. De Wilde et W. Wolff, Effects of an oil spill in outdoor model tital flat
1984   ecosystems. <u>Mar.Pollut.Bull.</u>, 15(3):102-06

Lebart, L., A. Morineau et J.P. Fenelon, Traitement des données statistiques. Paris,
1982   Dunod, 510 p.

Legendre, L. et P. Legendre, Ecologie numérique. Tome II: La structure des données
1979   écologiques. Paris, Masson Collection d'écologie, 13:254 p.

Laboratoire d'Hydrologie et de Molysmologie Aquatique. Evolution et repeuplement de
1983   sédiments pollués. <u>Rapp.Minist.Environ.Fr.</u>, Contrat "Mer" 82215:75 p.

MacCall, P.L., Community patterns and adaptative strategies of the infaunal benthos of
1977   Long Island Sound. <u>J.Mar.Res.</u>, 35(2):221-66

Peres, J.M. et J. Picard, Nouveau manual de bionomie benthique de la mer Méditerranée.
1964   Rec.Trav.Stn.Mar.Endoume, 31:5-137

Plante, R. et al., Stratégies d'échantillonages et fluctuations naturelles dans des
1983   ecosystèmes benthiques. Deux exemples en milieu tropical. In Actes 17ème symposium Européen de biologie marine, Brest. 27 sept.-1 oct. 1982 Oceanol.Acta., n° SP:pp.153-7

Reish, D.J., The use of sediment bottle collector for monitoring polluted marine waters.
1961   Calif.Fish Game, 47:261-72

Reys, J.P., Les peuplements benthiques (Zoobenthos) de la région de Marseille (France):
1976   aspects méthodologiques de la délimitation des peuplements par les méthodes mathématiques. Mar.Biol., 36(2):123-34

Sanders, H.L., 3-Structure of the soft bottom community. Limnol.Oceanogr., 5(2):138-53
1960

Sanders, H.L. et al., Anatomy of an oil spill: long-term effects from the grounding of
1982   barge Florida off West Falmouth, Massachusetts. J.Mar.Res., 38(2):265-380

Shannon, C.E. et W. Weaver, The mathematical theory of communication. Urbana Illinois,
1963   University of Illinois Press, 117 p.

Stirn, J., Manual des méthodes de recherche sur l'environnement aquatique. Huitième
1982   partie. Evaluation des modifications des écosystèmes marins dues à la pollution (Directives destinées au projet commun coordonné FAO(CGPM)/PNUE sur la pollution en Méditerranée). FAO Doc.Tech.Pêches, (209):75 p. Publié aussi en anglais

Stora, G., Recherches de bionomie descriptive et expérimentale (in vivo et in vitro)
1982   dans quelques biotopes littoraux soumis à des variations naturelles ou artificielles des conditions du milieu (notamment dans le golfe de Fos et l'étang de Berre). Thèse Doctorat d'Etat. Université de Aix-Marseille II, 327 p.

_____, Note préliminaire sur l'étude de la colonisation expérimentale d'un
1983   substrat dépeuplé par un peuplement lagunaire euryhalin et eurytherme. Rapp.P.-V.Réun.CIESM, 28(6):239-41

_____, Etude expérimentale in situ de la colonisation de substrats meubles par
1984   des populations annélidiennes. Oceanis, 10(7):761-73

Stora, G. et A. Arnoux, Effects of large freshwater diversions on benthos of a
1983   Mediterranean lagoon. Estuaries, 6(2):115-25

Tagatz, M.E. et Deans, Alterations in composition of field-and-laboratory-developed
1983   estuarine benthic communities exposed to di-n-butyl phtalene. Aquat.Toxicol., 3(3):239-48

Zajac, R.N. et R.B. Whitlatch, Responses of estuarine infauna to disturbance. 1. Spatial
1982   and temporal variation of initial recolonization. Mar.Ecol.(Prog.Ser.), 10:1-14

ACTION DES POLLUANTS SUR LES ECOSYSTEMES LITTORAUX MEDITERRANEENS

par

N. VICENTE
Laboratoire de Biologie Marine,
Faculté des Sciences et Techniques,
Saint Jérôme, 13397 Marseille Cedex 13

1. INTRODUCTION

La pollution est devenue de nos jours un lieu commun qui affecte toutes les mers du globe avec plus ou moins de gravité. Pour la Méditerranée, mer semi fermée dont le volume se renouvelle tous les 80 à 90 ans environ, la pollution pose des problèmes graves. Ses effets se font surtout sentir dans le bassin occidental beaucoup plus industrialisé, cependant le bassin oriental n'échappe pas à cette pollution bien que son niveau ne soit pas encore comparable à celui des côtes méditerranéennes italiennes, françaises ou espagnoles. Mais il est à craindre que le phénomène n'aille en s'amplifiant dans les années à venir avec l'industrialisation croissante de cette partie de la Méditerranée.

La gravité de cette pollution réside dans le fait qu'elle agit en priorité sur la frange littorale immédiate. Ce littoral marin est de plus en plus agressé de nos jours par toutes les nuisances du monde civilisé dans lequel nous évoluons: activités industrielles, tourisme intensif et urbanisation massive avec comme corollaire une ampleur sans cesse croissante des pollutions domestiques et industrielles.

Le milieu marin est devenu, en certains points, un véritable exutoire de notre civilisation. C'est une véritable poubelle, elle l'a toujours été, l'homme estimant que son pouvoir absorbant et diluant est infini.

S'il est vrai que la masse des océans présente de grandes possibilités de dilution, par contre sur le littoral marin où s'effectuent les principaux rejets, les dégats vont être considérables.

La frange littorale représente, au moins pour les cent premiers mètres de profondeur (c'est-à-dire le plateau continental, particulièrement étroit en Méditerranée), la zone où se manifeste le maximum de développement et d'exubérance de la vie. C'est aussi la zone où joue à plein l'effet de la pollution puisque c'est là qu'aboutissent presque tous les rejets de la vie terrestre: émissaires, eaux usées, transportées par les cours d'eau, rejets industriels, rejets agricoles, qui agressent chaque jour davantage le milieu marin.

Or ce milieu ne peut être de nos jours à la fois une poubelle et une nourrice. Si l'on veut préserver le "garde manger" que constituent les biocénoses littorales, il est indispensable et urgent sinon d'éviter les rejets, du moins de procéder à leur épuration et de mettre en place une gestion rationnelle des milieux littoraux autrement dit de parvenir à une planification écologique du littoral méditerranéen (Augier et Vicente, 1979).

Les aménagements littoraux: endigages, ports de plaisance agrandis et gagnés sur la mer, plages artificielles, représentent une forme de nuisance non négligeable pour la santé des biocénoses littorales (Meinesz et al., 1981).

Les terre-pleins et les plages artificielles notamment constituent une menace pour l'équilibre écologique du bord de mer. Ils détruisent irréversiblement, par recouvrement, la partie de la mer la plus riche en flore et en faune. En effet, la frange côtière entre 0 et 20 m est une zone où la lumière pénètre suffisamment pour permettre le développement des herbiers de Posidonies et de la plupart des algues qui contribuent à oxygéner les eaux littorales et à nourrir des milliers d'espèces animales ne pouvant se développer à plus grande profondeur.

De même l'aménagement d'un port entraîne pour le milieu marin des modifications de structure non négligeable du point de vue de la dynamique des eaux et de l'évolution biologique pour la zone marine où il est implanté. D'autre part, il est à noter que le rassemblement de navires va entraîner obligatoirement un certain degré de pollution organique, bactérienne et chimique qui de là va atteindre les rivages voisins.

Toutes ces considérations doivent être prises en compte dans la conception d'un ouvrage à créer, au même titre que les autres impératifs classiques à respecter dans ce genre de réalisation: voies d'accès, protection contre les houles, abris des vents dominants, etc.

C'est la somme de toutes ces pollutions et nuisances qui va finir par être fatale à l'équilibre écologique des biotopes littoraux et quelquefois d'une manière irréversible.

2. EFFETS DES POLLUTIONS SUR LES ECOSYSTEMES MARINS

Tous les écosystèmes ne vont pas réagir de la même manière face à l'agression des polluants. Quelquefois la rupture de l'équilibre va faire place à un nouvel équilibre avec redistribution des espèces dominantes, mais jamais l'écosystème ne présentera un aspect aussi florissant qu'à l'origine. Ce sont divers aspects, divers faciés de dégradation qui seront alors observés. Il en est ainsi pour les grandes prairies de zostères qui caractérisent le littoral méditerranéen: Posidonia oceanica, Cymodocea nodosa et Zostera noltii pour le bassin occidental et Halophila stipulacea pour le bassin oriental. Cette dernière espèce pouvant se trouver mélangée aux deux précédentes. Il a été observé cependant que Halophila se localise le plus souvent dans des zones où la compétition naturelle avec les autres phanérogames méditerranéennes est pratiquement inexistante et en particulier dans les fonds vaseux à grande profondeur où elle constitue des peuplements caractéristiques.

L'agression de ces grandes frayères par les diverses pollutions va entraîner des dommages considérables avec leur disparition quelquefois, de même que la raréfaction et la disparition de nombreuses espèces animales, notamment crustacés et poissons.

Ces phanérogames constituent les peuplements les plus sensibles du littoral qui paient un lourd tribut à la pollution des mers.

De même, les peuplements photophiles et sciaphiles de l'infralittoral présentent une sensibilité importante face aux pollutions (ex: la biocénose à Cystoseira stricta). Les zones riches en plancton d'eau pure sont également atteintes par les pollutions de même que les fonds coralligènes caractérisés par les tombants rocheux où vivent des multitudes d'invertébrés fixés particulièrement sensibles (gorgones, corail rouge, vers sédentaires, etc.).

Les peuplements des étages supra et médiolittoral qui se trouvent sous l'action directe des polluants de surface sont les plus atteints par les marées noires (algues photophiles et la faune qui leur est inféodée).

Effets sur les herbiers de phanérogames marines

Les Posidonies constituent un test pour la pollution de la Méditerranée. Ce sont les poumons du littoral qui disparaissent d'année en année. Ils disparaissent de plus en plus à cause surtout des aménagements aberrants qui les étouffent, des ancres des bâteaux qui les arrachent et des produits chimiques comme les détergents qui les brûlent (Augier et Boudouresque, 1970; Astier, 1972).

Cela est grave quand on songe qu'un mètre carré d'herbier de posidonies fabrique environ 10 l d'oxygène par 24 h et qu'un mètre linéaire de rhizome met un siècle à pousser dans des conditions naturelles. Par ailleurs, ces herbiers abritent une faune considérable qui trouve là un refuge et un lieu de reproduction. La faune vagile y est importante, représentée par de nombreux groupes zoologiques et elle sert de nourriture aux grands prédateurs que sont mollusques, crustacés et poissons (Ledoyer, 1962).

Rares sont les secteurs de la côte méditerranéenne du littoral Provence Côte d'Azur où l'on peut rencontrer des herbiers en bon état. Dans les Alpes maritimes sur 190 km de côte on compte 130 km de béton constitués par ports de plaisance, marinas, endigages et plages artificielles (Meinesz et Lefevre, 1978). Les secteurs les plus protégés sont situés autour des îles: Iles de Lérins, Archipel des Embiez près de Toulon et celui des Iles d'Hyères où se trouve le Parc national sous-marin de Port-Cros, seul parc sous-marin français qui possède encore l'un des plus beaux herbiers de posidonies, cependant très menacé par les ancres des plaisanciers (Augier et Boudouresque, 1970a).

La biocénose de l'herbier de posidonies est caractérisée par le groupe écologique des espèces vivant sur les feuilles; elle est également caractérisée par la combinaison de groupes écologiques qui réunit le groupe des espèces photophiles des frondaisons, le groupe des espèces sciaphiles des rhizomes et le groupe des espèces endogènes des mattes.

En l'absence de pollution, l'évolution de l'herbier est remarquable. Les rhizomes de Posidonia oceanica croissent non seulement sur le plan horizontal, mais aussi sur le plan vertical; cela leur permet d'échapper au lent ennoyage par les sédiments. Ces sédiments proviennent en partie des organismes ayant vécu sur les feuilles et surtout des particules transportées par les courants et la houle dont l'énergie cinétique est brisée au contact des feuilles jouant le rôle de "piège à sédiments". Les posidonies réagissent donc à ce lent ennoyage progressif par une croissance verticale des rhizomes, plus rapide. Il en résulte l'édification de véritables terrasses sous-marines constituées par un lacis de rhizomes dont les espaces sont colmatés par les sédiments grossiers, formations appelées "mattes" par les pêcheurs. La lente surélévation des mattes aboutit souvent à rapprocher celles-ci de la surface et leur destinée sera alors fonction de l'agitation des eaux.

En mode calme, c'est-à-dire essentiellement dans le fond de certaines baies, l'élévation des mattes aboutit à l'édification de ce que l'on a appelé un "récif barrière". Ces récifs barrières progressent lentement vers le large tandis que s'isole en arrière une lagune. La lagune aux eaux très calmes, est peu à peu comblée par les sédiments très fins et légers qui seuls franchissent le barrage des feuilles de posidonies. Pour pallier ce comblement, l'existence d'une circulation d'eau correcte est indispensable et les courants existants ne doivent pas être freinés par des aménagements aberrants tels que obstructions de passes, endigages ou enrochements.

Sur le fond de la lagune se constituent très souvent des pelouses à Cymodocea nodosa qui poussent sur les sédiments meubles dont les particules fines ont été filtrées par les rondaisons de l'herbier de Posidonies (Fig. 1).

D'autre part, les pelouses de C. nodosa ont une grande importance dans le cycle biologique de l'herbier de posidonies. Elles constituent le stade intermédiaire précédant l'installation des posidonies sur le sable sans végétation mais aussi elles se réinstallent là où les posidonies pour des raisons de modifications des conditions de milieu, ne peuvent plus subsister.

Des études de l'évolution de l'herbier peuvent être réalisées en balisant les limites supérieures et inférieures afin de suivre au cours du temps les fluctuations de ces limites.

Jadis, en eau claire, la limite supérieure de l'herbier débutait tout près du rivage, et la limite inférieure s'étendait jusqu'à 35 à 40 m maximum. De nos jours, avec la restructuration des rivages par l'homme, la limite supérieure a régressé, notamment au droit des petits ports jusqu'à - 15 mètres environ, et elle continue à reculer. C'est ainsi que l'on a pu constater en 1979 un recul de 80 cm en un an près d'un petit port de la côte marseillaise, alors que les conditions portuaires n'ont pas varié depuis plusieurs années (Christiani, 1980).

De nouveaux aménagements vont être responsables d'une dégradation rapide de l'herbier, dans les prochaines années. Il en est ainsi lors de l'installation des plages artificielles. Un exemple a été suivi par les biologistes dans la rade de Toulon. La construction d'un kilomètre de digue a provoqué la disparition de 25 hectares de posidonies (Nodot et al., 1978). L'apport de gravats, de décombres de

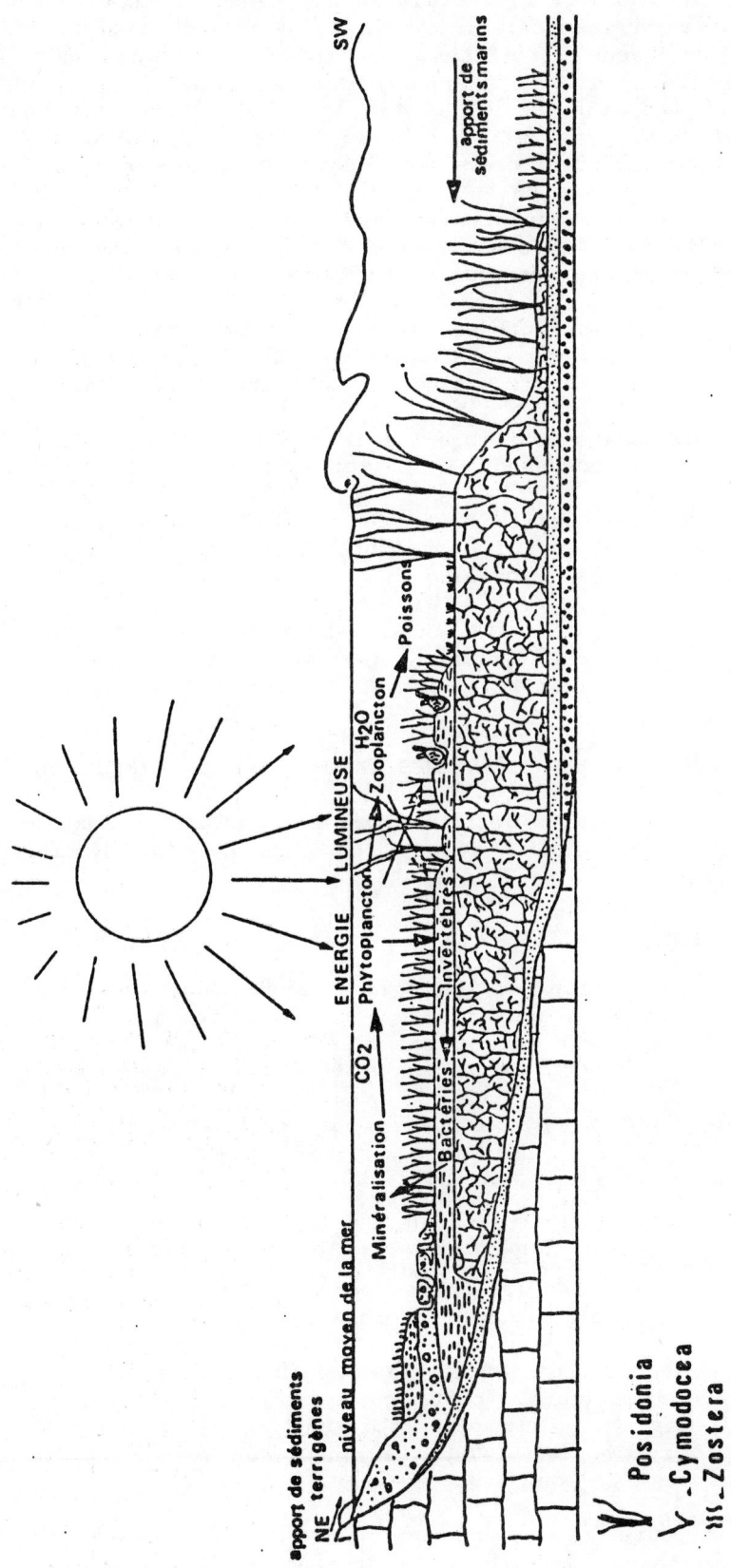

Fig. 1 Cycle de la matière dans la lagune du Brusc

toutes sortes, a entraîné une turbidité considérable des eaux. La photosynthèse s'en est trouvée perturbée, l'herbier voisin des plages artificielles s'est trouvé ennoyé et étouffé par les particules en suspension. Ainsi 20 hectares supplémentaires ont été détruits. Les eaux étant devenues turbides, les rayons de soleil ne pouvaient plus pénétrer suffisamment et la limite inférieure s'est rapprochée de la surface, de la sorte 22 hectares de plus ont disparu. En conclusion, pour construire un kilomètre de plages artificielles, ce sont près de 70 hectares d'herbier qui ont été sacrifiés. Cette destruction est particulièrement grave pour quatre raisons essentielles:

- Elle anéantit une surface vivante infiniment plus grande (15 fois dans le cas de Toulon) que celle des terrains "récupérés" qui, quelle que soit la valeur de leur aménagement, n'auront jamais l'intérêt écologique de l'herbier perdu.

- La vie sous-marine qui est intense dans les petits fonds est détruite avec la disparition de l'herbier, car la plupart des organismes de cette zone ont un besoin vital de la présence des posidonies pour leur reproduction.

- Les herbiers de posidonies contribuent à la stabilité des fonds marins et de la côte contre d'éventuels mouvements de la mer.

- Les eaux des plages alvéolaires sont peu renouvelées et mal oxygénées surtout en été ce qui les rend parfois insalubres. Elles sont souvent troubles et propices à un développement microbien.

Les herbiers sont également détruits au débouché des égouts et des émissaires urbains. Leur impact est toujours identique, quel que soit le secteur côtier. Les effets se manifestent à tous les niveaux (Astier et Tailliez, 1978).

3. ACTIONS DES POLLUANTS SUR LES ORGANISMES ET LES CHAINES ALIMENTAIRES MARINES

Les diverses biocénoses littorales affectées par les pollutions sont en temps normal le réceptable, le refuge d'organismes vivants les plus divers qui s'intègrent dans diverses chaînes alimentaires.

3.1 Mode d'action des polluants chimiques

Tous les polluants et en particulier les polluants chimiques vont avoir des effets délétères sur la plupart des organismes et affecter les divers maillons de ces chaînes alimentaires. Ainsi la faune et la flore sont-elles contaminées par divers polluants chimiques si nombreux qu'il est difficile de les passer tous en revue.

- Certaines fractions solubles d'hydrocarbures ont des propriétés cancérigènes, il s'agit des benzo 3-4 pyrènes. De nombreux poissons les concentrent, de même que le plancton végétal d'ailleurs (60 µg/g dans le plancton du Golfe de Gênes) (Ramade, 1977).

- Les détergents sont captés par les particules en suspension, de même que par les algues. Utilisés pour combattre les marées noires, ils constituent un remède pire que le mal, détruisant flore et faune littorales.

- Les pesticides comme le DDT, le lindane, les PCB (Phénochlor, Arochlor) agissent sur les systèmes enzymatiques, contaminent les coquillages, perturbent le développement des oeufs des mollusques, poissons et oiseaux marins (Escoubet et al., 1976; Chabert et Vicente, 1978). Ainsi, une dose de 0,01 mg/kg de DDT représente une dose mortelle pour les crevettes.

- Les métaux lourds: mercure, plomb, cadmium, nickel, cuivre, cobalt, etc. ralentissent la croissance, perturbent le développement et altèrent les organes: branchies, foie, rein. Ils s'accumulent dans les organismes et dans les sédiments marins.

Des dosages effectués par le laboratoire Central d'Hygiène Alimentaire de Paris en 1974 montrent que les poissons de Méditerranée ont des teneurs en mercure supérieures à ceux de l'Atlantique, en particulier le thon rouge, dont 70% des individus dépassent les normes internationales (0,7 mg/kg). Les concentrations sont de 0,485 mg/kg en Atlantique et 1,185 mg/kg en Méditerranée (Tableau I). Le mercure affecte le développement des oeufs de mollusques bivalves, s'intégrant aux membranes cellulaires et aux protéines pour former des mercaptides (Escoubet et Vicente, 1976; Vicente, 1978). Le plomb provoque la rupture des membranes cellulaires, le cadmium le ramollissement des os, car il perturbe le métabolisme du calcium, il a d'ailleurs des effets cancérigènes sur les poissons (Saxena et Yadav, 1981).

Tableau I.

Concentration en mercure dans quelques espèces de Méditerranée

Poissons	Mercure total en mg/kg Poids sec	Poissons	Hg total en mg/kg Poids sec
Sar	0,40	Pageot	0,31
Perche	0,99	Merlan	0,87
Sardine	0,46	Bogue	0,48
Vive	0,38	Thon	1,18

3.2   Action des polluants sur les chaînes alimentaires marines

Les divers polluants agissant directement sur les organismes qui peuplent les masses marines (plancton, poissons pélagiques) et le fond des mers (benthos) et ils peuvent se concentrer dans les chaînes alimentaires.

Les exemples de concentrations d'éléments chimiques dans les divers maillons des chaînes alimentaires marines sont nombreux.

3.2.1 Cas du mercure

L'exemple funeste de Minamata (Japon) pourrait se reproduire en Méditerranée si l'on n'y prend garde. Le mercure utilisé comme électrode dans la préparation de la soude caustique, sert de catalyseur pour la fabrication de fibres textiles. La synthèse d'une tonne d'acétaldéhyde rejette 30 à 100 g de mercure dans les eaux littorales.

A partir de doses infinitésimales de méthylmercure dans les eaux (0,1 µg/l), par le jeu des concentrations dans chaque maillon de la chaîne alimentaire on en retrouve 50 mg/kg chez les poissons, soit un coefficient de concentration de 500.000, et chez les pêcheurs contaminés 528 mg/kg dans les cheveux. Une telle contamination a entraîné à Minamata la mort de 48 personnes et l'invalidité totale de plusieurs centaines d'autres. Les mêmes phénomènes d'accumulation se rencontrent au sein de diverses chaînes trophiques.

Il convient d'être vigilants quand on constate les teneurs en mercure rencontrées chez de nombreux organismes en Méditerranée et surtout dans les sédiments (jusqu'à 20 mg/kg en certains secteurs au voisinage des égouts) où le mercure minéral peut être transformé en mercure organique beaucoup plus toxique.

3.2.2 Cas du Cadmium

Cet élément, bien que moins connu que le mercure pose d'énormes problèmes car on le retrouve dans les eaux méditerranéennes à des quantités non négligeables. Il va s'accumuler dans les coquillages notamment. Loin des côtes en plein océan, les concentrations en Cd sont très faibles (de 0,02 µg/l à 0,10 µg/l). Par contre dans les eaux côtières ou près des estuaires, les concentrations en Cd dissous croissent de manière significative (9 à 30 µg/l). Dans l'eau de mer plus de 90% du Cd est sous forme de chlorures. Les sédiments côtiers et ceux des estuaires en renferment des quantités non négligeables (1,8 à 6,8 ppm).

Une baisse de salinité augmente la prise du Cd par les bivalves et les crustacés par exemple, il en résulte qu'à faible salinité, le taux en calcium décroît et le captage du cadmium par les organismes en est facilité. De même une augmentation de la température entraîne une augmentation de Cd chez les organismes.

Il va agir à de faibles concentrations au niveau de phytoplancton premier maillon de la chaîne alimentaire et il va s'accumuler ensuite dans les animaux marins (mollusques, poissons), qui se nourrissent de plancton. Ainsi, il a été trouvé des facteurs de concentration:

$$\frac{Cd/g \text{ matière sèche}}{Cd/g \text{ eau filtrée}}$$

de 10^3 pour le sédiment, 10^4 pour le plancton, $10^2 - 10^3$ pour les algues, $10^3 - 10^5$ pour les mollusques, 10^3 pour les crustacés, 10^2 pour les poissons.

Le cadmium par ailleurs entraîne des troubles osseux chez les poissons et des perturbations dans la formation de la coquille des larves véligères de Mollusques car il modifie le métabolisme du calcium. C'est ainsi qu'on obtient de véritables monstres larvaires pour les coques et les palourdes qui ne parviennent pas à se fixer et voient leur développement interrompu.

Le cadmium affecte également la reproduction des poissons, le nombre d'oeufs pondus parvenant à éclosion diminue et entraîne des lésions au niveau des branchies (perturbations de l'activité respiratoire), des changements de la composition sanguine, des lésions de la colonne vertébrale. Il entraîne des pertes de poids, il va s'accumuler en priorité au niveau du rein.

Il modifie également le métabolisme du Zn au niveau des systèmes enzymatiques. Les concentrations létales se situent entre 0,01 mg/l et 10 mg/l dans l'eau de mer. Les oeufs et les stades larvaires sont les plus sensibles.

Le cadmium est considéré comme un élément non essentiel à l'homme et présente pour lui des effets toxiques très sérieux à faible dose. Les trois principales sources de contamination par le cadmium pour l'espèce humaine sont l'air, l'eau et surtout les aliments.

Parmi les aliments il faut citer notamment les mollusques et les crustacés marins qui peuvent concentrer de 10.000 à 1.000.000 fois le cadmium présent dans l'eau de mer: exemple, concentration dans les huîtres: 3,7 mg/kg.

3.2.3 Les composés organochlorés

Dans leur cas, on note aussi des phénomènes très importants d'accumulation et de concentration dans les chaînes alimentaires. Pour ce qui concerne le DDT par exemple, le quart de la production mondiale retourne à la mer par voie fluviale, les eaux de ruissellement et les divers rejets. Or, il suffit d'une concentration de 1 µg/l de DDT dans l'eau de mer pour qu'on en retrouve 70 µg/kg (ppb) dans le plancton, 15 mg/kg dans les poissons (15 ppm) et enfin 800 mg/kg dans les mammifères marins comme le dauphin, soit 1 million de fois environ la concentration de départ.

Ces mammifères marins (baleines, dauphins, phoques moines) paient un lourd tribut à la pollution de la Méditerranée et ils disparaissent de plus en plus. Se situant au sommet des chaînes alimentaires marines, ils se contaminent plus que n'importe quel autre organisme, et en particulier par les pesticides qui provoquent chez eux des troubles nerveux, des ulcérations de la peau, car souvent ils s'accumulent dans le lard sous-jacent. D'autres animaux comme les oiseaux marins peuvent être contaminés en se nourrissant de poissons. Cette contamination peut entraîner la mort et le plus souvent ce sont leurs oeufs qui renferment de grandes quantités d'organochlorés et dont la coquille devient très friable (Fig. 2).

Les PCB font eux aussi courir un grand danger à la faune littorale. Ils s'accumulent notamment dans les coquillages et les sédiments. Sur le littoral méditerranéen les travaux montrent des concentrations assez importantes dans les moules, en particulier au voisinage des émissaires (Escoubet et al., 1976).

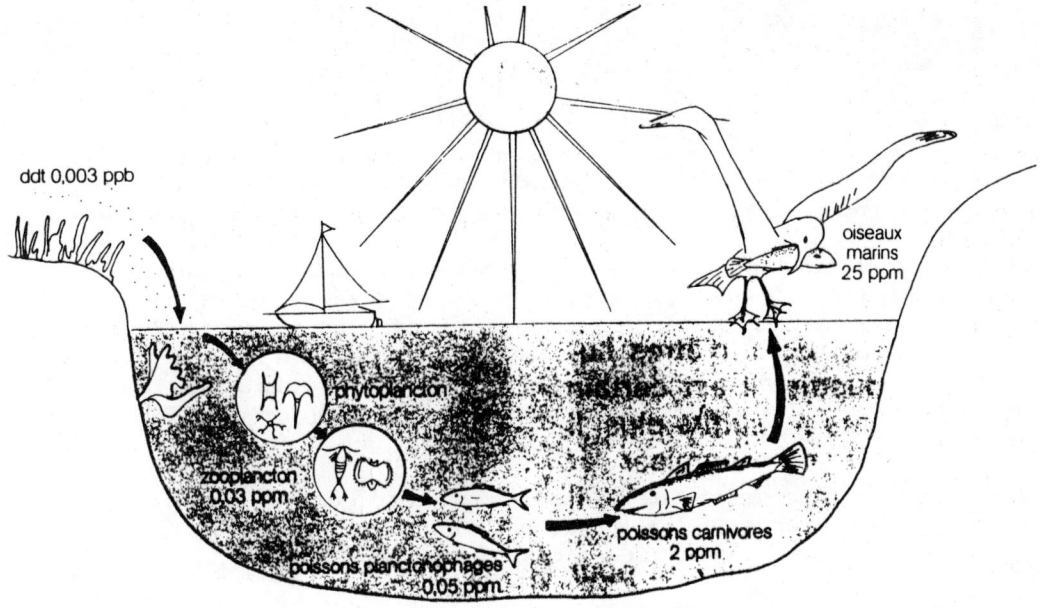

Figure 2 - Concentration de DDT dans une chaîne alimentaire marine.
Par lessivage des terrains cultivés, le DDT pulvérisé sur les végétaux à une concentration de 0,003 ppb arrive dans la mer et se retrouve dans le plancton à une concentration de 0,03 ppm (10 000 fois supérieure). Ce plancton sert de nourriture aux poissons planctonivores (sardines, harengs) dans l'organisme desquels le DDT se retrouve à 0,05 ppm. Ces poissons sont eux-mêmes dévorés par des poissons carnivores (cabillauds, thons) où la concentration en DDT est de 2 ppm. En bout de chaîne les oiseaux marins (goelands, cormorans) se nourrissent de poissons et se contaminent. La concentration en DDT chez eux est de 25 ppm.

4. CONCLUSIONS

Les solutions, les remèdes

Comme nous venons de la voir, les pollutions, sous des aspects multiples, affectent notre littoral méditerranéen. Atteignant la faune et la flore à tous les niveaux des chaînes alimentaires, par un effet de boomerang, la pollution finit par atteindre son responsable: l'homme qui se trouve au sommet de ces chaînes alimentaires puis qu'il consomme les produits de la mer contaminée. Mais cet homme qui d'Homo sapiens est devenu l'Home faber ou mieux l'Homo industrialis, prendra-t-il enfin les mesures qui s'imposent pour sauver cette nature qu'il agresse chaque jour davantage? Mettra-t-il tout en oeuvre pour sauver d'une mort inéluctable le berceau de multiples civilisations: La Méditerranée?

Pour lutter contre toutes ces nuisances, il existe pourtant des moyens, des remèdes. Bien que ce ne soit pas la panacée, il convient tout d'abord d'épurer les effluents usés à la source. Le traitement des eaux résiduaires dans les stations d'épuration est une solution à préconiser. Après un tel traitement, rien n'empêchera alors d'utiliser les émissaires pour rejeter une eau propre ou polluée à un degré tel que la diffusion des polluants encore existants ne représente qu'un aspect naturel de la vie humaine sur terre. Mieux vaut une épuration si minime soit-elle que pas d'épuration du tout. De nombreux travaux ont eu lieu et se poursuivent sur les problèmes de l'épuration des eaux domestiques et industrielles. Il faut distinguer des études destinées à trouver des procédés de traitement nouveaux (procédés physico-chimiques par exemple) ou des études destinées à améliorer des procédés déjà existants (miniaturisation, baisse du coût, etc.). De toutes façons, en l'état actuel des techniques, il faut distinguer deux grandes classes d'épuration: celles concernant les polluants biodégradables et celles concernant ceux qui ne le sont pas, autrement dit, l'épuration biologique et l'épuration physico-chimique. Le Lagunage est une option à préconiser quand celà est possible.

Plusieurs procédés d'épuration biologique sont applicables et varient seulement par la technologie mise en oeuvre: boues activées; percolation, disques biologiques, chenaux d'oxydation, etc. Cependant, dans tous les cas, le principe est le même: des bactéries et des microorganismes (phytoplanctontes) vont se développer en consommant comme nourriture les impuretés organiques contenues dans les eaux. Une aération est nécessaire, elle est réalisée dans le but de fournir de l'oxygène aux microorganismes soit par injection d'air sous pression, soit par contact avec l'atmosphère.

Il y aura ainsi diminution des impuretés et développement des microorganismes qui se retrouveront sous forme de boues qui devront être soutirées périodiquement. L'épuration industrielle doit être effectuée à la source même des rejets et il faut bien se garder de mélanger aux eaux urbaines des eaux industrielles, car, dans certains cas, les procédés d'épuration ne sont pas compatibles. En outre, certaines eaux industrielles par leurs caractéristiques, empêcheraient le dévéloppement des microorganismes épurateurs.

Les eaux résiduaires renfermant des toxiques comme les métaux lourds, les détergents, les hydrocarbures, les organochlorés sont traitées par divers procédés connus: filtrations poussées, centrifugation, flottation, osmose, inverse, etc.

Si ces procédés sont appliqués à la sortie de chaque usine, pour chaque effluent urbain, l'expérience démontre que, si l'on cesse de rejeter dans le milieu aquatique, quel qu'il soit, d'eau douce ou marin, des substances toxiques, la vie reprend toujours le dessus.

Les systèmes de mouillage des bateaux doivent être bien adaptés. Afin d'éviter les dégâts provoqués par les ancres, il est nécessaire de mettre en place des systèmes de mouillage réduisant l'impact de chaque bateau de plaisance sur les fonds (corps morts, bouées flottantes, etc.). Il faut pour cela tout d'abord apporter une bonne information au plaisancier qui n'est pas toujours au courant des dégradations qu'il peut occasionner.

Il est indispensable aussi de créer d'autres réserves naturelles afin de mettre à l'abri certains biotopes précieux du littoral méditerranéen (herbiers de posidonies, lagunes, marais, etc.) mais on ne pourra protéger des parcelles de ce littoral si on ne le protège pas dans son ensemble et par conséquent des mesures doivent être prises au niveau scientifique et juridique.

Il convient de mettre en place une gestion rationnelle des ressources naturelles sur tout le littoral méditerranéen.

Au niveau scientifique et technique, à l'heure où se mettent en place de nombreuses opérations d'aquaculture qui doivent constituer un complément appréciable pour la pêche artisanale, laquelle rencontre d'énormes difficultés à survivre, il est important de veiller à la sauvegarde du littoral en contrôlant l'impact des pollutions et en proposant des solutions. Pour ce faire, les études doivent être poursuivies sur l'évolution des divers écosystèmes. On devrait aboutir, à moyen terme, à la définition d'une méthodologie permettant d'estimer les flux polluants de toutes régions, parvenant aux façades maritimes. Cette action permettrait de traduire en charges polluantes les déversements directs ou indirects dans le milieu marin, et permettrait de mieux appréhender la notion d'acceptabilité à travers une quantification des données recueillies. Il convient de réduire les pollutions à la source en fixant des normes d'émission et en ayant recours à des technologies performantes et évolutives.

La notion "d'acceptabilité du milieu marin" repose sur un double concept: maîtrise des flux polluants à la source et connaissance fine du milieu récepteur. Une telle connaissance implique d'entreprendre ou de développer des recherches sur le devenir des polluants, sur leurs mécanismes d'action et sur leurs effets dans le milieu et les écosystèmes.

La richesse des écosystèmes marins fait que la contamination se produit le plus souvent à travers un grand nombre d'espèces et que les phénomènes de transfert de contaminants et de bioaccumulation sont importants à connaître, afin d'évaluer la toxicité globale d'un ou plusieurs polluants, en un site déterminé. Il est également nécessaire d'élaborer une méthodologie basée sur un certain nombre de tests biologiques adaptés aux divers sites et pouvant rendre compte de l'"état de santé" du milieu. La recherche de nouveaux indicateurs biologiques de pollution s'impose en ne retenant pas seulement des espèces résistantes (moules, balanes, oursins) mais aussi et surtout des espèces sensibles qui sont les plus vulnérables, et dans ce domaine, ce sont souvent les jeunes stades larvaires qui sont touchés en priorité. La connaissance des seuils de toxicité est également d'une grande importance et suppose une expérimentation approfondie en laboratoire.

Au plan juridique il est indispensable de parvenir enfin à une entente de tous les pays riverains de la Méditerranée pour une législation circumméditerranéenne. Trop de réunions et conférences n'en sont restées jusqu'alors, qu'au stade des voeux pieux, il faut à présent être fermes et résolus; puisqu'enfin, en 1980 a été paraphé à Athènes, par 17 des 18 pays riverains de la Méditerranée (seule l'Albanie s'est récusée) un protocole technique destiné à remédier à la pollution tellurique. Un pas important vient d'être franchi. En signant ce protocole les pays concernés s'engagent à équiper leurs villes côtières et leurs industries, loin en amont des cours d'usines d'épuration. Par ailleurs, le protocole indique en annexe une liste noire de produits interdits ou réglementés (mercure, cadmium, organochlorés, substances radioactives et cancérigènes).

Cette réunion d'Athènes marque donc certainement une étape décisive dans le trop lent processus de lutte anti-pollution amorcé pour sauver la Méditerranée.

Mais tout en envisageant les mesures de protection et d'épuration, il importe que chacun prenne conscience de la gravité du problème. Cette prise de conscience est actuellement de plus en plus nette chez les jeunes qui participent à des actions de sauvegarde de l'environnement: clubs de protection d'espèces animales en voie de disparition, nettoyage des plages, etc. Il est nécessaire que chaque enseignant, que chaque chercheur donne un peu de son temps à de telles oeuvres socio-éducatives.

Grâce à ces actions, le milieu marin dans sa zone la plus sensible, c'est-à-dire le littoral immédiat, pourra être sauvé et ses ressources préservées. Le choix, actuellement, est entre les déserts tels qu'on peut les voir à la sortie des égouts ou bien la fécondité qu'on peut observer dans les eaux poissonneuses d'un parc sous-marin comme celui de Port-Cros.

Si à court terme, c'est le problème de la salubrité des rivages qui se pose, à moyen et à long terme, compte tenu de l'évolution démographique galopante de la planète, ce sont les réserves d'eaux océaniques qu'il convient de conserver pures pour éviter d'avoir un jour à payer très cher le prix d'une irrigation par une eau qu'il faudra à la fois dessaler et dépolluer, et c'est la biomasse marine toute entière qu'il s'agit de préserver sachant qu'elle représente pour l'avenir 70% de l'évaluation de nos réserves nutritives. Si l'on veut exploiter de manière rationnelle toutes ces réserves, il est indispensable de prendre les mesures qui s'imposent pour enrayer la pollution.

Il n'existe qu'une alternative pour l'homme: sauver la nature ou périr avec elle. Nous laisserons le dernier mot au grand biologiste Jean Rostand aujourd'hui disparu: "l'idée ne pouvait venir aux hommes qu'ils auraient un jour à épargner cette géante (la nature), cette énorme, qu'ils auraient à en prendre soin, à la ménager, qu'il leur faudrait veiller à n'en pas gaspiller les ressources, à ne la pas endommager ou meurtrir, à ne la pas souiller en y déversant les excréments de leurs techniques".

5. REFERENCES BIBLIOGRAPHIQUES

Astier, J.M., Régression de l'herbier de Posidonies en rade des vignettes à Toulon.
1972 Ann.Soc.Sci.Archéol.Toulon Var, 24:97-103

Astier J.M. et P.H. Tailliez, Impact des effluents du grand collecteur du Cap Sicié sur
   1978   la vie des fonds marins. Bull.Fond.Océanogr.Ricard., Marseille 3:13-23

Augier, H. et C.F. Boudouresque, Végétation marine de l'île de Port-Cros (Parc National).
   1970   5. La baie de Port Man et le problème de la régression de l'herbier de posidonies. Bull.Mus.Hist.Nat.Marseille, 30:135-64

_____, Végétation de l'île de Port-Cros (Parc National). 6. Le Récif barrière.
   1970a   Bull.Mus.Hist.Nat.Marseille, 30:221-8

Augier, H. et N. Vicente, Essai de planification écologique en milieu marin. Bull.Fond.
   1979   Océanogr.Ricard, Marseille 3 Suppl. 8:1-40

Chabert, D. et N. Vicente, Contamination de Mollusques méditerranéens par un biocide
   1978   organochloré. Rev.Int.Oceanogr.Med., 49(3):45-8

Cristiani, G., Biomasse et répartition de l'herbier de Posidonia oceanica de la côte
   1980   Bleue (B. du Rh. France) et pollution marine par les métaux lourds. Thèse 3ème cycle. Université de Droit d'Economie et des Sciences Aix-Marseille, 150 p.

Escoubet, P. et N. Vicente, Effets de la contamination de Cerastoderma glaucum (Mollusque
   1976   Bivalve) par un sel de mercure. Haliotis, 7:151-5

Escoubet, P., J.L. Monod et N. Vicente, Concentration en pesticides (PCB et résidus de
   1976   DDT) dans l'eau les Mollusques de la zone des Embiez (Var-France). Haliotis, 7:141-3

Ledoyer, M., Etude de la faune vagile des herbiers superficiels de Zosteracés et de
   1962   quelques d'algues littorales. Thèse de 3ème cycle. Rec.Trav.Stn.Mar.Endoume, 25(39):117-235

Meinesz, A. et J.R. Lefevre, Destruction de l'étage infralittoral des Alpes Maritimes
   1978   (France) et de Monaco par les restructurations du rivage. Bull.Ecol., 9(3):259-76

Meinesz, A., J.M. Astier et J.R. Lefevre, Impact de l'aménagement du domaine maritime
   1981   sur l'étage infralittoral du Var - France (Méditerranée occidentale). Ann.Inst.Océanogr.Paris (Nouv.Sér.), 57(2):65-77

Nodot, C. et al., Etude d'impact des aménagements littoraux du Mourillon sur l'herbier
   1978   de Posidonia oceanica de la rade des Vignettes (Toulon-Var). Ann.Soc.Sci.Nat. Archéol. Toulon Var, 30:118-33

Ramade, F., Ecotoxicologie. Paris, Masson, Collection d'écologie, 9:205 p.
   1977

Saxena, O.P. et R.S. Yadav, Spontaneous epidermal neoplasm in freshwater aquarium fish.
   1981   Xephophorus maculatus (Gunther). Curr.Sci., 20(2):101-2

Vicente, N., Intégration du mercure aux structures cellulaires d'un Bivalve (Cerastoderma
   1978   glaucum, 1789). Haliotis, 9(1):35-40

REPORT ON BOTTOM FAUNA IN TWO NORTHERN ADRIATIC AREAS PRESUMED
TO BE INFLUENCED BY INPUTS

by

D. ZAVODNIK and J. VIDAKOVIC
Centre for Marine Research Rovinj
"Ruder Boskovic" Institute
52210 Rovinj, Yugoslavia

1. INTRODUCTION

Previous studies on general oceanographic conditions and dynamics of pelagic ecosystems of the northern Adriatic revealed the wide scale influence of the Po River discharge (Stirn, 1969; Revelante and Gilmartin, 1976; Degobbis et al., 1979; Smodlaka and Revelante, 1984). It was supposed that the continuous input of organic and inorganic seston, and perhaps some toxicants, could have an influence on benthos, at least in the areas facing the very mouth of the river.

On the other hand, a relatively closed marine ecosystem in an 11 km long fjord-like bay named Limski canal could be subject to alterations due to input of organic and mineral materials of terrestial origin and to the results of intensive shellfish and fish farming, as noted elsewhere (Millamena and Platon, 1977; Mattson and Linden, 1983).

The sediment-living macrobenthos in Limski canal is fairly well known (Vatova, 1931, 1935, 1943; Gamulin-Brida et al., 1968; Zavodnik, 1971), but only few quantitative data are available for the coastal and offshore area of the Po River mouth (Vatova, 1949; Ambrogi et al., 1983). The sediment meiofauna is also much better known in Limski canal (Daniels, 1970; Uffenorde, 1972; Vidakovic, 1984) if compared to offshore areas (Cita and Chierici, 1962; Ascoli, 1965; Vidakovic and Zavodnik, 1985).

To establish possible alterations of benthic communities in both areas, with regard to previous data, and to check the suppositions on input effects on the sediment living macro- and meiofauna, a repeat survey was undertaken as a part of the MED POL V project.

2. AREAS INVESTIGATED

Three offshore stations (ciphered SJ) were established in international waters just offshore from the Po River mouth, along a permanent transect for continuous oceanographic surveys between Rovinj and the Po River (Fig. 1, Table I). The stations were located in the area occupied by Vatova's (1949) zoocoenosis _Turritella_. Being about 20 Nm from the coast, Station SJ-7 served as a control station.

The inshore stations in Limski canal (ciphered LK) were situated in Vatova's (1935) zoocoenosis _Schizaster turritella_. Station LK-44, located at the entrance of the bay, was chosen as a control station.

Both areas are characterized by seasonal fluctuations in temperature, salinity and dissolved oxygen, and an annual change from a stratified to a vertically well mixed water column (Vatova and Milo di Villagrazia, 1950; Gilmartin et al., 1972). Although the alterations of salinity usually are minor, the sea water bottom layer suffers periodic and sometimes fairly sharp changes in some other parameters (Degobbis et al., 1979; Smodlaka, 1985). Especially important for bottom life can be anoxic conditions which were observed several times in the investigated offshore area of the northern Adriatic (Gilmartin et al., 1972; Smodlaka, 1985). In addition, and of prime importance for benthos in both areas, are variations in sedimentation rates, especially of pelitic material (Schreiber et al., 1968; Paul, 1970; Brambati et al., 1973).

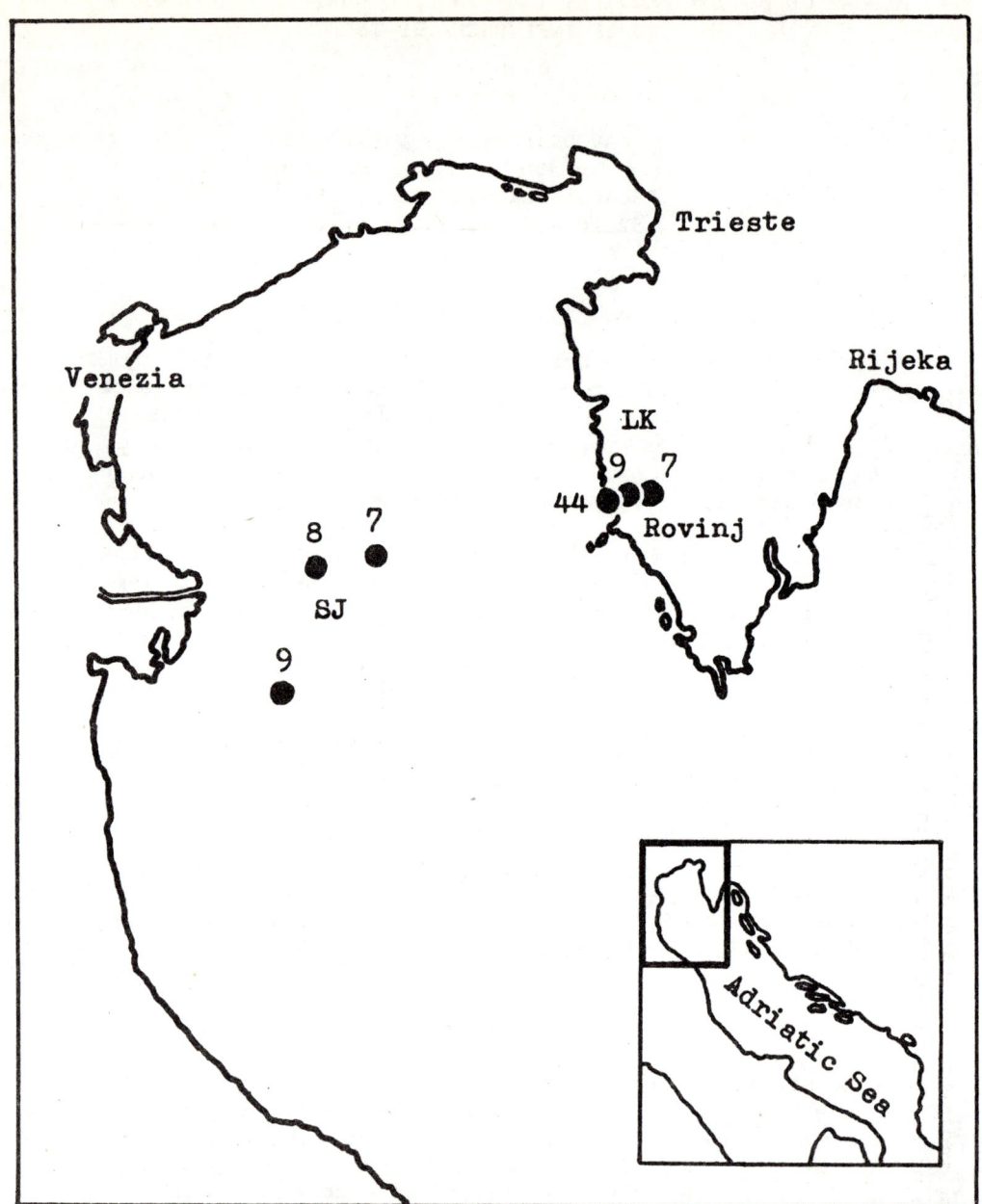

Figure 1. Research areas (SJ, LK) and stations.

3. METHODOLOGY

At offshore stations (SJ-7, SJ-8, SJ-9) sampling and measurements were performed at about two month intervals, from November 1982 to October 1983. Six surveys were completed. In Limski canal (stations LK-7, LK-9, LK-44) field research was performed usually once a month. Thus each station was surveyed 10-12 times.

Basic oceanographic, sedimentological and bacteriological parameters were studied using standard sampling and analytical methods (Oppenheimer and Zobell, 1952; Shepard, 1954; Strickland and Parsons, 1972; Dale, 1978). Benthic samples were taken by a van Veen $0.1 \, m^2$ grab. For macrobenthos usually 5 replicate samples (i.e. total from $0.5 \, m^2$ per station) were taken at each survey. The sediment was washed by overlow elutration through a 2mm sieve, and preserved in 4% neutralized formol. The extraction of organisms and further processing was done in the laboratory. The results were calculated in relation to the surface explored rather than to sediment volume to permit comparisons with previous data.

Table I

Basic data on locations, sediments and ranges of selected oceanographic parameters in bottom layers of sea water at stations investigated in 1982-1983.

Area Stations	Offshore (Bay of Venice)			Inshore (Limski canal)		
	SJ-7	SJ-8	SJ-9	LK-7	LK-9	LK-44
Latitude N	45°01.0'	44°59.8'	44°45.4'	45°08.2'	45°07.9'	45°07.6'
Longitude E	12°59.7'	12°49.8'	12°45.0'	13°42.8'	13°40.5'	13°36.8'
Depth (m)	33	31	31	8	18	30
Sediment type	clayey-sandy silt	clayey-sandy silt	clayey silt	clayey silt	clayey silt	sandy silt
Silt content (%)	43.30	52.53	62.84	—	—	—
Organic matter content (%)	12.45	11.30	15.55	12.58	16.23	17.90
Temperature (°C)	9.29-19.42	9.17-16.01	9.60-16.54	8.80-22.70	8.80-21.00	8.50-21.00
Salinity (Sx10^{-3})	38.14-38.33	38.03-38.42	37.99-38.42	35.89-38.76	35.70-38.48	34.79-38.55
Oxygen saturation (%)	49.1-99.0	31.2-100.3	24.3-99.7	78.0-118.5	68.7-106.8	65.5-108.7

At each survey four sub-samples per station were taken for meiofauna studies from an undisturbed van Veen grab sample with a plastic hand corer (inner diameter of 3.5 cm) (Vidakovic, 1984). Samples were fixed with 4% neutralized formol and stained with Rose Bengal. Sediment samples were processed by the elutration method. Special attention was paid to Nematoda, for which indices of diversity (H' and J), a mean dominance and a similarity index, and the interspecific encounter probability (PIE) were calculated (Gray, 1981).

4. RESULTS

4.1 Environmental research

At stations investigated during the period of this research, basic oceanographic parameters showed seasonal fluctuations within the ranges previously noted for these parts of the Adriatic Sea (Vatova and Milo di Villagrazia, 1950; Bozic et al., 1964; Gilmartin et al., 1972) (Table I). It should be emphasized, however, that offshore in the bottom layer of sea water oxygen concentrations clearly indicated deficiencies in November 1982 and from August to October 1983. Dissolved oxygen concentrations dropped patchily to only 25%, and rarely reached the saturation (100%) value. Obviously, these fluctuations were related to explicit stratification and reduced circulation of the water body, and to phytoplankton blooms (Smodlaka, 1985). It should be noted that high phytoplankton biomass (measured as chlorophyll a) and production were concentrated in surface and/or intermediate layers of sea water, and never reached a maximum near the bottom. In the same period, the bottom layer at LK stations was well aerated.

Analyses of the sediment indicated their heterogeneity by origin. They are poorly sorted and thus typical for low wave and current action (Gray, 1981).

4.2 Macrofauna

In both areas investigated, and at all stations, macrofauna was rather diverse. With regard to abundancies, major taxonomic groupings at offshore stations were Polychaeta, Echinodermata and Sipuncula, while in Limski canal sediments, Polychaeta, Mollusca and Crustacea were most characteristic (Table II). At a species level, in the area offshore the Po River mouth, the polychaetes Lumbrineris gracilis, Notomastus latericeus and Sternaspis scutata, the molluscs Turritella communis and Corbula gibba, the sipunculan Aspidosiphon sp. (in all probability A. kovalevskii) and the ophiuroids Amphiura chiajei and A. filiformis were most abundant. On average, in the period investigated, about 1000 specimens of macrofauna per square metre were extracted (Table III). Local and temporal variations were noted, but the constant dominance of Notomastus latericeus and Amphiura filiformis was evident.

In Limski canal, although the taxonomic data for Polychaeta are not yet available, the abundancies of macrofauna are lower in comparison with the results from offshore stations (Table II and III). For example, the only species at LK clayey silts, for which the mean abundancies calculated were more than 5 specimens per square metre, were Lunatia alderi, Dentalium dentalis, Sternaspis scutata, Gourretia minor and Upogebia tipica. None of the species identified until now showed such an extreme dominance as noted at the offshore stations.

4.3 Meiofauna

At all stations surveyed, the sediment meiofauna was characterized by Nematoda and Copepoda (Table IV). Several temporal variations in abundancies of each group were noted. In general, it seems that the sediments at offshore stations are inhabited by more numerous mobile meiofauna than are the stations located in Limski canal (Table III and IV).

Special attention was paid to analyses on Nematoda assemblages which, in respect to the total number of meiofauna, amount to 63-85%. At both areas, the chromadorid non-selective deposit feeders Dorylaimopsis mediterranea, Sabatieria ornata and S. punctata were very abundant (Table V). At some offshore stations, the highest abundance was noted for an omnivore Rhabdodemania sp. while at all inshore (LK) stations most abundant was an epigrowth feeder Hypodontolaimus sp. (Table V). Everywhere, relatively high species and dominance diversities of sediment nematodes populations were established (Table IV).

Table II

Mean abundancies (No. spm/m^2) of macrofauna at stations surveyed.

	Area	SJ			LK		
	Station	7	8	9	7	9	44
CNIDARIA							
Perigonimus sp.					0.3		
Cerianthus lloydii (Gosse)		2.1	5.0	3.5			
Epizoanthus arenaceus (D. Chiaje)							0.7
Epizoanthus sp.							0.2
Edwardsia claparedii (Panceri)		1.0	0.8		0.3	0.8	
Funiculina quadrangularis (Pallas)			0.4				
Virgularia mirabilis (O.F. Müller)		0.4					
MOLLUSCA							
Acanthochiton communis (Risso)					0.5		
Turritella communis Risso		40.4	14.6	5.5	0.3	0.8	3.3
Gourmya vulgata (Bruguière)					0.3		
Leiostraca subulata (Donovan)		0.7	6.9	1.0			3.7
Aporrhais pespelecani (Linnaeus)					0.3		0.9
Lunatia alderi (Forbes)			0.8		6.8	3.7	1.6
Lunatia guillemini (Payraudeau)							0.2
Murex brandaris Linnaeus						0.2	
Fusinus rostratus (Olivi)		0.7	0.4				
Cylichna cylindracea (Pennant)			1.5	0.5			0.2
Philine aperta Linnaeus		0.4		1.0			
Acera bullata Müller							0.2
Dentalium dentalis Linnaeus			0.4		23.2	5.0	1.9
Nucula nucleus (Linnaeus)				0.5	1.3		
Nucula turgida nitidosa (Winckworth)					1.8	1.0	3.3
Nuculana fragilis (Chemnitz)							0.2
Striarca lactea (Linnaeus)					1.6		
Modiolus barbatus (Linnaeus)					1.0		
Monia patelliformis (Linnaeus)		0.7			0.3		
Diplodonta brochii Deshayes						1.0	0.2
Thyasira flexuosa (Montagu)			0.8	1.0		0.4	0.2
Myrtea spinifera (Montagu)		1.8		4.5		0.4	0.9

Table II (Continued)

	Area Station	SJ 7	8	9	LK 7	9	44
Loripinus fragilis (Philippi)					0.8		0.9
Chama gryphoides Linnaeus					0.3		
Parvicardium ovale (Sowerby)					0.3		
Sphaerocardium paucicostatum (Sowerby)					2.1	0.6	
Acanthocardia echinata (Linnaeus)			0.4				
Pitar rude (Poli)		0.4	8.8		1.0	0.4	1.6
Dosinia lupinus (Poli)					0.3		
Dosinia sp.			0.4				
Chione ovata (Pennant)		0.4	12.7	3.5			
Venerupis rhomboides (Pennant)			1.1				
Venerupis pullastra (Montagu)					0.3		
Mysia undata (Pennant)		2.1	0.4				1.6
Gari fervensis (Gmelin)			0.4				0.2
Azorinus chamasolen (Da Costa)				0.5			
Quadrans serratus (Brocchi)		0.4			0.3		0.2
Moerella donacina (Linnaeus)			0.4				
Tellinella pulchella (Lamarck)		0.7	0.4	2.0	3.7	1.5	3.9
Abra alba (Wood)					1.8		2.8
Abra nitida (Müller)				0.5			0.2
Abra pellucida (Brocchi)		3.2	4.6	18.0			
Cultrensis adriaticus (Coen)		1.4	4.6	1.0	0.8		2.3
Solenidae gen. sp.		0.4					
Hiatella arctica (Linnaeus)		0.7	0.4		1.0		1.6
Saxicavella plicata (Linnaeus)						0.2	
Corbula gibba (Olivi)		3.6	58.5	86.0	2.6	2.9	
Rocellaria dubia (Pennant)					5.8		
Thracia pubescens (Pulteney)			0.4	0.5		0.4	
Cuspidaria cuspidata (Olivi)							0.5
SIPUNCULA							
Aspidosiphon kovalevskii (Murina)					0.5		6.5
Aspidosiphon sp.		55.0	48.8	114.0			
Phascolion strombi (Montagu)		0.4	0.4		0.3		
Golfingia elongata (Keferstein)		1.1					
Golfingia vulgaris (Blainville)			0.4	0.5			
Onchnesoma steenstrupii Koren & Danielssen						0.2	
Sipuncula indet.		2.5	4.2			0.2	

Table II (Continued)

Area Station	SJ 7	SJ 8	9	LK 7	LK 9	44
ANNELIDA						
Aphrodita aculeata Linnaeus	0.4		0.5			0.2
Evarnella cf. impar. (Johnston)	0.4					
Harmothoe cf. lunulata (Delle Chiaje)	6.1	0.4				
Harmothoe sp.			2.5			
Harmothoinae gen. sp.			0.5			
Leanira yhleni (Malmgren)		5.0	3.5			
Sthenelais boa (Johnston)	0.7					
Paranaitis kosteriensis (Malmgren)	0.7	1.5	0.5			
Eulalia sp.		2.3				
Mysta picta (Quatrefages)		0.8				
Phyllodoce sp.	1.1	0.8				
Paralacydonia paradoxa Fauvel		0.4				
Hesionidae gen. sp.	0.4					
Ophiodromus flexuosus (Delle Chiaje)		0.4				
Ancistrosyllis groenlandica McIntosh	0.7	1.5	1.0			
Pilargis verrucosa Saint-Joseph	0.7	0.4	0.5			
Sigambra tentaculata (Treadwell)	0.7					
Langerhansia cornuta (Rathke)	0.7	1.5				0.2
Eunereis longissima (Johnston)	0.7	1.9	2.0			
Neanthes lamellosa Ehlers	3.6	4.2	6.0			
Nephthys hombergi Savignyi		0.4				
Nephthys hystricis McIntosh	3.6	4.2	4.5			
Glycera convoluta Keferstein		3.8	4.5			
Glycera unicornis Savignyi	0.7	2.7	1.0			
Glycera rouxii Audouin & Milne Edwards	1.8					
Glycera sp.	1.4					
Goniada maculata Oersted	1.1	2.3				
Goniada norvegica Oersted			0.5			
Hyalinoecia bilineata Baird		3.5	1.0			
Hyalinoecia fauveli Rioja	1.4	0.4				
Eunice vittata (Delle Chiaje)	0.7	1.5				
Eunice sp.			0.5			
Marphysa bellii Audouin & Milne Edwards		6.1				
Marphysa kinbergi McIntosh		1.1				
Lumbrineris gracilis (Ehlers)	5.0	2.7	22.0			

Table II (Continued)

	Area	SJ			LK	
Station	7	8	9	7	9	44
Lumbrineris rovignensis (Fauvel)	0.4					
Drilonereis filum (Claparède)	0.4					
Spiophanes kroyeri Grube	0.7	1.1	1.5			
Polydora ciliata (Johnston)		0.8				
Polydora sp.	0.4					
Magelona alleni Wilson	2.1	1.1				
Poecilochaetus serpens Allen	1.1	0.4	1.0			
Chaetopteridae gen. sp.	1.8	0.8	1.5			
Aricidea fauveli Hartman	0.7	1.1	0.5			
Tauberia gracilis (Tauber)	0.7		1.0			
Cirrophorus branchiatus Ehlers		0.4				
Chaetozone setosa Malmgren	3.6	2.3	2.0			
Tharyx cf. marioni (Saint-Joseph)	2.1	1.1	0.5			
Tharyx heterochaeta (Laubier)		0.8				
Dodecaceria concharum Oersted	0.4	1.5				
Notomastus latericeus Sars	121.1	299.6	371.5			
Maldane glebifex Grube	5.0	0.4	0.5			
Asychis gotoi Izuka			0.5			
Owenia fusiformis Delle Chiaje	7.1	3.8	1.0			
Sternaspis scutata (Ranzani)	17.5	21.1	12.5	2.4	3.7	11.4
Amphictene auricoma (Müller)	0.7	1.1	1.5		0.2	
Lagis koreni Malmgren	0.4	1.5	0.5			
Ampharete acutifrons Grube	3.2	3.5	2.5			
Ampharetidae gen. sp.		0.8				
Amphicteis gunneri (Sars)	0.4	2.7				
Sosane sulcata Malmgren	0.4					
Melinna palmata Grube	2.1	23.8	3.0			
Lanice conchylega (Pallas)	0.4	0.8				
Polycirrus sp.		0.4	0.5			
Amphitritinae gen. sp.	2.1	0.4				
Thelepinae gen. sp.	0.4					
Terebellides stroemi Sars	0.7	0.4				
Sabella pavonina Savignyi			1.0			
Sabellinae gen. sp.	0.4					
Chone cf. duneri Malmgren		0.4				
Chone sp.	1.1					
Euchone rubrocincta (Sars)		0.4				

Table II (Continued)

	Area Station	SJ 7	8	9	LK 7	9	44
Ditrupa arietina (Müller)			1.1				
Pomatoceros triqueter (Linnaeus)							0.5
CRUSTACEA							
Processa nouveli nouveli Al Adhub & Williamson					0.3	3.3	0.9
Processa sp.		0.4	2.3		0.3	1.0	1.6
Philocheras bispinosus (Hailstone)						0.2	
Philocheras monacanthus (Holthuis)					0.3		
Philocheras sp.							0.2
Gourretia minor Gourret					6.0	4.4	
Upogebia tipica (Nardo)		0.4			3.4	12.1	
Upogebia sp.		0.4			0.3		
Callianassa sp.		0.4				0.2	0.2
Paguristes oculatus (Fabricius)							0.2
Pagurus cuanensis Bell		0.4					0.2
Anapagurus laevis (Bell)		1.1		1.5			
Anapagurus bicorniger A. Milne Edwards & Bouvier							0.2
Anapagurus sp.						0.2	0.7
Galathea intermedia Lilljeborg					0.5		0.2
Pisidia longimana (Risso)					0.3		
Liocarcinus maculatus (Risso)							0.2
Liocarcinus depurator (Linnaeus)		0.4		1.5			
Pilumnus sp. (juv.)					0.3		
Decapoda indet.		0.4		0.5	0.3	0.2	
Mysidacea indet.						0.4	
Cumacea indet.							0.7
Apseudes latreillei (H. Milne Edwards)						0.2	3.5
Isopoda indet.						0.2	2.6
Amphipoda indet.		3.9	3.8	2.5		1.9	5.6
TENTACULATA							
Scrupocellaria reptans (Linnaeus)		0.7					
Cribrillina radiata (Moll)					0.3		
Schizoporella linearis (Hincks)							0.5

Table II (Continued)

	Area	SJ			LK		
	Station	7	8	9	7	9	44
Schizoporella sp.					0.5		
Sertella beaniana King		0.4					
Lichenopora radiata Audouin			0.4				
ECHINODERMATA							
Ocnus planci (Brandt)		0.7					
Trachythyone elongata (Düben & Koren)						0.2	1.6
Labidoplax digitata (Montagu)						1.0	8.8
Leptosynapta inhaerens (O.F. Müller)						0.4	0.5
Astropecten irregularis pentacanthus (Delle Chiaje)					2.1	0.4	
Astropecten sp. (juv.)						0.2	
Amphiura chiajei Forbes		11.8	15.0	9.0	1.6	0.4	2.3
Amphiura cherbonnieri Guille		0.4	1.5				
Amphiura filiformis (O.F. Müller)		658.6	538.5	127.0		1.5	14.6
Amphiura brachiata (Montagu)		0.7					
Ophiothrix fragilis (Abildgaard)						0.4	
Ophiura ophiura (Linnaeus)				1.0			
Ophiura albida Forbes							0.2
Ophiura grubei Heller			0.4	0.5			0.7
Schizaster canaliferus (Lamarck)					0.3		0.5
VARIA							
Chondrilla nucula (Schmidt)					0.3		
Nemertina indet.		1.4	0.8	2.0	0.5	0.8	0.5
Echiura indet.						0.4	
Loxosomella atkinsae (Bobin & Prenant)							0.2
Loxosomella phascolosomata (Vogt)				0.5			
Phoronidea indet.		+	+			+	+
Ascidiella aspersa (Müller)						0.4	
Styela plicata Lesueur						0.4	
Polycarpa sp.		0.4					

Table III

Basic data on fauna and sediment heterotrophic bacteria.

Station	SJ-7	SJ-8	SJ-9	LK-7	LK-9	LK-44
No. of surveys	6	6	5	10	12	11
Macrofauna surveyed:						
Surface (m^2)	2.8	2.6	2.0	3.8	4.8	4.3
No. of species (*)	91	89	62	47**	44**	57**
No. of specimens ($\bar{x}$ n/1 m^2)	1013	1170	844	80**	45**	100**
Meiofauna surveyed:						
Surface (cm^2)	10	10	10	40	40	40
No. of species (Nematoda only)	22	24	20	55	45	73
No. of specimens ($\bar{x}$ n/10 cm^2)	1327	1245	612	736	312	767
Heterotrophs: ($\bar{x}$ n/1 g sediment)	2.08×10^5	2.42×10^5	4.03×10^5	1.68×10^5	1.43×10^5	1.90×10^5

* Unidentified taxonomic groups are considered as one species
** Polychaetes (unidentified) not included.

Table IV

Mean abundancies of sediment meiofauna ($\bar{x}$ per 10 cm^2) and analysis of nematode population.

Station	SJ-7	SJ-8	SJ-9	LK-7	LK-9	LK-44
Nematoda	1129	1024	467	474	197	571
Copepoda	135	149	96	174	86	140
Varia	63	72	49	88	29	56
Total meiofauna	1327	1245	612	736	312	767
Nematode population:						
Species diversity (H')	3.388	3.525	3.012	3.806	3.250	4.733
Dominance diversity (J)	0.760	0.769	0.697	0.746	0.678	0.839
Interspecific encounter probability (PIE)	0.871	0.867	0.796	0.850	0.807	0.946

Table V

Preliminary calculations on mean abundancies (No. spm/10 cm^2) of sediment nematodes at stations surveyed. One sample per SJ station and four samples per LK station were analyzed.

Area / Station	SJ 7	SJ 8	SJ 9	LK 7	LK 9	LK 44
ARAEOLAIMIDA						
Setoplectus sp.				0.3		
Halaphanolaimus sp.				2.0	0.5	
Leptolaimoides sp.				0.5		
Southerniella sp.				0.3		1.5
Axonolaimus arcuatus Steckhoven						0.3
Paradontophora sp.		6	5			4.5
Diplopletula sp.						0.5
DESMOSCOLECIDA						
Desmoscolex sp. 1		1		1.3	0.3	4.5
Desmoscolex sp. 2				0.3	0.3	0.8
Desmoscolex sp. 3						0.5
Desmoscolex sp. 4				0.5		
MONHYSTERIDA						
Siphonolaimus sp.						0.3
Coninckia sp.				0.3		
Desmolaimus sp.						0.8
Metalinchomeus sp. 1	3			0.8		2.5
Metalinchomeus sp. 2						1.3
Metalinchomeus sp. 3						0.8
Terschellingia longicaudata DeMan			7	3.0	2.0	12.0
Disconema sp.					0.3	
Paralinchomeus sp.	4		2		3.0	1.8
Monhystera sp.				0.3	0.5	
Steineria sp.	1					3.5
Theristus sp. 1	1	1		2.5	5.0	20.0
Theristus sp. 2				5.0	4.0	7.0
Theristus sp. 3					0.5	3.0
Theristus sp. 4				13.0	6.0	3.0
Theristus sp. (5)					0.3	
Theristus sp. 7				2.0		
Sphaerolaimus dispar Filipjev		1		1.0	4.0	4.8

Table V (Continued)

	Area Station	7	SJ 8	9	7	LK 9	44
Sphaerolaimus cf. macrocirculus Filipjev		3	1	4	0.5	0.8	1.0
Sphaerolaimus macrocirculus Filipjev					2.3	2.0	2.5
Sphaerolaimus minutus Vitiello		2	7	1	1.8	2.8	2.5
DESMODORIDA							
Microlaimus sp. 1				1			1.3
Microlaimus sp. 2					1.0		1.8
Metachromadora sp.					0.3	2.3	3.5
Pselionema sp. 1					0.5		2.3
Pselionema sp. 2							1.3
Desmodora pontica Filipjev					23.0		6.5
Desmodora conica Vitiello							1.3
Desmodora sp. 2							1.8
Croconema sp.		1	1	2			
Richtersia kreisi Boucher					0.3		0.8
Richtersia imparis Gerlach							0.3
Richtersia sp. 2							0.3
CHROMADORIDA							
Dorylaimopsis mediterranea DeZio		20	13	76	52.0	46.0	16.8
Hopperia sp.		2	2		5.0	50.8	2.5
Laimella sp.							1.3
Sabatieria hilarula DeMan		13	8		4.0		4.0
Sabatieria ornata (Ditlevsen)		32	18	30	11.5	1.0	18.3
Sabatieria punctata Stekhoven		32	26	20	24.0	8.8	26.0
Sabatieria sp.			4	11	4.5	2.3	0.5
Prochromadorella attenuata (Gerlach)					6.0	6.3	0.5
Prochromadorella ditlevseni (DeMan)				1	4.0		0.3
Actinomena sp. 1		9	16	1	12.0	0.8	14.0
Actinonema sp. 2		2	9	4			
Graphonema sp.		1			10.3	4.0	6.0
Hypodontolaimus sp.					79.8	68.8	31.8
Neochromadora sp.						0.5	0.8
Nygmatonchus sp.					0.3		
Spilophorella sp. 1					0.8	9.0	14.0
Spilophorella sp. 2					4.0	3.3	

Table V (Continued)

	SJ	SJ	SJ	LK	LK	LK
Area / Station	7	8	9	7	9	44
Neotonchoides warwicki Platt				1.3	0.8	4.5
Neotonchoides sp. 15		2	1		0.3	
Paracanthonchus sp.				0.5		
Longicyatholaimus complexus Warwick	2	18	2	32.3		29.5
Longicyatholaimus longicaudatus (DeMan)				1.0	4.0	2.5
Longicyatholaimus sp.						0.3
Marylinnia sp.				6.0		3.3
Metacyatholaimus sp.	4			7.0		26.0
Halichoanolaimus sp.		2	2		0.5	1.3
ENOPLIDA						
Parrironus sp.						1.5
Rhabdocoma sp.	1			0.3		0.3
Oxystomina sp. 1		1		0.8	1.0	2.8
Oxystomina sp. 2		2				1.3
Thalassoalaimus brevicaudatus Vitiello				0.3	0.3	
Thalassoalaimus sp.						0.8
Halalaimus filicorpus Vitiello				2.5	0.8	2.0
Halalaimus monstrocaudatus Vitiello	2	1	5		0.5	1.5
Anticoma sp. 1		1				1.0
Anticoma sp. 2					3.0	
Paramesacanthion sp.				0.8	0.8	1.3
Rhabdodemania mediteranea Boucher	35	58	15	2.5	2.3	10.5
Mononcholaimus sp. 1				0.3	0.3	0.3
Viscosia sp. 1		1	1	7.5	1.5	2.0
Viscosia sp. 2						2.0
Metoncholaimus sp.				0.5		0.3
Oncholaimus sp.						0.8
Prooncholaimus sp.	2			0.5		
Bolbella sp.				0.3	0.3	
Polygastrophora sp.						3.0
Bathyeuristomina sp.					0.5	1.0
Enoplidaea gen. sp.					0.3	
Species total	21	24	20	55	45	73
Specimens total	172	200	191	344	251	338

5. DISCUSSION AND CONCLUSIONS

The faunal lists (Table II and V) prove a great diversity at all offshore and inshore stations surveyed. Spatial variations, however, were noted, which partly may be attributed to variations in granulometric composition of the sediment (Table I). Thus, a calculated index of similarity indicated that nematodes from the two areas investigated belonged to different assemblages. It should be specially noted that, until now, at no station was any indicator species of pollution, in the sense of Picard(1965) and Stirn(1970), identified. Indicators of community instability were also rarely noted. Consequently, contrary to our initial expectations, it seems that no station surveyed was under the stress conditions of pollution. This supposition is supported by results of bacteriological survey and by calculations of PIE index for the sediment Nematoda assemblages (Table III and IV).

On the other hand, a clear dominance of some species indicates that basic oceanographic conditions in both areas studied were perhaps modified by combined effects of various factors. It seems that the temporal population of Notomastus latericeus had little impact on benthic communities at offshore stations, a phenomenon which is reminiscent of an appearance of Spiophanes bombyx in an area of the New York Bight (Swartz, 1976). Most of the dominant species are suspension or detritus film feeders, suggesting a high organic seston input. This can be of twofold origin: at first, sources of particulate organic materials can be Po River discharges, in Limski canal terrestial and fish farming inputs. The second source of organic seston input can be phytoplankton and microphytobenthos, especially through periodic blooms of unicellular algae. The secondary effects of algal blooms in the northern Adriatic can be deficiency in dissolved oxygen (Degobbis et al., 1979; Smodlaka, 1985) which sometimes have led to mass mortalities and possible local extinctions of non-tolerant benthic species (Zavodnik, 1977; Stachowitsch, 1983).

During the time of our research, in the offshore area anoxic conditions in the bottom layer of sea water were noted several times (Smodlaka, 1985) but no mass mortality of benthos was evidenced. Perhaps it occurred previously and initiated the diminution of Turritella communis population in the area, which probably gave an increase in abundance of a shell-living sipunculan Aspidosiphon kovalevskii (Vidakovic and Zavodnik, 1986). We suppose that the background for all these events, and the present faunal composition offshore the Po River mouth, are very intensive and almost continuous phytoplankton (and microphytobenthos) blooms favoured by the special temperature/salinity regime and by nutrient inputs from the Po River (Revelante and Gilmartin, 1976; Degobbis et al., 1979; Smodlaka and Revelante, 1984).

In the Limski canal inshore area, however, minor alterations in benthic macrofaunal assemblages occurred in comparison with data of about 50 years ago (Vatova, 1935). The abundancies of the macro- and sediment meiofauna were less than noted at the offshore Po River area, a fact which can be related to high input of terrestial inorganic seston (Paul, 1970) and the low input of particulate organic materials. It seems that silty bottom communities in Limski canal are not yet modified by pollution.

6. ACKNOWLEDGEMENTS

The authors are grateful to Mr. M. Juracic, Zagreb, for the results on granulometric analysis, to Miss D. Fuks, Rovinj, for providing bacteriological data, and to specialists for help in taxonomic studies of biological material: Dr. M. Hrs-Brenko, Rovinj (Bivalvia), Prof. Dr. L. Amoureux, Rennes (Polychaeta) and Dr. Z. Stevcic, Rovinj (Decapoda).

Thanks are also due to the technicians of the Centre for Marine Research at Rovinj: Mrs. E. Bozic-Rabak, MM. I. Korenic and R. Rabak for analyses of oceanographic parameters, and to Mrs. R. Sankovic for help at processing the biological material. We are much obliged to Dr. B.L. Bayne, Plymouth, for editorial comments on the manuscript.

The support of the Self-management Community of Interest for Scientific Research of S.R. Croatia, and of the FAO/UNEP (MED POL - Phase II) is acknowledged.

7. REFERENCES

Ambrogi, R., L. Amoureux and D. Bedulli, Contribution a l'étude des peuplements infra-
1983   littoraux face au delta du Po. Rapp.P.-V.Réun.CIESM, 28(3):189-90

Ascoli, P., Crociera talassografica adriatica, 1955. 6. Richerche ecologiche sugli
1965   Ostracodi contenuti in 16 carote prelevate sul fondo del Mare Adriatico.
Arch.Oceanogr.Limnol., 14(1):69-137

Bozic, E., M. Hrs-Brenko and M. Marinkovic-Roje, Oceanographical observations in the
1964   areas of Rovinj and Limski canal during 1963 and 1964 Hidrogr.God., (in Croatian).

Brambati, A., et al., Transport and sedimentation in the Adriatic Sea. Pubbl.Mus.Friulano
1973   Stor.Nat.Udine, (20):1-60

Cita, M.B. and M.A. Chierici, Crociera talassografica adriatica, 1955. 5. Ricerche sui
1962   Foraminiferi contenuti in 18 carote prelevate sul fondo del Mare Adriatico.
Arch.Oceanogr.Limnol., 12(3):297-359

Dale, T., Total chemical and biological oxygen consumption of the sediments in
1978   Lindaspollene, Western Norway. Mar.Biol., 49(4):333-41

Daniels, C.H., Quantitative ökologische Analyse der zeitlichen und räumlichen Verteilung
1970   rezenter Foraminiferen im Limski Kanal bei Rovinj (nördliche Adria).
Göttinger Arb.Geol.Paläont.Inst., 8:109 p.

Degobbis, D. et al., Increased eutrophication of the northern Adriatic Sea.
1979   Mar.Pollut.Bull., 10:298-301

Gamulin-Brida, H., A. Pozar and D. Zavodnik, Contribution aux recherches sur la bionomie
1968   des fonds meubles de l'Adriatique du Nord. 2. Biol.Glas., 21:157-201

Gilmartin, M. et al., Hydrographic data collected during approximately bi-weekly cruises
1972   in the northern Adriatic. Thalassia Jugosl., 8:149-213, 331-438

Gray, J.S., The ecology of marine sediments. An introduction to the structure and
1981   function of benthic communities. Cambridge, U.K., Cambridge University
Press, Cambridge studies in modern biology, 2:185 p.

Mattson, J. and O. Linden, Benthic macrofauna succession under mussels, Mytilus edulis
1983   L. (Bivalvia), cultured on hanging longlines. Sarsia, 68:97-102

Millamena, O.M. and R.P. Platon, Organic pollution in culture water resulting from excess
1977   feed and metabolic building. SEAFDEC Q.Res.Rep., 1(3):1-4

Oppenheimer, C.H. and C.E. ZoBell, The growth and variability of sixtythree species of
1952   marine bacteria as influenced by hydrostatic pressure. J.Mar.Res., 11:10-8

Paul, J., Sedimentologische Untersuchungen im Limski Kanal und vor der istrischen Küste
1970   (nördliche Adria). Göttinger Arb.Geol.Paläont.Inst., 7:75 p

Picard, J., Recherches qualitatives sur les biocoenoses marines des substrats meubles
1965   dragables de la region marseillaise. Rec.Trav.Stn.Mar.Endoume 36(52):1-160

Revelante, N. and M. Gilmartin, The effect of Po River discharge on phytoplankton
1976   dynamics in the northern Adriatic Sea. Mar.Biol., 34:259-71

Schreiber, B. et al., Gross beta radioactivity in sediments of the north Adriatic Sea:
1968   A possibility of evaluation the sedimentation rate. Arch.Oceanogr.Limnol.,
16:45-62

Shepard, F.P., Nomenclature based on sand-silty-clay ratios. J.Sed.Petrol.,
1954   24(3):151-8

Smodlaka, N., Primary production of the organic matter as an indicator of the
1985   eutrophication in the northern Adriatic Sea. Ph.D. Thesis, University of
        Zagreb, 259 p. (in Croatian).

Smodlaka, N. and N. Revelante, The trends of phytoplankton production in the northern
1984   Adriatic Sea: A twelve year survey. Rapp.P.-V.Réun.CIESM, 28(9):89-90

Stachowitsch, M., Mass mortality in the Gulf of Trieste: the course of community
1983   destruction. Mar.Ecol., 5(3):243-64

Strickland, J.D.H. and T.R. Parsons, A practical handbook of sea water analysis.
1972   Bull.Fish.Res.Board Can., (167):310 p.

Swartz, R.C., Research needs concerning pollution of the marine benthos. In
1976   Proceedings International Symposium on Marine Pollution Research, edited by
        S.P. Meyers. Gulf Breeze, pp. 60-8

Stirn, J., The northern Adriatic pelagial. Razpr.-Dissert.Slov.Akad.Znan.Umet.,
1969   (Cl. IV), 12(2):1-132 (in Slovenian)

_____, Biocoenological methods for assessments of marine pollution and problems
1970   of indicator species. Rome, FAO, MP/70/E-41:9 p. (mimeo)

Uffenorde, H., Ökologie und jahreszeitliche Verteilung rezenter benthonischer Ostracoden
1972   des Limski Kanal bei Rovinj (nördliche Adria). Göttinger
        Arb.Geol.Paläont.Inst., 13:121. p.

Vatova, A., La fauna bentonica del Canal di Leme in Istria. Mem.R.Com.Talassogr.Ital.,
1931   181:10 p.

_____, Richerche preliminari sulle biocenosi del Golfo di Rovigno. Thalassia,
1935   2(2):30 p.

_____, Le zoocenosi dell'alto Adriatico presso Rovigno e loro variazioni nello
1943   spazio e nel tempo. Thalassia, 5(6):61 p.

_____, La fauna bentonica dell'alto e medio Adriatico. Nova Thalassia, 1(3):110 p.
1949

Vatova, A. and P. Milo di Villagrazia, Sulle condizioni idrografiche del Canal di Leme
1950   in Istria. Nova Thalassia, 1(8):67 p.

Vidakovic, J., Meiofauna of silty sediments in the coastal area of the north Adriatic,
1984   with special reference to sampling methods. Hydrobiologia, 118:67-72

Vidakovic, J. and D. Zavodnik, Benthos off the Po River mouth - subject to stress
1985   conditions?. Journ.Etud.Pollut.CIESM, 7(1984):805-9

Zavodnik, D., Contribution to the dynamics of benthic communities in the region of Rovinj
1971   (northern Adriatic). Thalassia Jugosl., 7(2):447-514

_____, Benthic communities in the Adriatic Sea: reflects of pollution. Thalassia
1977   Jugosl., 13:414-22